NONEQUILIBRIUM PROCESSES IN THE PLANETARY AND COMETARY ATMOSPHERES: THEORY AND APPLICATIONS

ASTROPHYSICS AND SPACE SCIENCE LIBRARY

VOLUME 217

Executive Committee

W. B. BURTON, *Sterrewacht, Leiden, The Netherlands*
J. M. E. KUIJPERS, *Faculty of Science, Nijmegen, The Netherlands*
E. P. J. VAN DEN HEUVEL, *Astronomical Institute, University of Amsterdam,*
The Netherlands
H. VAN DER LAAN, *Astronomical Institute, University of Utrecht,*
The Netherlands

Editorial Board

I. APPENZELLER, *Landessternwarte Heidelberg-Königstuhl, Germany*
J. N. BAHCALL, *The Institute for Advanced Study, Princeton, U.S.A.*
F. BERTOLA, *Università di Padova, Italy*
W. B. BURTON, *Sterrewacht, Leiden, The Netherlands*
J. P. CASSINELLI, *University of Wisconsin, Madison, U.S.A.*
C. J. CESARSKY, *Centre d'Etudes de Saclay, Gif-sur-Yvette Cedex, France*
O. ENGVOLD, *Institute of Theoretical Astrophysics, University of Oslo, Norway*
J. M. E. KUIJPERS, *Faculty of Science, Nijmegen, The Netherlands*
R. McCRAY, *University of Colorado, JILA, Boulder, U.S.A.*
P. G. MURDIN, *Royal Greenwich Observatory, Cambridge, U.K.*
F. PACINI, *Istituto Astronomia Arcetri, Firenze, Italy*
V. RADHAKRISHNAN, *Raman Research Institute, Bangalore, India*
K. SATO, *School of Science, The University of Tokyo, Japan*
F. H. SHU, *University of California, Berkeley, U.S.A.*
B. V. SOMOV, *Astronomical Institute, Moscow State University, Russia*
R. A. SUNYAEV, *Space Research Institute, Moscow, Russia*
S. TREMAINE, *CITA, University of Toronto, Canada*
Y. TANAKA, *Institute of Space & Astronautical Science, Kanagawa, Japan*
E. P. J. VAN DEN HEUVEL, *Astronomical Institute, University of Amsterdam,*
The Netherlands
H. VAN DER LAAN, *Astronomical Institute, University of Utrecht,*
The Netherlands
N. O. WEISS, *University of Cambridge, U.K.*

NONEQUILIBRIUM PROCESSES IN THE PLANETARY AND COMETARY ATMOSPHERES: THEORY AND APPLICATIONS

by

MIKHAIL YA. MAROV

Keldysh Institute of Applied Mathematics,
Moscow, Russia

VALERY I. SHEMATOVICH

Institute of Astronomy,
Moscow, Russia

DMITRY V. BISIKALO

Institute of Astronomy,
Moscow, Russia

and

JEAN-CLAUDE GÉRARD

University of Liège, Belgium

KLUWER ACADEMIC PUBLISHERS
DORDRECHT / BOSTON / LONDON

A C.I.P. Catalogue record for this book is available from the Library of Congress

ISBN-13: 978-1-4020-0378-3   e-ISBN-13: 978-94-010-9555-6
DOI: 10.1007/978-94-010-9555-6

Published by Kluwer Academic Publishers,
P.O. Box 17, 3300 AA Dordrecht, The Netherlands.

Sold and distributed in the U.S.A. and Canada
by Kluwer Academic Publishers,
101 Philip Drive, Norwell, MA 02061, U.S.A.

In all other countries, sold and distributed
by Kluwer Academic Publishers,
P.O. Box 322, 3300 AH Dordrecht, The Netherlands.

Printed on acid-free paper

CONTENTS

CHAPTER 1

INTRODUCTION

This book reviews the approach to the kinetic simulation of nonequilibrium processes in the planetary atmospheres which the authors developed and dealt with since the 1970s. The results of this study, which are focused on the nonequilibrium collisional processes in the atmospheres of planets and comets, are thoroughly reviewed and discussed. Many specific problems of atmospheric modeling, involving numerical evaluation of aeronomic processes, are addressed and compared with the available experimental data.

The kinetic approach proved to be especially effective to model the interaction of the incident shortwave solar radiation with the rarefied gas of planetary upper atmospheres. It involves various processes of photolysis, energetic electron impacts, and accompanying numerous chemical reactions, as well as processes occurring in the intermediate ("transition") zones of planetary and cometary gas envelopes. The underlying mathematical treatment is based on the stochastic approach for the solution of the Boltzmann-type equation and implies the development of the efficient algorithms for its computer simulation. Some results of this study were previously summarized in the monograph issued in Russian (Marov *et al.*, 1990) and later in the review paper published in *Space Science Reviews* (Marov *et al.*, 1996).

The basic principles of stochastic simulation were first developed in the field of rarefied gas dynamics and were successfully applied to the solution of some engineering problems of aerodynamics and heat transfer. It is worth noting that, from the viewpoint of the fundamental problems of physics and mechanics involved, aeronomy can be considered as a new branch of rarefied and partially ionized gas dynamics. Hence basic principles and some modeling approaches used in these fields were incorporated and applied. However the broader scope of aeronomy interests and many specifics of the related problems, including the wide energy range of incident photons and electrons impacting thermodynamically open gas medium such as the upper atmosphere of a planet or cometary coma, required significant modification of the original concepts. New or upgraded methods and algorithms to address the relevant problems of non-equilibrium gas kinetics were developed and successfully implemented for the simulation of a wide range of aeronomical processes.

Qualitatively, the upper atmosphere of a planet can be defined as the atmospheric region stretching outward from some pressure level (starting from the stratopause though sometimes referred to as located from higher levels above the mesopause) where most of the incident short-wave (photolytically active) solar radiation is absorbed. It is essentially not bounded at the top. However, the density is too low for substantial energy absorption above the level, generally associated with exobase, from which escape of atmospheric particles into outer space becomes important. The rarefied partially ionized multicomponent gas that fills up the region between the mesopause and the exobase, referred to as the thermosphere, is one of the core elements of the general system of solar-planetary interactions, which is controlled by the gravitational and/or magnetic fields. Its nonequilibrium state is principally maintained by the incident solar radiation and partially by precipitation of energetic charged particles from the magnetosphere or directly supplied by the solar wind, depending on whether the planet possesses an intrinsic magnetic field and hence on the acceleration mechanism involved.

Solar extreme ultraviolet (EUV), soft X-ray and corpuscular radiation including energetic charged particle fluxes induced by the solar wind forcing, are the external sources responsible for the excitation, dissociation, and ionization of atmospheric molecules and atoms in this atmospheric region. These sources also result in the formation of neutral and ionized particles in translationally and/or internally excited states. The atoms, molecules, and their ions participating in the numerous exo- and endothermic chemical and ion-molecular reactions strongly influence the structure, energy balance, and dynamics of the thermosphere. The processes of molecular diffusion and thermal conductivity occur in the atmospheric gas, which is also involved in different transport phenomena, including horizontal and vertical motions with a broad range of spatial scales, wave transfer and turbulence, as well as ionospheric plasma drifts stipulated by the magnetospheric-ionospheric coupling. Evidently, the overall scenario is extremely complex, and only limited problems can be addressed for in-depth study. Many of these problems are intimately related with non-equilibrium processes in near-planetary space, and their analysis and modeling are the objectives of this book.

The following reasons encourage us to focus our attention on the non-equilibrium processes in the rarefied gas envelopes of the planets and other celestial bodies.

First, a significant part of atoms, molecules, and their ions is formed as radiatively and chemically active particles in translationally and internally excited states. They are produced by EUV photons absorption and by electron impact collisions as well as in the following chemical reactions induced by these external sources. These particles play a very important role in the chemistry of the planetary upper atmospheres. Indeed, chemical and ion-molecular reactions are enhanced if the atoms, molecules or ions participating in reaction are in translationally ("hot") and/or internally excited states. Besides, atmospheric particles can be excited to metastable states corresponding to forbidden electronic transitions. These metastable states have long lifetimes compared to the species excited through the permitted transitions. Because of the long lifetime and high inner energy of metastable excited species, chemical reactions in which they participate usually have much larger reaction rate coefficients than the corresponding reactions with species in the ground state. For example, the "hot" ground state $N(^4S)$ and metastable state $N(^2D)$ of atomic nitrogen are the dominant sources of thermospheric NO in the Earth's upper atmosphere. Nonetheless, while the NO molecules are only a minor neutral species, NO^+ ions are the major component in the E- and lower F-regions of the ionosphere as the result of ion-molecular reactions.

Second, the interaction of solar EUV photons and energetic charged particles with the upper atmosphere is a source of heat for the neutral and ionized atmospheric species. Atoms and ions emerging from the photolytic and electron impact processes with an excess kinetic energy and/or in internally excited states, are not only chemically converted to different species but may also recombine. Consequently, although these external sources are the main carriers of the deposited energy, direct collisional transfer of energy to atmospheric gas is usually small compared to the indirect heating by exothermic chemical and ion-molecular reactions. These reactions result in the conversion of the original photon and electron energy into kinetic energy of the reaction products. Besides, because the translationally excited products of these exothermic reactions are chemically active, they induce subsequent atmospheric reactions with activation energies, thus heating the ambient atmospheric gas by elastic collisions.

Third, the role of translationally and/or internally excited (nonthermal) particles in the atmospheric photochemistry and energetics is defined by the relaxation properties of the rarefied atmospheric gas. In particular, these properties define what fraction of the solar incident energy is lost through the various atmospheric emissions and thus is not converted into heat.

In a simplified qualitative mode, the relaxation properties of atmospheric gas could be estimated using the ratio of characteristic microscopic and macroscopic parameters, such as the mean free path between collisions and scale height, which is known as Knudsen number. In a simplified form, this number characterizes the possible deviation of gas thermal state from the local thermodynamic equilibrium (LTE). It follows from the kinetic theory of rarefied gases that if $Kn < 1$ the gas is in a near-LTE state and can be adequately described by the set of macroscopic (hydrodynamic) equations. There is a transition region for gas flow when $0.1 < Kn < 1$ and deviations from LTE are possible. Finally, the gas state is entirely defined by the dynamical parameters of the particles if $Kn > 1$. In this latter case the gas is referred to as nonequilibrium and its essentially arbitrary state is principally dependent on the initial and boundary conditions. All this means that for $Kn > 0.1$ the hydrodynamic approach starts failing and a set of Boltzmann type kinetic equations must be used in order to strictly describe the rarefied gas flows.

Formally, the upper atmosphere of a planet can be divided into specific regions in terms of their relaxation properties and respective Knudsen numbers). For example, in the Earth's lower thermosphere (from the mesopause up to about 150-200 km), the thermalization rate due to elastic collisions is high and relaxation is a local process. Thus one can assume that nonthermal particle transport is negligible. However it was found that, although a high thermalization rate occurs, the role of nonthermal particles in the processes of photochemistry remains significant even in this atmospheric region. In contrast, in the exosphere (above about 500 km) the thermalization rate is very small and nonthermal particles formed at these heights essentially do not interact with the ambient atmospheric gas. The most important role is played by nonthermal particles in the transition region between the thermosphere and the exosphere, or near the exobase, where the production rates of nonthermal particles are still significant while the elastic thermalization rate strongly decreases. It results in the production of a significant amount of nonthermal ("hot") particles in this region, which mainly contribute to the formation of a hot planetary corona.

As we already mentioned, only a microscopic kinetic approach ensures an adequate evaluation of the processes of thermalization for these "hot" particles taking also into account the nonequilibrium energy exchange between their translational and internal degrees of freedom. Recourse to a set of nonlinear Boltzmann equations with the source term allows to reach this goal. The left-hand side of such an equation describes the particle transport in the gravitational field while the right-hand side describes the collisional evolution of the atmospheric gas. The source term is the first part of the right-hand side expression, which represents the particle production rates in photolytic and impact reactions generated by the solar incident radiation. Incidentially, deposition of the external energy into rarefied atmospheric gas is considered to be the multichannel process accompanied by the production of superthermal, excited (in particular in the dissociation processes) and ionized atmospheric particles. In turn, the second part of the right-hand side expression is the set of nonlinear collisional integrals which describe the changes of the gas state through the following chemical and physical transformations in binary collisions of the atmospheric particles. This multichannel process involves both the elastic and inelastic collisions, chemical, ion-molecular and recombination reactions and therefore characterizes the relaxation of excited states of the atmospheric gas.

Let us note that when photochemically-produced "hot" particles are only small admixtures (and consequently the thermalization processes for these particles weakly disturb the local thermal state of a medium) their interaction with the ambient atmospheric gas may be described by a linearized form of the kinetic equation. This is the situation occurring in the lower thermosphere where the main atmospheric species are considered to be locally thermalized. But in the upper thermosphere and transition region between thermosphere and exosphere the system becomes nonlinear because the processes of relaxation occurring in the atmospheric gas significantly influence its thermodynamic state. This situation is clearly prevailing in the Earth's thermosphere where, due to elastic interaction with superthermal O (or N) atoms, the thermal oxygen atoms (the main atmospheric component at these heights) are persistently disturbed, resulting in the appearance of nonthermal atomic oxygen of secondary origin.

The solution of Boltzmann type equations describing aeronomical processes faces a number of complex computational problems. Advantage was found in the development of specific discrete mathematical models, based on the probabilistic approach to numerical processing of collisions within an ensemble of modeling particles. This means that the kinetic process under consideration is represented by its physical-probabilistic analogue. Intrinsically, this approach makes it necessary to develop adequate models for a stochastic simulation and efficient algorithms. Monte-Carlo algorithms totally meet these requirements because they allow to get multiple running of the simulation code instead of a direct numerical solution of the stochastic equations. Therefore, all the necessary physical characteristics of the system become available through the computer realizations of such a stochastic model. In addition, a rather comprehensive analysis of the kinetic processes involved becomes possible.

This approach adjoins a broader class of methods commonly used in applied mathematics and referred to as numerical experiments. In our particular case, it is mostly related to an assessment of the accuracy of the Boltzmann-type equation approximation and potential advantages that the stochastic simulation ensures in the solution of its numerical analogue. It is well known that the main difficulties in the solution of non-linear integro-differential Boltzmann equation are caused by both the large number of independent variables and the complicated structure of the collisional integrals which prevents resort to standard numerical techniques. Methods of direct statistical modeling involving the splitting technique for the processes of molecular collisions and collisionless displacements of gas particles with utilization of Monte-Carlo algorithms proved to be a powerful tool in the effective evaluation of nonstationary nonlinear problems of rarefied gas dynamics and in particular, aeronomical problems.

Although the advantage of this statistical approach was clearly demonstrated, the question still remains unanswered if it is associated with any distortion of the basic Boltzman equation. One proceeds from the physically justified assumption that time between collisions is asymptotically equivalent to collisional statistics in the ideal monoatomic gas and that the evolution of the model strongly follows a Markovian process. The precise jump-like Markovian process is thought to closely adapt the Boltzmann collisional statistics. Therefore, the evolution of the numerical stochastic model closely approximates the Boltzmann equation provided that the hypothesis of molecular chaos (random collisions without correlations) is additionally postulated. In other words, one may argue that this model serves as an asymptotic equivalent of the Boltzmann equation, best fitted to a spatially-uniform case (when the convective term is omitted).

It is important to note that the fact that the basic model equation is linear (in contrast to the original Boltzmann equation) strongly simplifies the solution of the problem which is represented by a finite number of particles trajectories. Through an averaging of a multitude of these trajectories, all the macroparameters of the gas flow can be calculated.

We can briefly review the remaining eleven chapters of the book as follows. The second chapter addresses the main properties of outer gaseous envelopes of planets and comets stretching outward into the interplanetary space and serves as a general introduction to the subject. The third chapter describes the rarefied gas of planetary and cometary atmospheres as physical and chemical systems and displays their peculiarities. We decided to give in both of these chapters a rather thorough review of the current knowledge in the field, to accommodate the readers not well familiar with the topic, being mainly focused on the problems of planetary aeronomy involved. The fourth chapter is the logical continuation of this approach where aeronomic processes, especially their nonequilibrium implications, are addressed in more detail. The following two chapters deal with the basic concepts of the kinetic approach to the mathematical modeling of a multicomponent (and generally nonequilibrium) rarefied gas system, involving the approximation of the Boltzman-type equation by stochastic numerical models and the study of their structures, as well as the development and realization of respective algorithms. The specifics of these numerical models for aeronomical applications, including the formation of nonthermal particles by solar electromagnetic and corpuscular radiation, their kinetics and relaxation processes in the planetary atmosphere are thoroughly discussed. The following five chapters are dedicated to various examples of kinetic simulation of nonequilibrium processes based on the developed technique. They include nonequilibrium chemistry of odd nitrogen in the Earth's thermosphere, kinetics of formation and maintenance of hot coronas of Earth, Venus and Mars; study of nonthermal particles in the upper atmosphere of Jupiter and their role in auroral emissions; transition regions of a cometary atmosphere involving near-surface (Knudsen) layer and processes of photochemistry in inner coma. The eleventh chapter also includes a brief review of the contemporary ideas on the chemical composition of comets, which provides a basis for their modeling. Concluding remarks are summarized in the last chapter.

The book is intended for specialists in the fields of aeronomy, planetary physics, space science and astronomy, but it will also be of interest to scientists dealing with rarefied gas dynamics, hydrodynamics, and computer modeling of astrophysical and atmospheric processes. It can be recommended to graduate and post-graduate students training in the above disciplines and related areas, as well as to qualified readers having a physical/mathematical background, curious to know more about processes occurring in the Earth's space environment and in the Solar system.

The writing of this book became possible due to multiyear fruitful contacts, stimulating discussions and colloborative efforts with our numerous colleagues. We are especially grateful to Oleg M. Belotserkovsky, Arne A. Pyarnpuu, Michael S. Ivanov, Galina I. Zmievskaya. Special thanks are given to Prof. Cornelius de Jager and the KLUWER Editorial Board for the encouraging offer to publish this book. The assistance of Angela Della Vecchia and Guy Munhoven in the preparation of the camera ready manuscript is gratefully acknowledged.

Any comment on the book contents and/or suggestion for its improvement would be highly appreciated.

M.Ya. Marov   V.I. Shematovich   D.V. Bisikalo   J.-C. Gérard

CHAPTER 2

PLANETARY AND COMETARY ATMOSPHERES: AN AERONOMY APPROACH

2.1. Planetary Atmospheres: General Characteristics

The atmosphere is a gaseous envelope surrounding a planet, held by gravity, having its maximum density just above the solid surface (if any) and becoming gradually thinner with distance from the ground, until it finally becomes indistinguishable from the interplanetary gas. As moving away from a planetary surface, different atmospheric regions are defined, with widely different properties, being the seats of a great variety of physical and chemical phenomena. Lower and middle atmospheres of the inner planets are thin compared to solid bodies and can be visualized as the skin of an apple, while atmospheres of the giant planetes are referred to as an outer shell of the gaseous-liquid body.

The properties of the planetary atmospheres vary widely even within the relatively small region of the solar system occupied by the planets and specifically inner planets. It is generally accepted that, unlike the atmospheres of the giant planets, the original reducing atmospheres of the terrestrial planets which included free volatiles, were lost during or at the end of the accretion process. The present-day secondary oxidizing atmospheres of the Earth, Venus and Mars were formed from condensed material (in which some of the volatiles were chemically bound or absorbed) during the thermal evolution of the planets by outgassing from the interior, mainly in volcanic eruptions. The basic data concerning the chemical composition, temperature, pressure, and density of the atmospheres of the planets are summarized in the Tables 2.1 and 2.2.

Atmospheres are largely responsible for the planetary environmental conditions and their dramatic difference. Their specific features are closely related with thermal regime and dynamics on the global and local scales and thus responsible for weather and climate conditions. Because an atmosphere is intimately linked with the processes in the interior and on the surface, its major neutral molecular constituents are to be considered as a legacy of the planetary evolution. Therefore atmospheric studies allow us to get more insight into solution of this fundamental problem. In turn, vastly extended rarefied upper atmospheres, being an element of circumplanetary space,

TABLE 2.1. Properties of the atmospheres of the inner planets.

Planet	Mercury	Venus	Earth	Mars
Z	He=20	CO_2=95	N_2=78	CO_2=95
	H_2=18	N_2=3-5	O_2=21	O_2=2-3
	Ne=40-60	Ar=0.01	Ar=0.93	Ar=1-2
	Ar=2	H_2O=0.01-0.1	H_2O=0.01-3	H_2=1(-3)-1(-1)
	CO_2=2	CO=3(-3)	CO_2=0.03	CO=4(-3)
		HCl=4(-5)	CO=1(-5)	O_2=0.1-0.4
		HF=1(-6)	CH_4=1(-4)	Ne < 1(-3)
		O_2 <2(-4)	H_2=5(-5)	O_2=0.1-0.4
		HCl=4(-5)	CO=1(-5)	Kr < 2(-3)
		SO_2=1(-5)	Ne=2(-3)	Xe < 1(-3)
		Kr=4(-5)	Kr=1(-4)	
		Xe=1(-6)-1(-5)	Xe=1(-6)	
μ		43.2	28.97	43.5
T_{max}	700	735	310	270
T_{min}	110	735	240	148
P	<2(-14)	90	1	6(-3)
ρ	1(-17)	6.1(-2)	1.27(-3)	1.2(-5)

a(-b)=a$\times 10^{-b}$. Z is the chemical composition (percent by volume); μ is the mean molecular mass; T is the surface temperature (K); P is the mean surface pressure (atm); ρ is the mean surface density ($g\,cm^{-3}$).

are involved in the multifacets complicated scenario of the solar-planetary interactions.

The gases making up the atmosphere can be divided into main and minor (trace) gases. In the first category are constant gases composing the bulk of the atmosphere which are relatively constant by volume and remain generally in the same proportion through time and from place to place throughout the lower and middle atmospheric regions due to efficient mixing. By definition it is homosphere, differing from above lying heterosphere. In contrast to the Earth's atmosphere composed mainly of nitrogen, oxygen, and argon, carbon dioxide is the main component of the atmospheres of Venus and Mars, with relative abundances of other gases only a few percent. Hydrogen and helium, with small abundances of hydrogen-bearing compounds, principally compose the atmospheres of the giant planets.

Although the minor (trace) constituents in the planetary atmosphere

TABLE 2.2. Properties of the atmospheres of the outer planets.

Planet	Jupiter	Saturn	Uranus	Neptune	Pluto
Z	$H_2=89$	$H_2=96$	$H_2=84$	$H_2=80$	CH_4
	$He=10$	$He=3$	$He=15$	$He=19$	
	$H_2O=1(-4)(?)$	$CH_4=2(-1)$	$CH_4=0.1-2$	$CH_4=0.1-2$	
	$CH_4=9(-2)$	$NH_3=2(-2)$	$C_2H_2=2(-5)$	$N_2=3(-1)$	
	$NH_3=2(-2)$	$C_2H_6=8(-4)$	$C_2H_6=<1(-4)$	$CO=1(-4)$	
	$HCN=2(-7)$	$C_2H_2=1(-5)$	$NH_3=?$	$HCN=1(-7)$	
	$C_2H_6=4(-3)$	$PH_3=3(-4)$		$NH_3=6(-5)$	
	$C_2H_2=8(-5)$	$GeH_4=4(-8)$		$C_2H_2=6(-6)$	
	$PH_3=4(-5)$	$C_3H_4=$trace		$C_2H_6=2(-4)$	
	$CO=3(-7)$	$C_3H_8=$trace			
	$GeH_4=6(-8)$				
	$C_3H_8=$trace				
μ	2.20	2.06	2.36	2.44	16
$T^{*)}$	135	105	51	52	32
$P^{*)}$	0.5	0.5	0.1	0.1	
$\rho^{*)}$	1(-4)	1.2(-4)	5.5(-5)	5.5(-5)	

$a(-b)=a\times10^{-b}$. Z is the chemical composition (percent by volume); μ is the mean molecular mass; T is the temperature (K); P is the pressure (atm); ρ is the density ($g\,cm^{-3}$).
$^{*)}$ At the reference pressure level in the atmosphere (except for Pluto).

amount to less than a small fraction of percent, they play a very important role in the atmospheric chemistry, especially in the chemistry of middle and upper atmosphere. The abundances of trace gases are strongly variable both in time and space. The main variable gases in the Earth's lower atmosphere are water vapor, carbon dioxide and ozone, and they play a considerable role in the heat balance, dynamics and weather formation, thus providing favorable conditions for life on the home planet. Sources of the minor constituents are volcanic eruptions (carbon-, sulfur-, and nitrogen-bearing compounds); combustion, both natural (e.g. CO_2) and anthropogenic (e.g. CO_2, SO_2, NO); biological processes (e.g. CH_4, N_2O, H_2, NH_3, H_2S, NO) and chemical reactions (e.g. HCl).

Temperature, pressure and density, together with gas chemical composition, define the state, structure and properties of the planetary atmosphere. Based upon a variety of changes which occur with height in the structure and properties, the atmosphere is commonly divided vertically into several zones, beginning with the lower densest *troposphere* (weather layer) and then up to *stratosphere, mesosphere* (both referred also as middle atmosphere) and very rarefied and essentially unconfined upper atmosphere (*thermosphere and exosphere*) at the top (see *Figs. 2.1, 2.2*). The Earth's troposphere contains about 80 % of the total air mass and its upper limit is defined by a sudden change in temperature, which stops decreasing and remains constant or starts increasing slightly from the *tropopause* located beween 7 and 17 km (depending on latitude) with a temperature -50 to -55^oC.

The *stratosphere* is defined as the region lying above the tropopause where the temperature lapse rate becomes zero, though the Earth's stratosphere shows a gradual increase of temperature due to the absorption of solar ultraviolet radiation by ozone, reaching a maximum of around 0^oC at 50 km (*stratopause*). In turn, the *mesosphere* is characterized by a positive lapse rate (decreasing temperature with increasing altitude) which extends to the *mesopause* (at about 85 km for the Earth with a temperature of about 180 K). Most of the mesosphere consists of neutral atoms and molecules in the same ratio as the lower atmosphere. The mesopause is the coldest atmospheric region. Both tropopause and mesopause are often referred to as the 'cold traps'.

The *thermosphere* is characterized by a negative lapse rate due to direct absorption of the solar extreme ultraviolet (EUV) radiation, diffusive-gravitational separation of molecules and atoms, steady change of the mixing ratio of atmospheric species with height, and auroral particles, electromagnetic fields and plasma motions influence. The *ionosphere* is essentially a part of the thermosphere extending upwards into the *exosphere*, which is defined as the region of nearly constant temperature where there are essentially no collisions for atoms moving upward. Therefore, beginning the base of the exosphere (*exobase, or critical level*) the atmospheric gas can dissipate into the outer space (for the Earth it is located at about 500 km). Depending on the phase in the 11-year cycle of solar activity, the Earth's exospheric temperature varies betwen 600 and 1400 K.

A simple mathematical form defined by the hydrostatic equation is widely used as the relationship between the main atmospheric parameters. It states that upward acceleration of gas molecules due to the pressure decrease must balance exactly the downward acceleration due to the force of gravity. When coupled with the equation of gas state, combining temperature, volume and pressure into one statement, this results in the barometric formula which defines the exponential decrease of pressure and density in the gravitational field of a planet towards outer space. The exponent in this formula is determined by the scale height depending on temperature, molecular weight of gas and acceleration due to gravity, and thus the rate of atmospheric thining with height. Since in the homosphere the average molecular weight remains constant, the same scale height is used in the barometric formula to describe the altitude distribution of all atmospheric species. In the heterosphere, the height distribution of all atmospheric species follows their own molecular or atomic weight due to molecular diffusion domain over turbulent mixing.

In the heterosphere, direct absorption of solar short-wave radiation as well as energetic particle impacts, most efficiently occur, giving rise to the numerous elementary processess of photochemistry, followed by chemical reactions, diffusion, airglow and auroral emissions, formation of a partially ionized plasma in the ionosphere, dynamic and electrodynamic interactions. Altogether, these processes involving energy/spectra transformations are referred to as planetary aeronomy, which conveniently integrates basic ideas of rarefied gas dynamics, chemical kinetics, plasma physics and electrodynamics and serves as the baseline to model of rarefied gaseous envelopes and space environment of the planets.

2.2. Planetary Upper Atmospheres and Ionospheres

2.2.1. EARTH

Neutral atmosphere. The structure and properties of the Earth's upper atmosphere — its temperature, density, and chemical composition — are determined by absorption of solar, electromagnetic and corpuscular radiation, mass, energy, and momentum transport from underlying regions in the lower atmosphere and processes involving magnetosphere-ionosphere interactions. Most of the energy exchange due to direct absorption of short-wave solar radiation in the far-ultraviolet and soft X-ray regions (from approximately 2000 Å to 10 Å) and solar corpuscular radiation occurs in the thermosphere. These absorptions cause dissociation, ionization and excitation of neutral species in the upper atmosphere, which partially emerge in nonequilibrium state. The longer-wave ultraviolet radiation (3200-2000 Å), which is largely responsible for the formation of the ozonosphere and in particular, screening the Earth from the energetic, harmful shortwave radiation, is absorbed lower in the middle atmosphere, along with the high-energy (up to 30 MeV) solar protons generated in flares and galactic cosmic rays. The magnetosphere-ionosphere interactions include the high-latitude precipitation of energetic particles, the excitation of current systems, leading to intense heating and the formation of large local density gradients.

Variations in the structure and composition of the upper and middle atmosphere are determined by perturbations of various kinds as well as by atmospheric dynamics. A complex set of physical, chemical, and biological processes associated with interactions between the atmosphere and

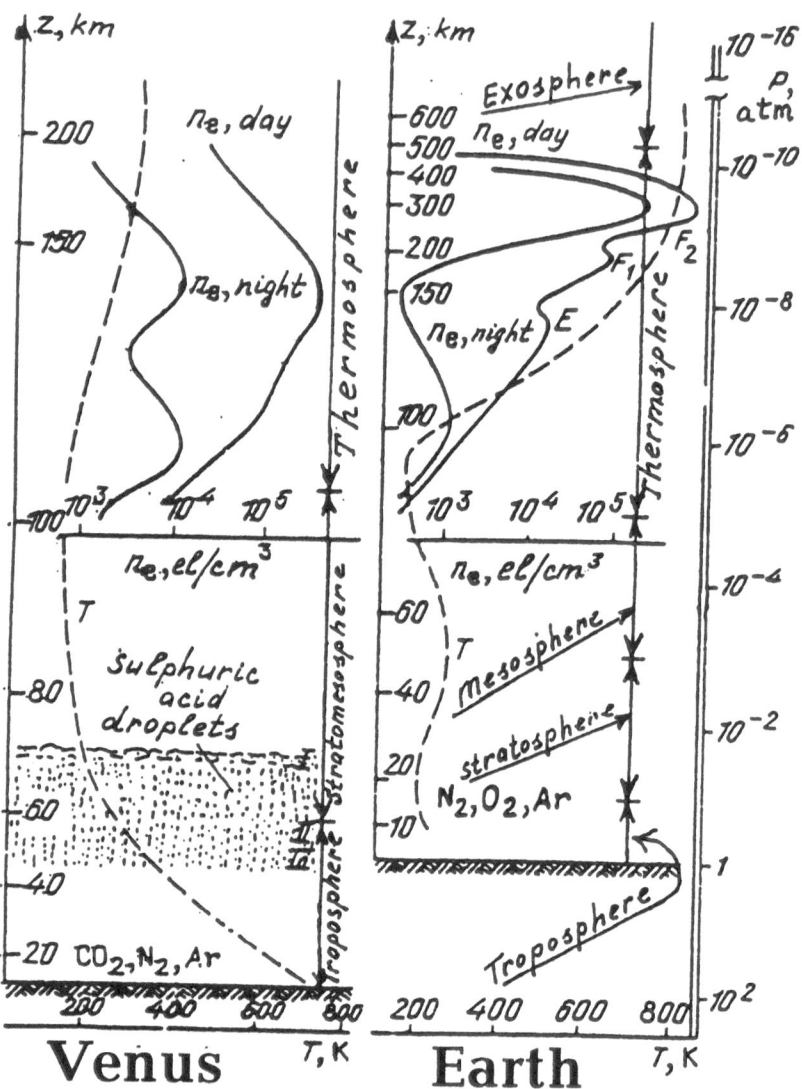

Figure 2.1. Structure of the atmospheres of Venus and Earth. For each planet the figure shows profiles of temperature and electron density in the ionosphere and cloud structure as a function of altitude and pressure.

the lithosphere, hydrosphere, and biosphere plays an important role in the formation and destruction of trace components (which have relative concentrations equal to 10^{-13}–10^{-5} of the major constituents N_2, O_2, and Ar) in addition to a variety of photochemical and chemical reactions, phase

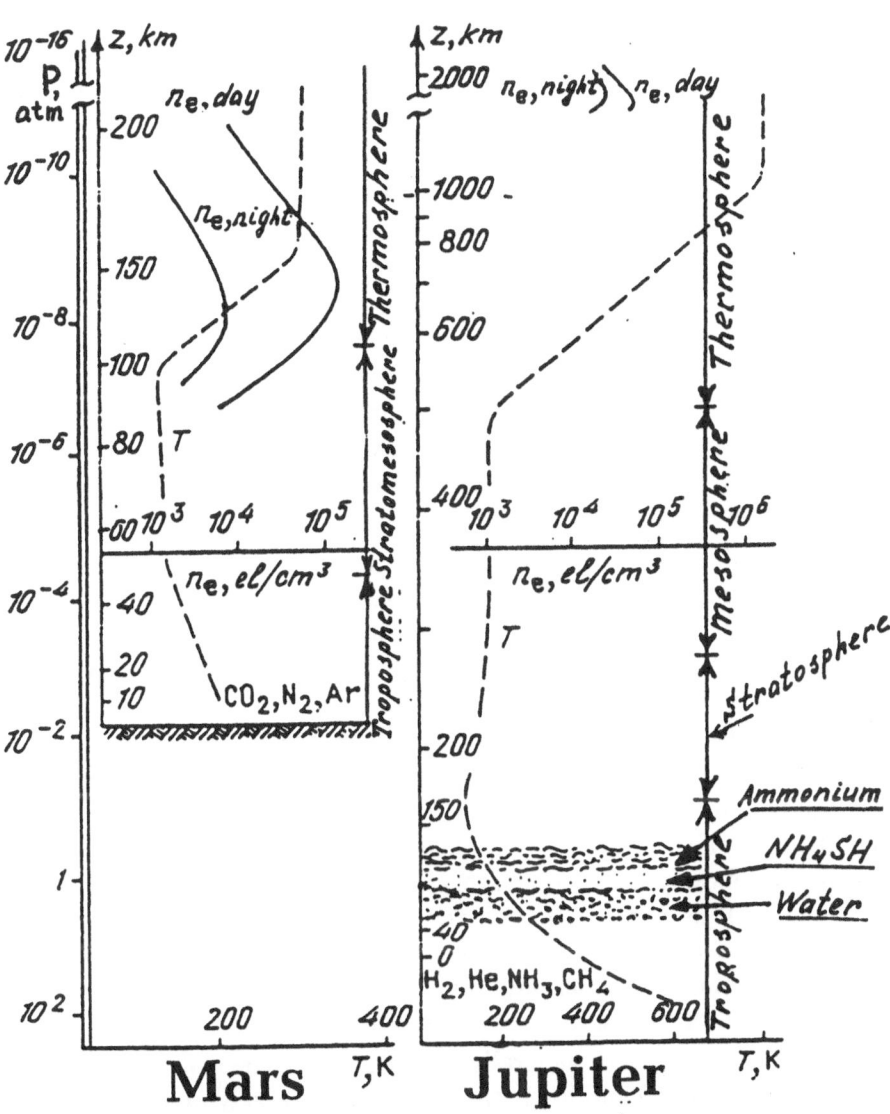

Figure 2.2. Structure of the atmospheres of Mars and Jupiter. For each planet the figure shows profiles of temperature and electron density in the ionosphere and cloud structure as a function of altitude and pressure.

transformations, and dynamic exchange processes.

The entire chemistry of the middle atmosphere (and, to a certain extent, the thermosphere) depends in some way on the formation of chemically active atomic oxygen. Its formation begins in the stratosphere through photodissociation of O_2. The fast reaction between O_2 and O leads to the formation of ozone (O_3), which itself is an effective absorber of ultraviolet radiation (at $\lambda < 3200$ Å). Ozone is rapidly destroyed in this photochemical reaction and the reaction between O_3 and O, and the oxygen atoms recombine. This sequence of processes is what makes up the well-known Chapman cycle.

Chemical radicals such as NO, NO_2, OH, HO_2, Cl, and ClO also have a substantial effect on the ozone concentration. The flow of these radicals from the troposphere into the ozonosphere is largely controlled by photochemical reactions involving halogen compounds, methane, nitrous oxides, and water. This leads to additional catalytic paths for the destruction of ozone. The anthropogenic factor is coming to play an increasing role in the observed increase in the concentration of the parent molecules for these radicals [primarily dichlorodifluoromethane (CCl_2F_2) and trichlorofluoromethane (CCl_3F) - freons, and the carbonated halogens ($C_2H_3Cl_3$, C_2Cl_4, etc.), as well as N_2O and CH_4].

Radical-radical reactions in the stratosphere lead to the formation of a number of intermediate compounds that have relatively low catalytic activity but readily decompose into the original radicals photochemically. These include radicals such as HNO_3, $ClONO_2$, H_2O_2, HOCl, HO_2NO_2, etc. The formation of hydrocarbonates (H_2CO, etc.) and CO is also important from the point of view of stratospheric chemistry. Other chemical reaction products serve as initial products for the formation of aerosols. These are primarily SO_2 and COS, which serve as a basis for the formation of droplets of sulfuric acid (with concentrations ranging from 0.5 cm^{-3} to 10 cm^{-3}) concentrated in a stratospheric layer at approximately 20 km altitude, as well as NH_3. Although the concentrations of all these minor gaseous constituents typically do not exceed millionths of a percent, they have a substantial effect not only on atmospheric chemistry but also on energy exchange processes in the middle atmosphere.

The upper atmosphere generally has a simpler chemistry. The chemistry of the mesosphere and lower thermosphere is largely driven by atomic oxygen formed through the photodissociation of O_2. The concentration of atomic oxygen begins to exceed that of molecular oxygen above an altitude of approximately 120–140 km. Another important component in this region is nitric oxide (NO), whose concentration depends on the concentration of atomic nitrogen. NO is most efficiently produced by collisions with O_2, and the main loss channels include photodissociation, photoionization, and vertical transport. Atomic nitrogen (like atomic hydrogen) is formed via ionization processes in the middle atmosphere accompanying the dissociation of molecular nitrogen and water vapor. N and H initiate a chain of catalyzed chemical reactions involving oxygen and ozone. The catalytic action of odd nitrogen and hydrogen compounds (in the form of NO_x and HO_x) significantly accelerates the destruction of ozone in reactions between O_3 and O.

On the average, N_2 remains the dominant component to 180 km altitude, because the efficiency of its dissociation is low. O becomes the dominant component higher in the thermosphere. At even higher altitudes, H and He begin to dominate. He and O concentrations are approximately equal between 500 and 700 km, and those of H and He between 900 and 1800 km, depending on the phase of the solar cycle, which affects the temperature of the upper atmosphere. Concentrations of N_2, O_2, and O in the thermosphere increase with increasing temperature, while the concentration of H_2 decreases (due to the increase in its rate of dissipation). The variations in He concentration are more complex. The concentration of He is highest at solar minimum (especially in the winter hemisphere-the so-called 'winter helium bulge'). Let us note that, although the hydrogen and hydrogen compounds (such as CH_4, H_2O, and OH) are minor constituents of the thermosphere, they may still play a significant role in chemical transformation and radiative heat exchange mechanisms.

Ionosphere. Photoionization and impact ionization by energetic electrons of major and minor components are responsible for the formation of the ionosphere. They also play an important role in the energy balance of the upper atmosphere, as well as in atmospheric chemistry, and lead to various important chains of aeronomic reactions that affect the structure and dynamics of individual regions in the atmosphere. However, the ionosphere itself exerts a strong influence over the macroscopic properties of the medium, even at levels of the atmosphere where the electron density is relatively low (no greater than $\sim 10^{-9}$ at 100 km). The density of the ionosphere increases with altitude, and plays an extremely important role in electrodynamic interactions and in the dynamics of the atmosphere as a whole. In dynamic terms, the Earth's atmosphere can generally be treated as a neutral medium below ≈ 150 km; at higher altitudes, it must be treated as a thermal plasma that is primarily controlled by the geomagnetic field.

Electron density profiles (ionospheric structure) are generally distinguished by the D, E, and F regions differing also in the ionic composition. The maximum electron density (up to 10^6 cm^{-3}) is observed during the daytime at approximately 280 km altitude (the F_2 layer). Less well-defined maxima occur at altitudes of 150–200 km (F_1 layer), at approximately 110 km (E layer), and 80 km (D layer), with electron densities ranging from 10^5 cm^{-3} in the F_1 layer to 10^3 cm^{-3} in the D layer. The density of each of these layers is substantially lower at night.

The lower layers of the ionosphere (D and E layers), from approximately 60 km to approximately 150 km, consist primarily of NO^+ ions, although the ratio of neutral NO molecules to N_2 molecules is no more than 10^{-8}. O_2^+ ions are also present in comparable quantities. The ionosphere from the F_1 layer (160–180 km) up to 600–800 km at solar minimum and 800–1000 km at solar maximum consists mainly of O^+ ions. Above 1000 km at mid-latitudes and low latitudes, the ionosphere is dominated by atomic hydrogen ions H^+; at latitudes corresponding to the outer plasmasphere, the O^+ ion begins to predominate up to altitudes of several thousand kilometers. It should be noted that ions of the metals Mg, Fe, Ca, and Si, apparently meteoritic in origin, have been observed at altitudes of 100–110 km. The drift of these long-lived ions in the large wind gradients at these altitudes leads to the appearance of the sporadic E_s layer.

We have given a rather simplified description of the ion component distribution that corresponds to a temporal and spatial average for undisturbed conditions. This structure undergoes major deviations as a result of variations of the initial neutral composition, and the rate of ion generation and recombination. These variations are due to changes in the flux intensity of ionizing radiation and the efficiency of dynamic exchange processes. At mesospheric altitudes, an important role is also played by cyclic variations in water vapor concentration, which is subject to photodissociations associated with changes in production of atomic oxygen. The concentration of NO and other minor components depends on this effect as well as the rate of vertical transport.

The increased downward flux of NO molecules out of the thermosphere and the decreased photolytic destruction rate during the winter in turn lead to an increase in the concentration of NO^+ ions. This increase, together with the seasonal restructuring of the wind system in the middle atmosphere, is associated with the anomalous winter increase in electron density in the D layer. Meteorological control over the lower ionosphere is one of the notable factors affecting variation in NO^+ concentration as well as of more complex formations, cluster ions. In the night ionosphere NO^+ ions may predominate throughout the F_1 region, up to 250–300 km near the equator. Under such conditions, the density of electrons and ions is almost invariant with altitude, which is conducive to the development of ionospheric irregularities and instabilities.

The situation in the F layer of the ionosphere is more complex, especially at high latitudes. Thus, the region of anomalously low electron density at the altitude of the F layer (the main ionospheric trough) on the nightside shows significant variations. This trough serves as a kind of boundary between the mid-latitude ionosphere and polar ionosphere. The position of the trough is highly dependent on solar and geomagnetic activity as well as time of day and season. The entire structure of the ionosphere undergoes dramatic changes during geomagnetic disturbances, largely due to dynamic processes in the atmosphere.

Energy balance and variations of macroparameters. The various atmospheric constituents are involved in a wide variety of complex chemical transformations. Products of these reactions have come to play a determining role both in the composition of the atmosphere and ionosphere and in heat exchange. The energy balance of the middle atmosphere is largely governed by the heating of the ozonosphere, while absorption of EUV radiation and the action of particles and fields from the magnetosphere play decisive roles in the energy balance of the thermosphere. The energy input to the atmosphere above 100 km from EUV radiation is approximately 10^{12} W [or 2 erg/(cm^2 s)]. The mean energy input from the magnetosphere due to solar-wind particles and the resulting current systems is 5×10^{10} W , or approximately 0.1 erg/(cm^2 s). However, during geomagnetic storms associated with solar flares, the energy input from this source may become comparable to, or even several times larger than, the energy input due to electromagnetic radiation, reaching tens of erg/(cm^2 s). A significant fraction of the energy from the charged particles injected into the magnetosphere is released in the auroral zones, while even more of the energy is trapped in the so-called ring current that encircles the planet at middle and low latitudes. Energy of $\sim$0.1 erg/cm^{2}s is released when the ring current decays.

Various dynamic processes, including wave processes, also make a significant contribution to the energy balance of the middle atmosphere and thermosphere. Dynamic processes are primarily responsible for energy redistribution on a global scale. However, tides, acoustic and internal gravity waves, and turbulence also play an important role in the thermal balance of various individual regions and in the observed spatial and temporal variations of structural parameters. For example, the bulk heat release rate due to the dissipation of internal gravity wave energy in the lower thermosphere is comparable to that of the other energy sources.

The chemical composition and related energy input determine the global structure of the upper atmosphere-primarily the vertical profiles of macroscopic parameters such as the temperature T and the density ρ. These parameters undergo substantial changes at high altitudes.

Temperature variations in the stratosphere and mesosphere are (like the temperature variations at the surface of the Earth) generally no greater than 50–60 K. Annual and semiannual components with altitude-dependent phase and amplitude are clearly present in the temperature and pressure variations. Semiannual variations in the atmospheric parameters are especially characteristic of the upper stratosphere at tropical latitudes, while quasibiennial variations are observed in the lower stratosphere near the equator.

Temperature variations in the thermosphere and exosphere reach 600–800 K in amplitude, which means that the variations in the density ρ are just as large. The basic variations in temperature and density may be divided into two major groups. The first group includes those due to changes in the influx of solar electromagnetic radiation, which depend on solar activity and time of day. This group is dominated by solar cycles, multiples of 11-year period, and also includes the 27-day cycle frequently observed in density variations, which is associated with the rotational period of the Sun and the existence of long-lived local active regions on its surface. These variations are generally well-correlated with the decimetric radio flux of the Sun ($F_{10.7}$) or Wolf numbers (W), which serve as a measure of the effectiveness with which the upper atmosphere is heated by EUV radiation. The second group of variations is generally subdivided into geomagnetic disturbances and semiannual effect comparable in magnitude to the diurnal effect. This group may be considered the overall response of the neutral atmosphere to magnetospheric-ionospheric interactions. These variations show the best correlation with the planetary geomagnetic disturbance indices (K_p or A_p), which are also serve as a measure of the effect of the solar plasma on the Earth's geomagnetic field. In addition to the variations mentioned above, relatively small variations in T and ρ as a function of latitude and season have been observed, along with variations in ρ as large as 25 percent in amplitude and periods ranging from several minutes to several hours (with 24-, 12-, and 18-hour periods — multiples of the solar and lunar days — being especially prominent).

The largest variations in the neutral upper atmosphere are those involving the 11-year solar cycle. One characteristic indicator of this variation is the temperature of the upper thermosphere, which is nearly isothermal. This temperature is virtually identical to the exospheric temperature (T_∞). Including the diurnal variations, the latter temperature varies from 500–700 K at solar minimum ($F_{10.7} \approx 70$–80) to 800–1300 K at solar maximum ($F_{10.7} \approx 200$). As far as density is concerned, the expansion and contraction of the upper atmosphere due to these temperature variations (i.e., the diurnal and semiannual variations) cause the mean density to vary severalfold near 300 km altitude and by over two orders of magnitude at 500–600 km altitude.

The diurnal variations in T and ρ are the most regular: the minima occur at approximately 0400–0500 LT (local time), while temperature reaches its maximum value at 1600 LT and density at 1400 LT. This is due to various special characteristics of the diurnal atmospheric expansion as well as the dynamics of the atmosphere. The largest diurnal variations in the thermosphere occur at 400–500 km altitude. The maximum of the diurnal variations shifts to lower altitude as solar activity decreases.

The variations associated with geomagnetic activity are particularly complex in nature. Even relatively weak magnetic disturbances cause global changes in T and ρ, with the relative effect of these variations being more apparent at solar minimum. During strong disturbances, temperatures of up to 3000 K have been recorded in auroral regions, whereas the density variations in the mid-latitude atmosphere (at 500 km altitude) have been as large as a factor of 8. The semiannual density variation has two peaks, in June-July and December-January, with the second peak having a slightly larger amplitude. This variation includes a larger range of altitudes than the diurnal density variations, from 90 to 1200 km. The largest variations (up to a factor of 3) are observed near 500 km, with virtually no dependence on phase of solar cycle. No direct source for the semiannual variations has as yet been identified; it has been suggested that they might be due to periodic variations in the heliographic latitude of sunspot activity centers.

These variations provide convincing evidence of the fact that the parameters of the upper atmosphere can only be adequately interpreted and utilized if accompanied by specification of the time of day, latitude, level of solar activity, level of geomagnetic activity, etc. Appropriate empirical models describing the structure of the atmosphere and the variations in its structure as a function of a limited number of parameters have been derived through the generalization and systematization of experimental results. This approach makes it possible to characterize the state of the upper atmosphere in a given region of space under certain geophysical conditions to high accuracy.

Dynamics. The fact that the variations in temperature and density are global is a reflection of the extremely dynamic nature of the middle and particularly upper atmosphere and their reaction to the external disturbances that affect energy and mass exchange. These variations lead to motions on various spatial scales all the way from planetary-scale circulation to local flows and wave processes.

The most important characteristic of middle atmosphere dynamics is the global circulation system. Its distinguishing feature is stable zonal flow, which periodically shifts direction over the course of the year from west to east during the winter to east to west in the summer. This flow is most prominent at altitudes of 60–70 km, and is due to the variation in insolation with latitude, which leads to a change in the amount of ultraviolet radiation absorbed by ozone and the subsequent reradiation of energy in bands of O_3, CO_2, and H_2O. The corresponding wind system, sometimes called the monsoon, is characterized by the stratospheric circulation index (SCI), which is the mean wind velocity in a layer 10 km thick at the level of the stratopause (~ 50 km). The SCI may reach values as high as 50-60 m/s.

Wave processes have a significant effect on the distribution of temperature, pressure, and density, as well as on wind velocity, degree of ionization, and atmospheric emissions. The wave field in the middle atmosphere reflects the close connection between the troposphere and the dynamics of overlying regions. The planetary waves, which have horizontal dimensions comparable to the radius of the Earth and periods on the order of a day or more, have an especially strong effect on mean zonal flow. The so-called sudden high-latitude stratospheric warmings in particular are due to these waves. Deviations from the regular restructuring of the mid-latitude circulation system (especially in the mesosphere) and the observed asymmetry in the middle-atmosphere circulation pattern relative to the poles during winter (in contrast to the symmetric pattern that occurs during summer) are also associated with these planetary waves.

Winds due to tidal oscillations are superposed on this wind system. Short-period variations-acoustic waves and internal gravity waves (caused by perturbations associated with the restructuring of meteorological processes, instabilities of various types, such as wind shear, heating in the auroral regions, etc.) — are also active in these same regions of the middle atmosphere. These waves have both vertical and horizontal (zonal and meridional) components. The periods of these acoustic-gravity waves range from several milliseconds to several hours, with wavelengths ranging from a few to hundreds of kilometers. Energy dissipation due to internal gravity wave modes makes a significant contribution to the thermal balance of the upper mesosphere and lower thermosphere.

Like the middle atmosphere, the thermosphere is characterized by a broad dynamic spectrum of motions, from large-scale wind systems to wave processes. The motions in the equatorial and mid-latitude thermosphere are largely determined by variations in the flux of EUV radiation, but are also strongly affected by the high-latitude regions most influenced by the magnetosphere-ionosphere interaction. The high-latitude neutral-gas circulation can also be stimulated by ions undergoing ionospheric convection, the effectiveness of which depends to a great extent on the amount of heating that occurs as energetic charged particles penetrate the auroral zones. The release of thermal energy in these more-or-less regularly occurring processes does make some contribution to the overall thermospheric circulation system. And since the redistribution of energy in these local heating processes is global in nature, the thermosphere, which as a whole has a stable structure and an effective energy dissipation mechanism in the form of viscosity and thermal conductivity, is essentially in a state of constant dynamic restructuring.

The system of zonal and meridional winds in the thermosphere at altitudes above approximately 100 km is comparatively stable. These winds, caused by density gradients due to diurnal and seasonal variations in solar energy flux, lead to meridional flow of air masses from the summer to the winter hemisphere. The stable nature of this circulation is strongly affected by increases in geomagnetic activity, leading to the formation of an additional system of large-scale eddies that may reach the equator after their formation in the winter auroral zone. One distinguishing feature of the zonal wind is that the 'evening' winds (from west to east) and 'morning' winds (from east to west) have different velocities. The terms 'morning' and 'evening' winds refer to the time periods 1800–2400 LT and 0400–1200 LT, respectively. As a result, the atmosphere at an altitude range of 200–400 km at latitudes of $30° - 40°$ rotates with velocity $V \approx 100$ m/s, which on the average is above the Earth's intrinsic rotation.

The wind system at high latitudes, on the other hand, is primarily directed poleward and to the west during the night, switching to equatorward and to the east in the morning. This wind system depends strongly on the level of geomagnetic disturbance. The morphology of the wind systems and waves excited during disturbances initiated by the full set of electrodynamic interactions is unusually complex. The global 'response' of the thermosphere to heating in the auroral regions largely involves changes in the dynamic regime of the atmosphere.

The dynamics of the thermosphere, under the influence of the ionospheric plasma, has a strong effect on the behavior of the ionosphere, which is the determining factor below ~ 150 km; above this level, the motion of the electrons and ions is largely controlled by the electrical and magnetic fields. At middle and low latitudes, the electrical fields are generated by winds associated with thermal tides, while the electrical fields in the high-latitude atmosphere are largely generated by interactions between the magnetosphere and ionosphere.

Magnetospheric-ionospheric interactions. The pattern of reletionships between the Earth's magnetosphere, ionosphere, and neutral upper atmosphere is complex and we only briefly emphasise here some important points relating to coupling of the processes in the upper atmosphere and circumplanetary space.

A general picture looks as follows. The solar wind deforms the geomagnetic field at distances on the order of 10 Earth radii (R_E) and brings lines of force originating in the polar regions over to the nightside. These lines of force and associated plasma form a magnetotail or plume behind the Earth consisting of two branches, a northen branch and a southern branch, separated by a neutral sheet with a maximum field strength of 1 nT (10^{-5} oersted). The magnetotail extends outward over a distance of at least 1000 R_E. A shock wave is formed as the supersonic solar plasma flows onto the magnetosphere, the characteristics of the flow changing radically with flow passage. The region between the shock wave and the boundary of the magnetosphere contains a turbulent plasma layer with a highly fluctuating magnetic field. The thickness of this layer — the magnetosheath — increases with distance from the frontal point, but is almost independent of the level of geomagnetic disturbance. Near the frontal point, the thickness of the transition zone averages 2–4 R_E. Immediately behind this region is the magnetopause (which is characterized by diffusion and large pressure gradients) and the polar cusp regions, which are constantly occupied by high-velocity plasmoids and very strong electric fields. This leads to constant replenishment of the magnetosphere with solar-wind particles. The neutral sheet, a plasma sheet containing plasma as hot as the solar wind, plays an important role in this process.

The topology of the magnetosphere and the position of its outer boundary on the side facing the Sun (the dayside magnetopause) are determined by the dynamic pressure of the solar wind flow and the orientation of the interplanetary magnetic field. When the flow of solar plasma is decelerated, the magnetosphere is compressed, the magnetopause moves closer to the Earth, and the strength of the magnetic field increases, marking the initial phase of a magnetic substorm. The electric current systems formed at the boundary of the magnetosphere cause the lines of force to deform. The deformation of the outer magnetosphere and the fact that some of the lines of force are swept back to the nightside cause the magnetosphere to be asymmetric. The magnetopause can be likened to a pulsating membrane through which some of the kinetic energy of the solar wind is transferred to the magnetosphere. Instabilities in the magnetopause and shock wave zone convert some fraction of this energy into energy of plasma oscillations and electromagnetic radiation radiated from the magnetosphere into interplanetary space with a broad wave spectrum.

The magnetosphere is divided into two main regions filled with plasma having different properties — the inner magnetosphere and the outer magnetosphere. The inner magnetosphere (out to radius $\sim 4R_E$ in the equatorial plane) — the plasmasphere — contains a low-temperature plasma of relatively high density. The plasmasphere is structurally and morphologically related to the ionosphere, and in fact appears to be a continuation of it. The outer magnetosphere is filled with both high-temperature and relatively low-temperature plasma formed by particles trapped by the geomagnetic field. These two plasmas are determined by the initial nature of the charged particles. The hot plasma is injected through the dayside magnetopause and from the magnetotail, whereas the thermal plasma comes from the ionosphere via hydrodynamic light-ion acceleration at high latitudes (the polar wind) and acceleration due to longitudinal electric currents in the auroral regions.

The polar wind, a relatively thin supersonic flow of low-energy protons and helium ions directed upward, is subject to strong spatial and temporal variations. The motion of the thermal plasma in the plasmasphere, as well as energetic protons and electrons in the outer magnetosphere associated with the radiation belts, is largely controlled by large-scale electric fields resulted from magnetospheric convection, which leads to systems of electric currents directed parallel to the magnetic lines of force, with the return current flowing through the polar ionosphere. The thermal ions experience a sharp 'cutoff' near the boundary of the plasmasphere (plasmapause), where the thermal plasma accumulates for long periods of time.

Particles in the magnetosphere move along complex trajectories determined by their rotation about the magnetic line of force, their oscillation between the northern and southern hemispheres, and their drift in longitude. If the angle formed by the velocity vector of the particle and the magnetic field strength (pitch angle) decreases below the critical value and the point at which the particle is reflected (mirror point) descends into the ionosphere, the particle will lose all its energy in collisions with atoms and molecules in the upper atmosphere.

The mean density of charged particles in the magnetosphere is controlled by the balance between sources and losses. Ionospheric particles are an important source in addition to the solar-wind particles (electrons, protons, and heavier ions) transported into the magnetosphere from its boundary by fluctuating electric and magnetic fields. Ionized hydrogen, helium, nitrogen, and oxygen formed in photochemical processes (as well as electrons) enter the magnetosphere from the ionosphere. The plasmasphere (which is trapped by mid-latitude lines of force and lies relatively close to the Earth) is mainly filled with protons formed in the ionosphere by charge-exchange reactions between oxygen ions and atomic hydrogen, as well as He^+ and O^+ ions, a significant fraction being in nonequilibrium state and responsible for the formation of the hot oxygen corona. During magnetic storms, the relative abundance and energy of O^+ ions increase, reaching several tenths of a percent of the total concentration. Another source of particles for the magnetosphere includes protons, helium and hydrogen ions heated in the auroral regions of the ionospheric plasma, where they are accelerated and flows are formed along magnetic field lines.

Magnetospheric loss mechanisms include, first, exchange with the solar plasma through the magnetopause and the ejection of magnetospheric plasma into the magnetotail, and second, scattering of high velocity particles due to cyclotron instability. Characteristic manifestations of the cyclotron instability mechanism accompanying particle precipitation into the upper atmosphere include pulsating aurorae, various types of low-frequency electromagnetic radiation of magnetospheric origin covering a wide frequency range, and the excitation of geomagnetic field pulsations. All of these phenomena are in some way associated with the development of electric fields and current systems in the magnetosphere and ionosphere, leading to energy and mass exchange.

Aurorae are a prominent manifestation of the powerful energy and mass exchange processes and the interaction between current systems in the high-latitude upper atmosphere at high levels of geomagnetic disturbance, with the numerous nonequilibrium aeronomy processes involved. Aurorae are among the most well-known phenomena in the entire chain of effects linking the Sun and Earth. They occur in some form almost continuously within the polar ovals located asymmetrically about the geomagnetic poles in the northern and southern hemispheres at the edges of the polar caps. The closest approach of the polar ovals to the geomagnetic poles occurs on the dayside, while the largest separation occurs on the nightside. Strong diffuse auroral emission is almost always present along the band of the oval, along with discrete aurorae of various form and intensity. Under quiet geomagnetic conditions, the auroral boundaries lie at 78° and 68° geomagnetic latitude. Intense aurorae generally occur on the nightside, but sometimes also occur on the dayside. During intense magnetic storms, bright discrete auroral forms encompass the entire polar oval for 360° around the magnetic pole.

When particles of magnetospheric origin (mainly electrons and protons) precipitate into the atmosphere, this leads to collisional excitation of atmospheric atoms and molecules and the emission of photons in several characteristic lines and bands - especially prominent are various molecular band systems and characteristic lines from oxygen and nitrogen, which give rise to the observed range of colors. The types and intensity of the aurorae depend on the energy spectrum of the electrons and/or protons and the spatial and temporal characteristics of the acceleration process. The complex morphology of this phenomenon is also determined by the structure and evolution of the current systems along the magnetic lines of force, the microstructure of the electric and magnetic fields within individual auroral arcs, and the overall dynamics of the polar atmosphere during the magnetic storm. Dissipation of energy from current systems provides the basic energy input to the high-latitude thermosphere.

2.2.2. VENUS AND MARS

Neutral Atmospheres. Unlike the Earth's upper atmosphere, where oxygen plays a dominant role, processes in the upper atmospheres of Venus and Mars are largely controlled by the photochemistry of carbon dioxide, which, with its dissociation products CO and O, remains the dominant component to altitudes of approximately 200 to 250 km. Above this altitude, helium and hydrogen gradually become the dominant components in the atmospheres of these planets, as in the Earth's atmosphere.

The processes of CO_2 dissociation and ionization are accompanied by the inverse processes of association and recombination under conditions of intense dynamic exchange (turbulent diffusion and circulation). Minor atmospheric constituents, primarily H_2O — and on Venus sulfur compounds — also play an important role in atmospheric chemistry, affecting the altitude distribution of various components and the thermal regime. Photochemistry and atmospheric dynamics explain the predominance of CO_2 to high altitudes in the thermospheres of both planets.

The bulk of the energy of EUV radiation is converted to heat in photolysis reactions, as well as in the thermalization of newly created photoelectrons. However, the association reactions involving CO and O have low efficiency at high altitudes, and only play an important role in the mesosphere and lower thermosphere, where photolysis products are removed most actively. The remainder of the EUV energy is scattered and reradiated by the atmosphere, providing important information on its properties. An analysis of the dayglow in the Martian atmosphere has provided an estimate of the temperature of the exosphere $T_\infty = 350$ K, and emission line data have been used to calculate the profiles of CO_2, CO, and O with altitude, which are in good agreement with mass-spectrometer data. Approximately the same value was obtained for the exospheric temperature from measurements of the characteristics of the upper-atmosphere dayglow on Venus: $T_\infty = 375 \pm 100$ K. The value of $T_\infty \simeq 400$ K on the dayside was confirmed by measurements of the drag on the Pioneer-Venus probe. The temperature on the nightside proved to be 100 K — much lower even than the temperature of the mesopause in the Earth's atmosphere.

The fact that the mean exospheric temperatures of Venus and Mars are much lower than that of the Earth can be explained by the large net radiation fluxes of energy in the infrared bands of CO_2. This also the reason why the base of the exosphere is approximately 200 km lower for these planets than for the Earth's atmosphere. As for the unusually low temperature of the nightside upper atmosphere (cryosphere) of Venus, it can probably be explained by turbulent heat conduction, in addition to the radiation.

Ionospheres. Venus and Mars have ionospheres, but they are less dense and closer to the surface than the Earth's.

One substantial difference between the ionospheres of Mars and Earth is that the former does not have the F_2 maximum formed in the Earth's ionosphere due to O^+ ions as a result of the relationships between the processes of ionization, recombination, and diffusion. This difference is apparently due to the fact that the charge-exchange reaction between O^+ and CO_2

is faster than that between O^+ and N_2, which continues to be the dominant atmospheric constituent in the Earth's atmosphere to relatively high altitudes. This prevents the accumulation of O^+ ions on Mars at altitudes less than approximately 200 km. The maximum electron density N_e in the dayside Martian ionosphere occurs at altitudes $h = 135$–140 km and is $N_e \approx 2 \times 10^5$ cm^{-3}, i.e., an order of magnitude lower than the density in the F_2 layer of the Earth's ionosphere. Another, less well-defined maximum (small inflection point) in the $N_e(h)$ profile occurs at an altitude of approximately 110 km, with an electron density of $N_e \approx 7 \times 10^4$ cm^{-3}. On the nightside, the peak N_e moves down to an altitude of 100–110 km, with a maximum density of $N_e \approx 5 \times 10^3$ cm^{-3}.

The main constituent of the Martian ionosphere is ion O_2^+, which is mainly formed in the charge-exchange reaction between CO_2^+ and O. A model of the photochemical equilibrium based on this reaction and several associated reactions is consistent, to the first approximation, with the measured distributions of various components up to an altitude of ~ 2000 km.

Above this level, the relative concentration of the lighter ions increases. The lower ionosphere of Venus also consists mainly of ions O_2^+, but with a significant quantity of O^+ as a secondary constituent. CO^+ and CO_2^+ ions are also present in significant quantities. O^+ ions dominate above 160 km on the nightside and 200 km on the dayside, except for during the predawn hours, when concentration of H^+ ions is approximately the same as that of O^+ ions. The dayside maximum (with density $(2 \div 5)10^5$ cm^{-3}) occurs at 140 km, while a sharp decrease in electron density is observed at the 250–300 km level. The ionopause — the boundary between the thermal ions in the ionosphere and streams of energetic particles from the solar wind — is located at approximately this same level (with an altitude ranging from 290 km on the nightside to 1000 km at the terminator). On the nightside, an extended region of moderate electron density $\sim 1 \times 10^3$ cm^{-3} (up to over 3000 km in altitude) forms. This region contains several local maxima, which have density 5-10 times higher, near 150 km altitude. The ion density and composition in the lower ionosphere are subject to substantial variations.

The dayside ionosphere of Venus generally satisfies the condition of photochemical equilibrium, i.e., is largely controlled by local processes involving the formation and destruction of ions, like the ionosphere of Mars and the E layer in the Earth's ionosphere. The situation on the nightside is more complex. The narrow maximum in N_e that forms at the same (or perhaps slightly higher) altitude as the maximum in the dayside ionosphere is probably due to processes involving the transport of ions from the dayside and bears a closer resemblance to the F_2 layer of the Earth's ionosphere in morphology and nature. This horizontal transport of ions to the nightside hemisphere is presumably due to the diurnal pressure gradient of the ionospheric plasma, which does not undergo any appreciable drag because the planet has no intrinsic magnetic field. Precipitation of energetic electrons due to electromagnetic interactions also contributes to the ionization of the nightside thermosphere, and is probably responsible for the secondary peak in N_e below the main peak.

One of the interesting characteristics of Venus ionosphere is the extremely high electron ($T_e \approx 5000$ K) and ion ($T_i \approx 1000$ K) temperatures, which are both much higher than the temperature of the neutrals. This is indicative of the low efficiency of thermal relaxation processes (in contrast to Earth, where there are no substantial differences between temperatures of electrons and ions and of neutrals up to an altitude of ≈ 500 km), and therefore, a persistance of nonequilibrium state of the ionospheric plasma. One should also note that these high values of T_e and T_i are maintained on the nightside, against the background of a neutral cryospheric temperature of ≈ 100 K, which confirms the high efficiency of horizontal transport and ionization processes on the nightside.

Circumplanetary space. The structure and thermal state of the upper atmospheres and ionospheres of Venus and Mars are closely linked with the processes resulted from the interactions between the planet and the solar wind.

The solar wind flow around Mars does not form anything analogous to the magnetosphere of Earth. The configuration of circumplanetary space around Mars is most similar to that observed for Venus, which implies that Mars has no intrinsic magnetic field (except for remnant magnetism in the crust). Although the problem of magnetic field of Mars did not yet meet an unambiguous solution, a number of features actually do support a gas-dynamic model for the solar plasma flow and direct exchange of mass between the solar plasma and the upper atmosphere in which the intrinsic magnetic field does not play a critical role. These features are as follows: the properties of the detached shock wave, whose position does not show any significant deviation from a mean altitude of ~ 1500 km, even when there are variations in solar-wind pressure (in contrast to the model magnetosphere); formation of a boundary layer (where a decrease in flow velocity and drops in electron and ion temperature are observed) on the dayside and nightside of the planet; and evidence suggesting the presence of streams of O^+ ions in the boundary region, moving away from the planet. Nevertheless, further research is needed on the origin of the structure of circumplanetary space around Mars, because the patterns observed are not completely congruent with solar plasma interaction with a non-magnetized body.

The topology of space in the neighborhood of Venus is completely determined by the direct interaction of the solar plasma and the ionosphere. The most characteristic feature of this interaction is the formation of a transition region — the ionopause — which forms on the dayside of the planet in a region behind the bow shock, at altitudes above ~ 300–500 km. Here the pressure of the solar wind is in approximate equilibrium with the kinetic pressure of atmospheric gas and the pressure of the magnetic field induced by the currents in the ionosphere. In addition to the formation of the transitional zone identified with the ionopause, the processes in the solar plasma flow area around Venus also include successive heating and (partial) thermalization of ions largely being in noneqilibrium state; the formation of a region of reduced density behind the shock wave (i.e., the ionosheath, roughly analogous to the magnetosheath in Earth's magnetosphere); the important role played by the dynamics of the ionosphere and the electric fields within the ionosphere; and the formation of an 'umbra' zone of reduced density behind the planet.

This model for the solar plasma flow around Venus is consistent with ideas concerning the important role played by ionospheric convection and the interaction between the ionospheric and interplanetary magnetic fields. In this model, the pressure in the dayside ionosphere is much lower than the thermal plasma pressure, and can be comparable to it only near the terminator, where it presumably has the greatest effect on the morphology of movement. The strongest magnetic fields are expected in the nightside ionosphere, primarily in the vicinity of the large-scale holes reflecting the interaction between the zone of reduced density (magnetotail) and the ionosphere. All of these processes taken together reflect the specific nature of the interaction between Venus and the solar wind, which is fundamentally different from that observed for Earth or any other planet with a magnetic field. By their configuration these patterns resemble the interaction of solar wind plasma with comets.

2.2.3. MERCURY

Mercury essentially has no atmosphere: the upper limit to the atmospheric pressure is $\sim 2 \times 10^{-14}$ atm (density 10^{-17} g cm^{-3}). This corresponds to conditions in the terrestrial atmosphere at approximately 800 km altitude and places Mercury in the same class as the Moon, which has an extremely rarefied gaseous envelope of similar density and composition. This implies that the exosphere is in direct contact with the surfaces of these bodies and very rarefied atmospheric gases can be mostly found in nonequilibrium state. The solar wind (which supplies protons and α-particles in addition to heavier elements such as Ne, C, O, etc.) obviously plays the dominant role in the creation and maintenance of the Mercurian atmosphere. Continuing evolution of volatiles (primarily helium from the radioactive decay of uranium and thorium) from the planetary crust may also make some contribution. Charged particle streams due to acceleration in the magnetosphere also play an important role. Because of the large difference between the dayside and nightside temperatures the Mercurian atmosphere is markedly asymmetric.

Mercury possesses a noticeable magnetic field of about 350 nT and its magnetosphere is small compared to that of the Earth. One of its distinguishing features is that the solar plasma approaches quite close to the planet's surface because of the rather weak magnetic field and the lack of an atmosphere. The bow shock (detected from the sharp increase in the magnetic field) is located at ≈ 1.5 R_M from the center of the planet. The boundary of the magnetosphere (i.e., the magnetopause, which carries the electrical currents responsible for sudden changes in the direction of the magnetic field) lies at approximately the same distance. This means that the planet itself occupies a significant fraction of the magnetosphere, so that there is no room for a magnetic trap. Thus, Mercury does not have a region where energetic charged particles are trapped within a regular magnetic field; there are essentially no radiation belts. Second, there are no longitudinal currents in Mercury's magnetosphere due to the lack of an atmosphere, since they cannot be closed by transverse conduction in the ionosphere. Solar-wind plasma is mainly present in the outer magnetosphere and subject to strong boundary fluctuations in the regions where the field is disturbed.

2.2.4. OUTER PLANETS AND SATELLITES

Unlike the small and dramatically evolved inner planets, the outer giant planets (Jupiter, Saturn, Uranus and Neptune) are large, mostly gaseous bodies which preserved essentially unchanged (primitive) since their origin. All of these bodies (except for Uranus) are radiating more energy than they receive from the Sun that dramatically impacts their atmospheric dynamics. Outer parts of the giant planets represent immensely deep reduced atmospheres dominated by hydrogen and helium, other elements forming compounds with hydrogen (methane, ammonia, water, hydrocarbons). Once the height profiles of the hydrocarbons and other minor species were obtained by *Voyager* and *Galileo* measurements, the validity of various chemical schemes proposed were checked through the development of photochemical models. Energetic charged particles precipitated from the magnetosphere tremendously exert on the atmospheric chemistry as well, including formation of complex compounds in clouds with subtle shades of colors. Of special interest is the fact that chemical reactions occurring in the atmospheres of these distant worlds today, involving beutifully colored features (chromophores) in the clouds, possibly resemble the chemistry that occurred in the earliest history of our Solar system.

Jupiter and Saturn offer many apparent similiarities of their atmospheric and specifically aeronomy characteristics, though there are also distinct differences in terms of quantative behavior of the dynamics, energy budget and the plasma processes. Middle atmospheres of both planets appear to be well mixed up to a homopause level at about several nbars, and termosphere and exosphere encompass a huge region above this level. Temperature increases from approximately 140K in the middle atmosphere on the order of 1100 K in the exosphere of Jupiter (variations from ~ 850 to ~ 1300 K) and 600–800 K in the exosphere of Saturn about the same time of day and latitude. The vertical transport of species in the upper atmosphere is governed by the eddy diffusion processes, but they are about two orders of magnitude more efficient on Saturn ($K \sim 10^8$ cm^2 s^{-1}) than on Jupiter. Although the reason of this descrepancy is not completely clear, it is thought it could result from a possible stratospheric/mesospheric thermal structure or more efficient turbulence in the Saturnian atmosphere. In turn, the higher exospheric temperature on Jupiter can be explained by the direct supply of sufficient energy by auroral particles precipitation in the thermosphere, though this source is not efficient on Saturn. Both Joule heating caused by the departure of the magnetospheric plasma from corotation with the planet like in the case of Jupiter, and inertia-gravity wave mechanisms were proposed for accounting for the observed exospheric temperature of Saturn.

The maximum electronic density in the ionosphere of Jupiter at the heights between 500 and 1000 km exceeds 10^5 cm^{-3} and about an order magnitude less in the ionosphere of Saturn. The electronic temperature generally agrees with that of neutral species in the same altitude range. In the Saturnian ionospheric chemistry significant role probably plays an efficient loss mechanism for topside protons by collisions with vibrationally excited molecular hydrogen, along with vertical ion drifts. The Jovian upper atmosphere is involved in the complicated processes of magnetospheric-ionospheric interactions. The most dramatic interactions occur between the Jovian upper atmosphere and Io (closest to Jupiter Galilean moon) torus which is assumed to be partially responsible for the Jovian aurorae. The energy powering these aurorae amounts to $(0.3 - 1.0)\ 10^{14}$ W, compared to 10^{11} W for the Earth's aurorae. This energy is assumed to come mostly from electrons and protons beamed down the magnetosphere, but also involves heavier ions of oxygen and sulfur.

The system of Saturnian moons behaves as a bandpass filter when moving in the magnetosphere. The icy satellites and the largest moon Titan are significant internal sources of plasma, which forms a plasma sheet resembling the Earth's plasmasphere but only equatorially confined. It is loaded with protons and heavier ions, such as N^+, N_2^+ or H_2CN^+ from Titan's exosphere. However, the mass-loading rate in this Saturnian plasma sheet is about two orders of magnitude less than the source of mass load by volcanically active Io and therefore a plasma-dominated Jovian-like magnetodisk dominated by sulfur and oxygen from Io does not form.

The beautiful aquamarine colors of Uranus and Neptune are caused by the absorption of red light by the methane gas in the atmospheres of these planets. Sunlight reflected by the planets is thus deficient in red rays, but the short wavelengths are scattered strongly, just as in the Earth's atmosphere. Middle atmospheres of Uranus are dominated by the photochemistry of CH_4 and its by-products C_2H_2 and C_2H_6. The products of methane photochemistry condense to form several discrete layers of aerosols as they diffuse downward in stratosphere, being possibly undergone by additional ultraviolet-induced polymerization followed by charged particle irradiation.

The composition and pressure levels of the major aerosol layers on Neptune are found to be generally similar to those in neighboring Uranus, despite the fact that Neptune's clouds and hazes display considerably more temporal and spatial variability and dynamical activity than observed on Uranus. Aerosol heating due to the absorption of sunlight is an important source of power warming and thus surprisingly warm stratospheric temperatures (~ 150 K) on these distant bodies. However, hydrocarbons, abundant in the middle atmospheres of both planets, are substantially absent from the upper atmosphere of Uranus, what is also contrasted to the upper atmospheres of Jupiter and Saturn. On the other hand, similar to these giants, trace disequilibrium species in the high stratosphere of Neptune support the idea about ammonium and/or hydrosulfide ammonium clouds deeper in its atmosphere.

The deficiency of hydrocarbons in the upper atmospheric regions of Uranus, in contrast to very dynamic atmosphere of Neptune with highly variable bright 'patchy' clouds and hazes well above the main cloud layer, can be explained by very slow upward diffusion of methane into the Uranian upper atmosphere, to supply the photochemical reactions that produce hydrocarbons and thus to keep up with their downward removal by condensation and following on sedimentation. Consequently, one may assume that, unlike Neptune with an array of thermodynamic and interrelated chemical processes with CH_4 involvement, the upper atmosphere of Uranus is dominated by the hydrogen photochemistry, what is confirmed by the fact that the H column abundance is consistent with photochemical production driven by solar ultraviolet. Energy deposition precipitation of high energy electrons from the magnetosphere can be the most efficient source responsible for the unexpectedly high exospheric temperature ($T_\infty \simeq 750$ K). It corresponds to the total heating rate 0.4 erg/(cm^2 s), several orders of magnitude larger than EUV [10^{-3} erg/(cm^2 s)] and provisional auroral [2×10^{-2} erg/(cm^2 s)] inputs. These energetic electrons would result in the nonequilibrium processes responsible for atmospheric particles excitation and an excess ultraviolet luminosity observed.

The most striking characteristic of both Uranus and Neptune is the orientation of their magnetic dipoles which are strongly deviated from the axes of rotation and offset from the center of figure. This creates peculiar magnetospheric configurations. Because the rotational axis of Uranus lies nearly in the plane of its orbit (such that the angle between the two axes is an astonishing 55°) this has the potential effect of causing the planet's auroral zone to occur near its equator, rather than at the pole. The shape of its plasma sheet in the magnetotail is generally similar to a geometric model of plasma sheet of the Earth and hence electrons are assumed to diffuse radially inwards from a source in the magnetotail.

The large icy satellites of the outer planets (first of all Io, Europa, Titan and Triton, as well as Pluto), which are referred to as closely related objects, much like the inner planets, are of special interest for the aeronomy. The most intriguing is Titan having a dense mostly nitrogen atmosphere that is assumed to be in the process of evolution from a primitive hydrogen-rich state. Possibly the most important point is that the study of Titan allows us to investigate an evolving, oxygen-poor atmosphere in which complex chemical reactions are occurring today that may resemble some of the reactions responsible for the first steps along the path from chemistry to biology on the early Earth. In turn, Pluto and Triton, satellites of Neptune, have many similarities and are considered to be closely related objects of essentially the same origin, having extremely tenuous atmospheres. The methane atmosphere of Pluto may periodically collapse depending on the incident solar radiation. The surface pressure of the Triton's mostly nitrogen atmosphere is only 16 x 10^{-6} bar and temperature is only 37 K. No other satellites have atmospheres except for a trace gas envelope for Io and Europa.

Voyager-2 could not obtain spectra of Triton, but ground-based observations show a number of as-yet-unidentified absorption features that must be caused by solid forms of various photochemical reaction products. They may be caused by dark material evidenced by isolated patches being blown by local winds. Some of these patches are the end points of vertical columns that are carrying this dark material up into the atmosphere. Regardless the origin of these columns (dust devils; low-temperature volcanism; thermal chimneys punching through an inversion layer) they provide a useful illustration of energetic reactions at very low temperatures.

Io is distinguished by the enormous contemporary volcanism that is the reason of the atmospheric maintenence, as well as the plasma torus formation along its orbit and the ions supply to the Jovian magnetosphere. According to the *Voyager* masurements, gaseous SO_2 is perhaps the dominant constituent of Io's atmosphere, being in equilibrium with frost SO_2 on the surface at $T = 130$ K. The SO_2 atmosphere surface pressure is estimated 1×10^{-7} bar, what corresponds to the number density 5×10^{12} cm^{-3}. Upper limits of common volatiles H_2O, CO_2, NH_3, and CH_4 are exceedingly small, while photochemistry of SO_2 could lead to O_2 as the major gas on the night side. SO_2^+, S^+, S_2^+, O^+, and O_2^+ are the main ion of Io plasma torus, though its inner part contains also sodium ions. Reasonable processes for loss to the torus include thermal escape, sputtering, and sweep-up of ions. Material may also escape from Io as ions directly into the plasma torus by 'magnetospheric pickup'. As it was mentioned earlier heavy ions projected from the torus along the Jovian magnetic field lines and accelerated by strong electric fields are assumed to contribute to the auroral activity in the polar regions of Jupiter.

The *Hubble Space Telescope* detected the spectral signature of molecular oxygen on Europa in UV spectra. It was identified as an extremely tenuous atmosphere of O_2, having a surface pressure $\sim 10^{-11}$ bar. The origin of this atmosphere is associated with water ice surface exposure to sunlight and dust impacts, the oxygen being the fragment of water molecules. Thus the aeronomy of Europa might resemble some reactions in the oxygen-rich upper atmosphere of the Earth. On the other hand, upper limits for oxygen (possibly diurnally varying) atmospheres of other Galilean satellites, also resulting from photolysis of water vapor, was established by *Voyager* at 10^{-9} bar for Ganymede and a higher value for Callisto.

We already emphasized the point that a really unique body in the outer Solar system is Titan. Unlike Jupiter and Saturn, Titan was certainly not expected to retain large amounts of hydrogen and helium, as these gases can easily escape from the satellite's relatively weak gravitational field. The generally lower temperatures in the Saturn system compared with Jupiter may explain, however, why Titan has an atmosphere while Ganymede and Callisto do not, although the similar densities of all three satellites suggest similar bulk compositions. One may also expect substances such as methane, ammonia, nitrogen, carbon monoxide, and argon to have been more prevalent in the materials that formed the more distant satellites around a mother planet which generated lower heat during its formation. Indeed, laboratory studies of the ability of water ice to trap and retain gases as it forms have demonstrated that this ability rapidly diminishes with increasing temperature. In particular, at temperatures above 135 K, gases that have either been trapped in the crystal structure of ice or merely absorbed on icy surfaces easily leave the ice.

In other words, even if the ice that now composes Ganymede and Callisto was formed at lower temperatures, once it had been heated above this temperature, its gas content would have been markedly depleted. This is apparently what happened to the materials that formed these satellites before their incorporation: they were in effect 'baked out'. In the cooler environment at Titan's distance from Saturn, however, temperatures evidently remained well below 135 K and the gases were not released until the satellite formed. In turn, heat liberated during the accretion of Titan would have been more than sufficient to liberate gases from the infalling ices. It could also have driven reactions among them, forming new species. These gases then became the satellite's earliest atmosphere. Additional heat was generated by the energy released from the decay of short-lived radioactive elements in the rocky component of Titan. Just as the inner rocky planets melted and differentiated, this distant satellite must have formed a core of dense (rocky) material that became surrounded by a mantle of ice. Thus the formation of a secondary, outgassed atmosphere on Titan followed the same general pathway that produced the early atmospheres on Mars, Venus, and Earth.

The atmosphere of Titan contains about 10 times as much nitrogen as our planet, giving a value for the surface pressure of 1.5 times the sea-level pressure on Earth. Among minor constituents there is a few percent of methane and a rich variety of other compounds, principally hydrocarbons such as ethane (C_2H_6) and acetylene (C_2H_2) and nitriles such as hydrogen cyanide (HCN). A tiny amount of carbon dioxide is present, as is carbon monoxide. Given the presence of CO from the beginning, the formation of a small amount of CO_2 is understandable through OH intermediate formation from CH_4, without the presence of water vapor. Present uncertainties in the determination of the mean molecular weight of the atmosphere allow as much as 10 to 15 percent of primordial argon ($^{36}Ar + ^{38}Ar$), but this gas may also be present only in tiny quantities.

Let us also note that nitrogen and methane molecules are assumed to be broken apart by cosmic rays, high energy ultraviolet photons from incident sunlight, and electrons bombarding the atmosphere from Saturn's magnetosphere, and the fragments are recombining to make some additional minor constituents. In this process, some hydrogen is produced; the amount of H_2 represents a steady state between production and escape.

At the very low temperature of the Titan's lower atmosphere, the vapor pressure of water is vanishingly small. Water vapor could be supplied to this low temperature atmosphere from the outside, as showers of ice particles formed by meteoritic and cometary impacts with icy satellite surfaces, or directly, when this debris gravitationally captured by the Saturn system is swept up by Titan as it orbits the planet, though it would still be very low. In this case, the small amounts of CO and CO_2 observed could be produced, nonetheless, from photochemical reactions between the resulting water vapor and the already present methane. If, however, one could magically move Titan closer to the Sun, say to the orbit of Mars, the character of the atmosphere would immediately change: the warmer surface temperatures would allow plenty of water vapor to enter the atmosphere, and the resulting supply of oxygen would rapidly convert methane and its byproducts to carbon dioxide, exactly the dominant carbon-carrier we find on Mars. It means that the Titan aeronomy would be essentially similar to what we have on Mars and Venus.

The atmospheric temperature initially decreases with altitude from 94 K at Titan's surface until the tropopause is reached at a height of about 60 km, where is a turn-around, and temperature begins to increase with height, eventually reaching a value of 175 K, some 80 K higher than the surface temperature. This structure is caused primarily by the absorption of ultraviolet light in Titan's upper atmosphere. Absorption occurs not only in the photochemical processes, but also in nondestructive excitation of the resulting molecules, and in the dark material making up the aerosol.

Evidently, some of this airborne material in the Titan's atmosphere that completely hide the surface is simply condensed forms of the atmospheric gases because, with the exception of hydrogen and carbon monoxide, everything will condense at the tropopause temperature. However, the aerosol layers we see are not just simple condensation products, for in that case, they would be not dark but white or gray. The dirty-orange color revealed at the *Voyager* images of Titan suggests that some additional chemistry is occurring, transforming the simple molecules into more complex substances. Some of these are probably polymers. Both hydrogen cyanide and acetylene are known to form dark polymers that could certainly contribute to the observed effects. Indeed, laboratory experiments starting with the principal ingredients of Titan's atmosphere and using a variety of energy sources to drive the reactions have little difficulty in producing dark organic material that may well offer a good analogue to Titan's aerosols. This aerosol formation responsible for 'ultraviolet haze' high in the Titan's atmosphere is a specific feature of the aeronomy processes on this distant body.

2.3. Cometary Atmospheres

Comets are asteroid-like icy bodies, usually of several to tens km across at highly-eccentricic orbits around the Sun. The main population is located in the Oort cloud (at $\sim 10^4$–10^5 a.u.) and Kuiper belt (beyond the orbit of Neptune up to $\sim 10^3$ a.u.). The main characteristics of several repre-

TABLE 2.3. Physical characteristics of well-studied comets.

Comet	Effective diameter D_{eff}, km	Albedo geomet. P_v	Rotation period, hr	H_2O production rate Q, s^{-1}	Perehilion distance r, AU
Halley	12.0	0.04	2.84	6×10^{29}	0.8
Arend-Rigout	10.4	0.03	13.5	2×10^{26}	1.58
Neujmin	20.8	0.02	12.7	2×10^{26}	1.68
Scwassmann-Wachmann	30.8/ 17.2	0.04/ 0.13	14.0/ 32.3		
Tempel 2	11.2	0.02	8.9	2×10^{27}	1.71
Encke	< 4.4/2?		22.4?	6×10^{28}	0.76
IRAS-Araki-Alcock	8.0			2×10^{28}	1.03

The data are summarized according to Weaver *et al.* (1987), where a more complete bibliography on the tabular data is given. $D_{eff} = 2\sqrt{ab}$, where a and b are the projections of the semiaxes at maximum illumination. The quantity Q is based on the vectorial model of Festou and Feldman (1981).

sentative comets are shown in Table 2.3 and the basic ideas about coma formation and its internal structure are schematically depicted in the diagram *Fig. 2.3.*

Comets provide a unique opportunity to investigate the most primitive, primordial matter in the protoplanetary nebula, from which the larger bodies were presumably formed in the process of accumulation. There is also enormous interest in studying the processes of gas dynamics, molecular kinetics, and heat and mass transfer that accompany the formation of a comet's gas-dust atmosphere and its evolution as the comet moves along its heliocentric orbit, as well as plasma effects during loading of the solar wind with cometary ions and the formation of a shock wave.

Ideas about the nature and properties of comets have broadened considerably in recent years, thanks to improved astronomical measurement methods and laboratory research and the space missions to comets Halley and Giacobini-Zinner. The most important results have been obtained dur-

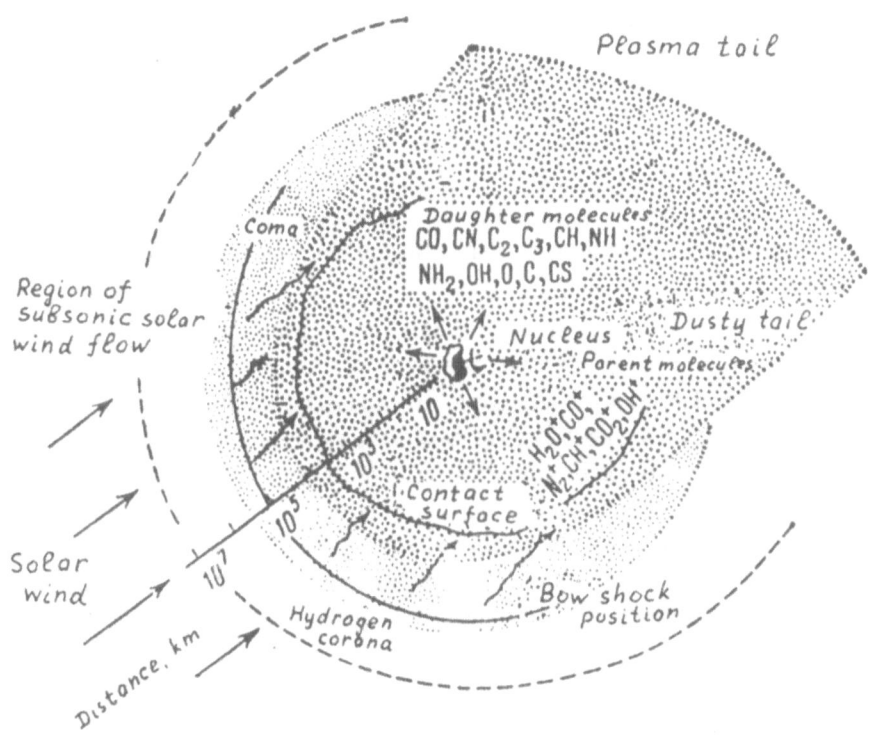

Figure 2.3. Diagram of the formation of a cometary atmosphere and its interaction with the solar wind (according to Marochnik, 1985).

ing the flights of the *Vega* and *Giotto* craft to comet Halley, with the latter also accomplishing an encounter with comet Grigg-Skjellerup in 1992 and measuring a number of its physical properties. It made it possible to improve significantly modeling approaches to answer some principal questions about the physical structure, dynamics, and chemistry of comets, including nonequilibrium processes in transition regions, such as the boundary near-surface layer and inner coma, with the chemical reactions involved.

Basic ideas on the physical structure of a cometary nucleus are based on the model of dirty snow ball that principally fits the results of ground-based observations and measurements from spacecraft. The absorption of solar radiation by a cometary nucleus leads to its heating and sublimation of volatile components, forming the comet's atmosphere (coma) and enormously extended tail. Both have a highly heterogeneous structure that varies with time. A characteristic element of the coma structure is the presence of jets and shells (halos). Their formation is due both to variations in the intensity of sublimation from the surface, screened by the mantle, of the rotating nucleus as the insolation varies, and to variations in outgassing through cracks and pores in the mantle (crust) from subsurface gas reservoirs under excess pressure. Both sources are responsible for the loss of dust, frozen into the ice fraction of the nucleus, entrained by the gas streams. This has also been associated with individual active centers, apparently with sizes of several hundred meters, possibly consisting of sections of dust-ice conglomerates with different degrees of exposure, porosity, or cracking.

The density of the escaping gas decreases in approximately inverse proportion to the square of the distance from the surface of the nucleus and is highest along the axis of the cone. Because the outflow occurs not into a vacuum but into a surrounding atmosphere, so-called Mach disks are formed at the bow shock in the transition from supersonic to subsonic velocity, and configurations resembling 'barrels' develop at the jet boundaries. These effects undergo certain changes when dust is present in the stream, due to the interaction of dust particles with gas molecules (frictional and heat-transfer processes). This considerably complicates the theoretical description of such a heterogeneous medium.

Chemical composition of the cometary atmosphere includes both parent molecules (H_2O, CO_2, HCN, CO, NH_3,...) and daughter molecules and radicals (H, OH, CH, NH, CN, C_n,...) produced under the influence of the solar EUV radiation through the processes of dissociation, ionization, and complex of chemical reactions in the inner and outer coma. Data of spectroscopic measurements provide a basis for the revealing of the abundances of various components, as well as the relationships between parent and daughter products during sublimation of the nucleus, with allowance for photolysis and subsequent chemical reactions, the efficiency of which varies with distance from the nucleus. The problems of chemical composition will be specially addressed in the Chapter 11 dealing with the cometary atmospheres.

It is evident that the aeronomy of the rarefied gas coma involving multiple species and their efficient transformations by the incident solar radiation, as well as nonequilibrium states of the reaction products, is extremely complicated. Its complexity dramatically increases due to the very close relationship of the coma formation and evolution with the processes within and on the surface of nucleus responsible for the outgasing of volatiles accompanied by production of dust component. Irregularities of sublimation and some specific properties of ejections are related to a certain heterogeneity in the physical structure of the nucleus and its chemical composition, as well as to nonuniform surface patterns and the presence of individual inclusions.

During the spacecraft encounters with comet Halley, gas-dust ejections were observed both in the continuum and in individual emission lines and they appear clearly in images sent by *Giotto*. Jets are usually some 500 m in diameter, but a fine structure is detected within them, which may be due either to the aforementioned heterogeneity in the topography of the active regions from which the gas and dust are emitted or to hydrodynamic interactions between the escaping gases and inactive regions located within them. It was found that large amounts of dust are being lost and the detected submicron particles have low densities, in the 0.1–1.0 $g\,cm^{-3}$ range i.e., they have a highly porous structure consisting of even smaller particles (or giant molecules?). The cumulative mass distribution $n(m)$ of numbers of particles in the $m > 10^{-10}$ g range corresponds to an m^{-s} power law, with $s \sim 1$, whereas $s \ll 1$ for particles near 10^{-17}–10^{-16} g. In other words, the particle mass distribution function rises with decreasing particle mass up to a cutoff threshold at the smallest detected masses 10^{-17} g, which correspond to a particle size of hundredths of a micron.

Besides dust particles, ice crystals (water clusters) of various sizes may be formed in the comet's expanding atmosphere due to recondensation of water vapor. This process has been estimated to involve perhaps up to 15% of the H_2O in the inner coma at distances within $\sim$ 500 radii of the nucleus, after which such particles disappear. Recondensation should result in some increase in the stream velocity in comparison with the model of a uniformly expanding gas. Attempts have been made to find indirect confirmation of this phenomenon by comparing with mass-spectrometric measurements on *Giotto*, although direct identifications of clusters of this kind are hindered mainly by the masking influence of dust in the continuum at all wavelengths.

Many efforts were paid to study the behavior of a gas-dust mixture in the region near the nucleus using methods of hydrodynamics and heterogeneous mechanics. It has been shown that the interaction between phases causes a number of qualitatively new effects that markedly influence the dynamics and energetics of the carrier and disperse phases. The radial distributions of the main parameters of the carrier and disperse phases calculated on this basis made possible a qualitative estimate of the contribution of dissipative processes that disrupt the adiabaticity of flow and also lead to changes in the parameters of the gas stream. A relative contribution of gas heating due to heat transfer from hot dust and chemical energy sources stipulated mainly by H_2O photodissociation and following on reactions, as well as the patterns of hydrodynamic velocity, stream density, and temperature distribution in the inner coma, were evaluated. While this approach ensures a general agreement with the observational data, it is considered as only a rather coarse approximation. A more complete study of evolution of the multicomponent gas of the inner coma in the solar radiation field with the involvement of nonequilibrium processes requires a more detailed analysis of the kinetics of sublimation and photolytic and elastic/inelastic collisions in the cometary atmosphere. The results of such analysis will be discussed in detail in Chapter 11.

In the conclusion of this review chapter, we suggest that even a brief survey of the contemporary knowledge of the planetary atmospheres would require to refer to enormous number of original publications. For this reason, because of space limitations and also the difficulty of objective citation of works of historic or primary scientific interest, we preferred to limit the list of references mainly to monographs and outstanding reviews. In these, the demanding reader interested in a deeper understanding of the discussed topics and the problems involved may find out more complete bibliography.

General references to Chapter 2.

Beaty, J.K., and Chaikin, A. (eds): 1990, *The New Solar System*, 3rd Edition, Sky Publ. Co., Cambridge Univ. Press.

Beebe, R.: 1994, *Jupiter. The Giant Planet*, Smithsonian Institution Press.

Chamberlain, J. W. and Hunten, D.: 1987, *Theory of Planetary Atmospheres. An Introduction to Their Physics and Chemistry*, 2-nd edition, Academic Press.

Comets (L.L. Wikeming ed.) : 1982, The University of Arizona Press, Tucson, AZ.

Gombosi, T. I., Nagy, A. F., and Cravens, T. F.: 1986, 'Dust and neutral gas modeling of the inner atmospheres of comets', *Rev. Geophys. Space Phys.* **24**, 667.

Herbert, R. : 1978, *Introduction to the Atmosphere*, 3-rd edition, McGraw-Hill Book Company, NY.

Houghton, J. T. : 1986, *The Physics of Atmospheres*, 2nd edition, Cambridge University Press.

Huebner, W. F. (ed.) : 1990, *Physics and Chemistry of Comets*, (Astronomy and Astrophysics Library), Springer-Verlag.

Kelley, M.C.: 1986, *The Earth's Ionosphere. Plasma Physics and Electrodynamics*, Academic Press, 1986.

Lewis, J. S.: 1995, *Physics and Chemistry of the Solar System*, Academic Press.

Lewis, J. S. and Prinn, R. G.: 1984, *Planets and Their Satellites. Origin and Evolution*, Academic Press, 1984.

Marochnik, L.: 1985, *Meeting with a comet*, Nauka, Moscow.

Marov, M. Ya.: 1986, *Planets of the Solar System*, second edition, Nauka, Moscow.

Marov, M. Ya. : 1994, 'Physical Properties and Models of Comets', *Astron. Herald (Solar System Research)*, **28**, 301.

Marov, M. Ya. and Kolesnichenko, A. V.: 1987, *Introduction to the Planetary Aeronomy*, Nauka, Moscow.

Marov, M. Ya. : 1994, 'Inner Planets', in *Space Biology and Medicine*, 2nd edition, NASA.

Marov, M. Ya., and Grinspoon, D.: 1997, *The Planet Venus*, Yale University Press, New Haven.

Mars (H.H. Kieffer et al. eds.): 1992, The University of Arizona Press, Tucson, AZ.

Morrison, D., and Owen, T.: 1988, *The Planetary System*, Addison Wesley.

Neptune and Triton (D.Cruikshank ed.): 1994, The University of Arizona Press, Tucson, AZ.

Owen, T.: 1994, 'Outer Planets', in *Space Biology and Medicine*, 2nd edition, NASA.

Origin and Evolution of Planetary and Satellite Atmospheres (S. K. Atreya et al. eds.): 1986, The University of Arizona Press, Tucson, AZ.

Parks, G.K.: 1991, *Physics of Space Plasmas. An Introduction*, Addison Wesley.

Rodgers, J. H.: 1995, *The Giant Planet Jupiter. Practical Astronomy Handbooks 6*, Cambridge University Press.

Satellites (J.A. Burns and M.S. Matthews eds.): 1986, The University of Arizona Press, Tucson, AZ.

Uranus (J.T. Bergstralh, E.D. Miner, and M.S. Matthews eds.): 1991, The University of Arizona Press, Tucson, AZ.

Venus (D.M. Hunten *et al.* eds.): 1983, The University of Arizona Press, Tucson, AZ.

Whipple, F. L.: 1985, *Mystery of Comets*, Smithsonian Institution Press, Washington.

CHAPTER 3

RAREFIED GASES OF PLANETARY AND COMETARY ATMOSPHERES AS A NONEQUILIBRIUM PHYSICAL AND CHEMICAL SYSTEM

3.1. Main Physical-Chemical Processes in the Upper Atmospheres

In the previous chapter we showed that the chemical composition, structure and dynamics of the atmosphere are defined by specific physical character-istics of the planet. Although gaseous envelopes of celestial bodies exhibit many differences, it is not our goal to address the peculiarities involved; in-stead we shall focus on the main physical-chemical processes in the Earth's atmosphere which is referred to as an example of the planetary upper at-mosphere that was most thoroughly studied. It is worth to emphasize that, despite the diversity of atmospheric chemical compositions, several common groups of the collisional processes can be distinguished which are specified by molecular mechanisms of absorption and transformation of electomag-netic and corpuscular solar radiation. It means that the processes described in this section generally well represent other planets and therefore this ap-proach can be also applied to their study.

The upper atmosphere, being partially controlled by the planetary mag-netic field, is one of the main elements of the Earth's space environment, which is involved in the complex system of solar-terrestrial interactions. Radiative transfer and spectral transformation of solar energy when pen-etrating the upper atmosphere, accompanied by the numerous aeronomic processes, compose fundamental problems of space physics.

It follows from multiyear experimental and theoretical studies of the solar-terrestrial interactions that the overall photochemical, plasma and dy-namic processes occurring throughout the thermosphere, ionosphere, and magnetosphere are responsible for the global energy budget, structure and all other properties in the near-Earth space (Akasofu and Chapman, 1972; Rees, 1989). We shall narrow our focus, however, to the aeronomy, namely to the elementary aeronomic processes in which neutral and charged par-ticles are formed and destroyed. The most abundant species of the Earth's upper atmosphere are neutral particles, molecules N_2, O_2 and atoms O, N,

He, and H dominating.

Thermal energy of neutral particles in the upper atmosphere varies within the limits $kT_\alpha \sim 0.01 \div 0.2$ eV, and they can be found in the rotationally, vibrationally, and electronically excited states. Due to the interaction between the rarefied atmospheric gas and EUV-photons with energies $E_{h\nu}$, exceeding the photolysis energy $E_\alpha^{(0)}$ (i.e., dissociation and ionization threshold energy) of the atmospheric particles, the numerous elementary processes occur. The following photochemical reactions are most important in the Earth's upper atmosphere:

$$O_2 + h\nu \to \begin{cases} O(^3P) + O(^3P, {}^1D, {}^1S) \\ O_2^+(X^2\Pi_g, \ldots, B^2\Sigma_g^-) + e_\nu \\ O(^3P, {}^1D, {}^1S) + O^+(^4S) + e_\nu \ , \end{cases} \quad (3.1)$$

$$N_2 + h\nu \to \begin{cases} N(^4S) + N(^4S, {}^2D, {}^2P) \\ N_2^+(X^2\Sigma_g^+, \ldots, B^2\Sigma_u^+) + e_\nu \\ N(^4S, {}^2D) + N^+(^3P) + e_\nu \ , \end{cases} \quad (3.2)$$

$$O + h\nu \to O^+(^4S, {}^2D, {}^2P) + e_\nu \ . \quad (3.3)$$

In the photolytic reactions (3.1) – (3.3) neutral and ionized particles in electronically excited states, as well as photoelectrons which are energetically active relative to the background atmospheric gas, are formed. Similar to these reactions of photolysis, impact processes also take place resulted from high-energy particle fluxes penetrating into the upper atmosphere (photoelectrons, precipitating charged particles of magnetospheric origin, etc.):

$$O_2 + e_\nu \to \begin{cases} O_2(v, a^1\Delta_g, \ldots, \Sigma_{Ry}) + e_\nu' \\ O(^3P) + O(^3P, {}^1D, {}^1S) + e_\nu' \\ O_2^+(X^2\Pi_g, \ldots, B^2\Sigma_g^-) + e_\nu' + e \\ O(^3P, {}^1D, {}^1S) + O^+(^4S) + e_\nu' + e \ , \end{cases} \quad (3.4)$$

$$N_2 + e_\nu \to \begin{cases} N_2(v, A^3\Sigma_u^+, \ldots, \Sigma_{Ry}) + e_\nu' \\ N(^4S) + N(^4S, {}^2D, {}^2P) + e_\nu' \\ N_2^+(X^2\Sigma_g^+, \ldots, C^2\Sigma_u^+) + e_\nu' + e \\ N(^4S, {}^2D) + N^+(^3P) + e_\nu' + e \ , \end{cases} \quad (3.5)$$

$$O + e_\nu \rightarrow \begin{cases} O(^1D, \ldots, \Sigma_{\Delta l, \Delta s}) + e'_\nu \\ O^+(^4S, ^2D, ^2P) + e'_\nu + e \ . \end{cases} \tag{3.6}$$

The photolytic and impact processes are most effective at altitudes of 100–300 km and consequently, the most dense thermal plasma layer ($n_e \sim 10^4 \div 10^6$ cm^{-3}) is formed there.

The ions produced in the reactions (3.1) – (3.6) have Maxwellian distribution with average energy $kT_{a+} = 0.03 \div 0.2$ eV, appropriate to the energy distribution of neutral particles. However, the photodissociation products carry the excess of kinetic energy (because the energy of the absorbed photon usually exceeds the dissociation threshold), which is transferred then into atmospheric heat by to elastic collisions with the dominant atmospheric components.

Free electrons produced in reactions of ionization (with energy $W_e < E_{h\nu} - E_a^{(+)}$ or $W'_e < W_e - E_a^{(+)}$), in contrast to ions, have an arbitrary initial energy distribution, dependent on the energy spectrum of EUV radiation or the energy of precipitating particle flux. By their properties, such electrons with energies $W_e = 10^1 \div 10^2$ eV occupy an intermediate position between the thermal collisional plasma of ionospheric origin ($W_e \leq 1$ eV) and the collisionless magnetospheric plasma ($W_e = 0.1 \div 100$ keV) and are refered to as superthermal electrons.

The mean free path for superthermal electrons can be more or less than characteristic macroscopic scales of near-Earth's space plasma, and their kinetic energy and pitch-angle distribution have a rather complicated shape, usually very different from Maxwellian distribution (Krinberg and Taschilin, 1984; Schunk and Nagy, 1978). However, during a very short period $\sim 10^{-3} \div 10^0$ s a large part of these superthermal electrons lose the initial energy $W = 10^1 \div 10^2$ eV (basically by to inelastic collisions with neutral particles) and they become thermalized, i.e. the equilibrium Maxwellian distribution with average energy $kT_e = 0.03 \div 0.3$ eV sets up (Krinberg, 1978).

Energetically active — superthermal, excited and ionized — particles produced in reactions (3.1) – (3.6) are locally thermalized due to binary collisions with particles of atmospheric gas. The main channels of thermalization of fast photo- and secondary electrons are the inelastic reactions of excitation, ionization and dissociation of the main atmospheric components, as well as the Coulomb interaction with thermal electrons.

In turn, for translationally and/or internal excited atomic and molecular reaction (3.1) – (3.6) products, all possible channels of binary interactions are important, including inelastic processes of excitation and quenching of metastable states, photochemical and ion-molecular reactions, etc., though elastic thermalization has a dominant role.

The ions O_2^+, N_2^+, and O^+ produced by ionization of neutral particles, can be further transformed into ions of other types in the following reactions of atmospheric ion-molecular chemistry (Rees, 1989; Torr and Torr, 1982):

$$O^+ + H \to H^+ + O \ , \tag{3.7}$$

$$O^+(^4S, {}^2D, {}^2P) + \begin{cases} O_2 \to O_2^+ + O \\ N_2 \to \begin{cases} N_2^+ + O \\ NO^+ + N(^4S) \\ O^+(^4S) + N_2 \end{cases} \\ O \to O^+(^4S, {}^2D) + O \ , \end{cases} \tag{3.8}$$

$$O_2^+(X^2\Pi_g, a^4\Pi) + \begin{cases} N_2 \to N_2^+ + O_2 \\ O \to O_2^+ + O \\ NO \to NO^+ + O_2 \\ N(^4S) \to NO^+ + O \ , \end{cases} \tag{3.9}$$

$$N_2^+ + \begin{cases} O_2 \to O_2^+ + N_2 \\ O \to \begin{cases} O^+ + N_2 \\ NO^+ + N(^2D) \end{cases} \\ NO \to NO^+ + N_2 \ , \end{cases} \tag{3.10}$$

$$N^+ + \begin{cases} O_2 \to \begin{cases} O_2^+ + N(^4S) \\ NO^+ + O \end{cases} \\ O \to O^+ + N \ . \end{cases} \tag{3.11}$$

Reactions (3.9) – (3.11) weakly influence the electron concentration and they are important only in the study of the ionic composition of thermal ionospheric plasma, whereas reactions (3.7, 3.8) exert a stronger influence on the structure of the upper atmosphere. The O^+ ion in reaction (3.8) is long-lived species with respect to the process of its transformation into molecular ions, which are rapidly destroyed in the following reactions of dissociative recombination with thermal electrons:

$$O_2^+ + e \to O(^3P) + O(^3P, {}^1D) \ , \tag{3.12}$$

$$N_2^+ + e \to N(^4S) + N(^4S, {}^2D) \ , \tag{3.13}$$

$$NO^+ + e \to N(^4S, {}^2D) + O(^3P) \ . \tag{3.14}$$

The ion-molecular reactions are very important in the analysis of the thermal state of the upper atmosphere, because they are accompanied by significant thermal energy outputs. In turn, the charge exchange reaction (3.7) is responsible for the formation of the plasmasphere, because it results in the production of protons which are experienced by the gravitational force, though 16 times weaker than for the initial ion O^+.

Neutral-neutral chemical reactions and processes of collisional quenching of metastable states play an important role in the energetics and chemistry of the upper atmosphere, namely:

$$O(^1D) + \begin{cases} N_2 \rightarrow O + N_2 \\ O_2 \rightarrow O + O_2 \ , \end{cases} \tag{3.15}$$

$$N(^4S) + \begin{cases} NO \rightarrow N_2 + O \\ O_2 \rightarrow O + NO \ , \end{cases} \tag{3.16}$$

$$N(^2D) + \begin{cases} O \rightarrow N + O \\ O_2 \rightarrow NO + O(^1D) \ , \end{cases} \tag{3.17}$$

$$N_2(A^3\Sigma) + O \rightarrow N_2 + O \ , \tag{3.18}$$

$$O + O + M \rightarrow O_2 + M \ . \tag{3.19}$$

It should be noted that the reaction of atomic oxygen recombination is efficient only in the lower thermosphere and mesosphere.

An important feature of photochemical reactions (3.7) - (3.19) should be mentioned: these reactions are predominantly exothermic, i.e. are accompanied by the release of thermal energy. It means that the reaction products are formed with excess kinetic energy therefore, they possess increased chemical activity and significantly contribute to the energetics of the upper atmosphere.

A considerable influence upon energetics of the upper atmosphere is also exerted by thermal electrons, because their elastic and inelastic interactions with the neutral and charged heavy particles of the upper atmosphere lead to the excitation of internal states of atmospheric particles and to the subsequent radiative cooling of atmospheric gas. For example, an excitation of the fine-structure levels of atomic oxygen may occur (Moffet, 1988)

$$O(^3P_j) + e \rightarrow O(^3P_{j'}) + e \ ,$$

where $J = 0, 1, 2$ are total moment of the oxygen atom in the ^3P state.

Finally, it is necessary to include in the set of aeronomic reactions the Coulomb interactions between charged atmospheric particles; resonant and nonresonant charge exchange processes between ions and neutral particles; and other collisional processes connected with the momentum and energy redistribution in the rarefied gas of the upper atmosphere (Rees, 1989; Krinberg, 1978; Schunk and Nagy, 1980).

Charged particles formed at the ionospheric heights are usually referred to as thermal (or cold) plasma, because the average kinetic energy of particles is considerably smaller than the energy of charged particles in the near-Earth's space plasma. Under action of a pressure gradient, this ionospheric plasma can move upward along geomagnetic field tubes and fill up the plasmasphere. Here the plasma temperature is increased to $kT_e = kT_{\alpha+} = 0.5 \div 1$ eV and the ion concentration amounts to $n_e = 10^2 \div 10^3$ cm^{-3}.

These values considerably differ from those for magnetospheric plasma mainly composed of electrons and protons with concentration $\sim 0.1 \div 10$ cm^{-3} and average energy $\sim 0.1 \div 100$ keV (hot plasma). The kinetic energy distribution of magnetospheric particles is non-Maxwellian, and pitch-angle distributions are usually anisotropic (see, e.g., Krinberg, 1978). Analysis of characteristic spatial-temporary scales in the magnetosphere allows us to consider the magnetospheric plasma as collisionless and hence to use the drift approach for its description (Akasofu and Chapman, 1972).

We conclude that the above discussed complex of aeronomic collisional processes reflects the transformation of solar electromagnetic and corpuscular radiation when penetrating rarefied gas of the Earth's upper atmosphere, accompanied by collisional transformation of the absorbed solar energy into heat. Essentially the same processes occur in the atmospheres of other celestial bodies while particular schemes of elementary aeronomical processes responsible for the absorption and transformation of external energy can be different (Chamberlain and Hunten, 1987; Atreya, 1986). Therefore this principal approach can be applied to other planets of the solar system as well.

3.2. Nonequilibrium Features of a Rarefied Atmospheric Gas

A rarefied, partially ionized, multicomponent gas (weakly ionized plasma) of the upper atmosphere is controled by gravitational and magnetic fields. Its nonequilibrium state is maintained by the incident solar radiation. Solar extreme ultraviolet (EUV), soft X-ray, and corpuscular radiations including fluxes of energetic charged particles induced by the solar wind forcing are the external sources responsible for the excitation, dissociation, and ionization of atmospheric molecules and atoms in the upper atmosphere.

A substantial fraction of atoms, molecules, and their ions are formed as radiatively and chemically active particles in translationally and/or internally excited states. They are produced by EUV photons and by charged particle impact collisions as well as in chemical reactions. In this section we consider the mathematical description of the formation and relaxation in an ambient of these nonthermal particle gas. The relative importance of these two processes define the gas kinetic state.

It is well known that the fundamental distinction between equilibrium and nonequilibrium physical-chemical systems with constant fluxes of mass and energy is their behavior relative to time reversibility. In an equilibrium system each influx is compensated by outflux, i.e. the system is invariant to time reversibility. This symmetry can be destroyed by fluxes through the system, which will disturb this equilibrium state. When being close to balance, the physical-chemical system is in steady-state and disturbances imposed on it decrease with time (Prigogine, 1962). Among all possible system states the stationary one is of especial interest. In this state the thermodynamic properties of the system do not change in time, but can change in space. The intensive properties of this system cannot be continuous on the boundary however, where the exchange of mass and energy between the system and medium can occur. If the system is in a stationary state, the flows of mass and energy are time independent. The equilibrum state is thus defined as the stationary state in which intensive properties of the system are continuous when transition through the system's boundary occurs.

Plasma (rarefied, partially ionized gas) of the upper atmosphere is in a stationary nonequilibrium state and hence, can be addressed as thermodynamically open physical-chemical system. If, however, there are some disturbances (e.g., sporadic changes of the solar activity), the atmospheric plasma will be in a non-stationary state. An example of this system is so-called relaxation plasma, i.e. the plasma arising when an external source influencing a stationary plasma changes or terminates.

The translational degrees of freedom of atmospheric particles, as well as some internal degrees of freedom (rotational, vibrational and electronic excitation) may be considered as separate subsystems of the general thermodynamic system - plasma as a whole. The parameters of these statistical ensembles of particles (subsystems) are characterized by distribution functions (DF) (Prigogine, 1962).

The velocity distribution function characterizes the part of particles of α-species with velocities in interval $(\mathbf{c}, \mathbf{c} + d\mathbf{c})$, located at the given point of space $(\mathbf{r}, \mathbf{r} + d\mathbf{r})$ at the given time t

$$dn_\alpha(\mathbf{c}, \mathbf{r}, t) = F_\alpha(\mathbf{c}, \mathbf{r}, t) \, d\mathbf{c} \, d\mathbf{r} = n_\alpha(\mathbf{r}, t) f_\alpha(\mathbf{c}, \mathbf{r}, t) \, d\mathbf{c} \, d\mathbf{r} \ . \qquad (3.20)$$

Here, for the one-particle distribution function f_α the following normalized expression is used

$$\int F_\alpha(\mathbf{c}, \mathbf{r}, t)\, d\mathbf{c} = n_\alpha(\mathbf{r}, t) \int f_\alpha(\mathbf{c}, \mathbf{r}, t)\, d\mathbf{c} = n_\alpha(\mathbf{r}, t) \ ,$$

where $n_\alpha(\mathbf{r}, t)$ is the local number density.

It is possible to describe the distribution of component α particles by discrete levels of inner excitation by functions of relative population of level i in given space point $\mathbf{r}$ at time t:

$$x_\alpha(\mathbf{i}, \mathbf{r}, t) = n_\alpha(\mathbf{i}, \mathbf{r}, t) \,/\, n_\alpha(\mathbf{r}, t) \tag{3.21}$$

with the following normalized expression

$$\sum_i x_\alpha(\mathbf{i}, \mathbf{r}, t) = 1.$$

Here the designation i of a set of quantum numbers describing the inner excitation of a particle (rotational, vibrational and electronic excitation) is introduced.

If the degrees of freedom of component α particles are considered as the isolated subsystems of the atmospheric plasma, then in each subsystem stationary equilibrium distributions is established (Hirschfelder *et al.*, 1954; Mitchner and Kruger, 1973; Hochstim (Ed.), 1969):

- for translational degrees of freedom it is the Maxwell distribution

$$f_\alpha^{(M)}(\mathbf{c}) = \left\{ \frac{m_\alpha}{2\pi k T_{\alpha,tr}} \right\}^{\frac{3}{2}} \exp \left\{ -\frac{m_\alpha c^2}{2k T_{\alpha,tr}} \right\} \ , \tag{3.22}$$

where k is the Boltzmann constant, and $T_{\alpha,tr}$ - the kinetic temperature;

- for inner degrees of freedom it is the Boltzmann distribution

$$x_\alpha(i) = \frac{n_\alpha(i)}{n_\alpha} = \frac{g_{\alpha,in}(i)}{Q_{\alpha,in}(T_{\alpha,in})} \exp \left\{ -\frac{E_\alpha(i)}{k T_{\alpha,in}} \right\} \ . \tag{3.23}$$

In this distribution $g_{\alpha,in}(i)$ is the statistical weight, $Q_{\alpha,in}$ - the statistical sum, and $T_{\alpha,in}$ - the temperature of the considered level of inner excitation.

Actually, an exchange of energy between particles of different components and various degrees of freedom occurs when they collide, as well as under the action of external factors.

In the general case it is possible to distinguish the following groups of physical-chemical processes in a gas system under consideration:

- collisional

$$\alpha_1(\mathbf{i}_1) + \alpha_2(\mathbf{i}_2) \to \alpha_3(\mathbf{i}_3) + \alpha_4(\mathbf{i}_4) \ , \tag{3.24}$$

including elastic collisions, inelastic collisions of excitation and de–excitation, chemical reactions, etc.;

- radiative

$$\alpha(i) + h\nu \rightarrow \begin{cases} \alpha(i') \\ \alpha_1(i_1) + \alpha_2(i_2) \ , \end{cases} \tag{3.25}$$

including: 1) photoprocesses with absorption in a line, such as resonant absorption, as well as spontaneous and induced radiation; 2) photoprocesses with absorption in continuous spectrum, such as photodissociation, photoionization, and radiative recombination. Photon scattering and free-free absorption do not influence directly the atmospheric gas kinetics.

Processes (3.24), (3.25) represent the complex of physical-chemical reactions in rarefied gas under the influence of UV radiation. The range of these reactions extends from excitation of rotational and vibrational levels to dissociation and ionization, and also includes the backward reactions.

We shall characterize the probability of reaction (3.24) realized by the dynamic scheme $I \rightarrow F : (\mathbf{c}_1, \mathbf{c}_2) \rightarrow (\mathbf{c}_3, \mathbf{c}_4)$ by the scattering function

$$I_{\alpha_I \rightarrow F(i_I \rightarrow F)}(\mathbf{c}_1, \mathbf{c}_2 \rightarrow \mathbf{c}_3, \mathbf{c}_4) = |M_{FI}|^2 \, \delta(E_I - E_F) \, \delta(\mathbf{P}_I - \mathbf{P}_F) \ , \tag{3.26}$$

where M_{FI} is the matrix element of transition between the initial and final states of the elementary process (3.24) (Blochintsev, 1961), and dynamic characteristics of particles are connected by conservation laws of mass, momentum, and energy

$$\begin{cases} m_{\alpha_1} + m_{\alpha_2} = m_{\alpha_3} + m_{\alpha_4} \\ m_{\alpha_1} \mathbf{c}_1 + m_{\alpha_2} \mathbf{c}_2 = m_{\alpha_3} \mathbf{c}_3 + m_{\alpha_4} \mathbf{c}_4 = (m_{\alpha_1} + m_{\alpha_2})\mathbf{c} \\ m_{\alpha_1} c_1^2 + m_{\alpha_2} c_2^2 = m_{\alpha_3} c_3^2 + m_{\alpha_4} c_4^2 + 2\Delta E(\alpha_I \rightarrow F(i_I \rightarrow F)) \\ \Delta E(\alpha_I \rightarrow F(i_I \rightarrow F)) = E_{\alpha_3}(i_3) + E_{\alpha_4}(i_4) - E_{\alpha_1}(i_1) - E_{\alpha_2}(i_2) \ . \end{cases} \tag{3.27}$$

Here $\mathbf{c}$ is the velocity of center of mass, and ΔE- energy output of the reaction. The scattering cross-section (the parameter usually used in kinetics) for reaction (3.24) can be easily obtained from (3.26)

$$\sigma(\alpha_I \rightarrow F(i_I \rightarrow F); \ \mathbf{c}_1, \mathbf{c}_2 \rightarrow \mathbf{c}_3, \mathbf{c}_4) = \frac{1}{|\mathbf{c}_1 - \mathbf{c}_2|} \int I(\alpha_I \rightarrow F(i_I \rightarrow F)) \, d\mathbf{c}_3 \, d\mathbf{c}_4 \ . \tag{3.28}$$

The values of (3.26), (3.28) depend on the interaction potential of the particles involved and can be calculated by methods of quantum mechanics or measured in laboratory experiments (Hochstim (Ed.), 1969; Mitchner and Kruger, 1973).

For processes of absorption (3.25), analogous scattering functions may be defined (Blochintsev, 1961). However, for radiative transitions in lines, the Einstein coefficients are generally used:

- for spontaneous radiation $\alpha(i) \rightarrow \alpha(i')+h\nu_{\alpha(i \rightarrow i')}$ the atomic coefficient $A_{\alpha(i \rightarrow i')}$ characterizes the probability of a given transition and may be written in the following form (Blochintsev, 1961; Grim, 1969)

$$A_{\alpha(i \rightarrow i')} = A_1 \frac{g_{\alpha(i')}}{g_{\alpha(i)}} |f_{\alpha(i \rightarrow i')}| \nu^2_{\alpha(i \rightarrow i')} , \quad i > i' , \qquad (3.29)$$

where $h\nu_{\alpha(i \rightarrow i')} = E_{\alpha(i)} - E_{\alpha(i')}$ is the average frequency of radiated photon, $f_{\alpha(i \rightarrow i')}$- the oscillator strength in the case of absorption at transition $\alpha(i \rightarrow i')$. It should be noted that the oscillator strength is an undimensional atomic constant varying in interval $0 - 1$ ($f_{\alpha(i' \rightarrow i)} = -f_{\alpha(i \rightarrow i')} g_{\alpha(i')}/g_{\alpha(i)}$), and $A_1 = 8\pi^2 e^2/m_e c^2$ is the atomic constant;

- for resonant absorption and induced radiation $\alpha(i) + h\nu \rightarrow \alpha(i')$ the following coefficient is used

$$B_{\alpha(i \rightarrow i')} = B_1 \frac{f_{\alpha(i \rightarrow i')}}{\nu_{\alpha(i \rightarrow i')}} , \quad B_1 = \frac{\pi e^2}{m_e h} . \qquad (3.30)$$

The interrelation between scattering functions (3.26) for direct and backward directions of reaction (3.24) results from the principle of microscopic reversibility (Blochintsev, 1961; Hochstim (Ed.), 1969; Mitchner and Kruger, 1973):

$$I(\alpha_{I \rightarrow F} | c_1, c_2 \rightarrow c_3, c_4) = I(\alpha_{F \rightarrow I} | c_3, c_4 \rightarrow c_1, c_2) , \qquad (3.31)$$

since $|M_{FI}|^2 = |M_{IF}|^2$ in agreement with the symmetry relative to the time reversibility of the Schredinger equation. A similar expression for the relation of the cross-sections of the different radiative processes may be used. For example, the interrelation between the Einstein coefficients has the following form

$$\frac{A_{\alpha(i \rightarrow i')}}{B_{\alpha(i' \rightarrow i)}} = \frac{8\pi h}{c^2} \nu^3_{\alpha(i \rightarrow i')} ; \quad g_{\alpha(i')} B_{\alpha(i' \rightarrow i)} = g_{\alpha(i)} B_{\alpha(i \rightarrow i')} , \qquad (3.32)$$

These expressions (ratios of reciprocity) permit to establish the relationship between the kinetic characteristics of direct and inverse directions for collisional (3.24) and radiative processes (3.25). In particular, in the case of a Maxwell-Boltzmann equilibrium distribution of particles, the reaction rates in both (direct and inverse) directions of reaction (3.24) are equal (principle of detailed equilibrium) (Hirschfelder et al., 1954; Mitchner and Kruger, 1973; Hochstim (Ed.), 1969).

To calculate the kinetic characteristics of nonequilibrium rarefied gas, one must know the scattering functions of the reactions involved, the distribution functions of gas particles by translational and inner degrees of freedom and the radiation field. The distribution functions of the gas particles may be obtained from the solution of the kinetic Boltzmann equation. This equation is usually written in the following general form

$$\frac{\partial}{\partial t}F_{\alpha(i)} + \mathbf{c}\frac{\partial}{\partial \mathbf{r}}F_{\alpha(i)} + \mathbf{S}_\alpha\frac{\partial}{\partial \mathbf{c}}F_{\alpha(i)} = \left\{\frac{\delta F_{\alpha(i)}}{\delta t}\right\}_{col} + \left\{\frac{\delta F_{\alpha(i)}}{\delta t}\right\}_{rad} , \quad (3.33)$$

where $\mathbf{S}_\alpha$ is the external force field. The right-hand side of the equation (3.33) describes changes of the component α particle number density due to collisional (3.24) and radiative (3.25) processes. The detailed expressions for these collision integrals will be given in the next chapter. The collisional term $(...)_{col}$ is defined by the distribution functions of gas particles participating in reactions (3.24), whereas the radiative term $(...)_{rad}$ is defined by the radiative external field and can be considered as the source of particles in the photoprocesses (3.25).

The ratio of characteristic time scales of collisional τ_{col} and radiative τ_{rad} processes defines the relaxation rate of the gas state to a stationary (and maybe local equilibrium) one. At $\tau_{col} \ll \tau_{rad}$ (relatively high density of gas) collisional processes determine the population of the excited levels. In rarefied gases characterized by a small density, radiative processes define the gas state. The total relaxation rate is defined by the gas density, the energetic parameters of radiation field and the cross-sections of collisional and radiative processes (see, e.g., Hochstim (Ed.), 1969; Mitchner and Kruger, 1973).

Under collisional relaxation, the exchange by translational energy is the fastest process and the average value of the energy exchange in the elementary act is $\delta_{el} = 2m_{\alpha_1}m_{\alpha_2}/(m_{\alpha_1} + m_{\alpha_2})^2$. It means that if masses of the gas particles are not very different, the averaged kinetic temperature will be reached very rapidly, after a few elastic collisions. The rotational energy exchange rate in collisions of particles of various types or one-species particles in various quantum states, usually requires $\sim 10^1 - 10^2$ collisions to reach the average temperature $T_{rot} = T_{tr}$.

In turn, the balance between translational and vibrational degrees of freedom is reached after a considerably greater number of collisions $\sim 10^3 - 10^8$ (depending on temperatures and nature of colliding particles). The vibrational energy exchange is considerably faster. If the deficiency of the vibrational quantums energy is small compared to their energy, the average value will be reached after $\sim 10^2 - 10^3$ collisions. Subsequently, when the vibrational temperature is settled out, the relaxation evolves to a common gas temperature.

The characteristic times for reaching equilibrium between the translational degrees of freedom and their electronic excitation and ionization are strongly dependent on features of the system. For example, for the upper atmosphere the condition $\tau_{el} \gg \tau_{vib}$ is usually valid. The characteristic times of different chemical reactions can also widely change. For impact reactions, it is close to the characteristic time of translational relaxation (at small values of activation energy); however, if the reactions proceed with the participation of excited or charged particles, their characteristic times can reach values of the relaxation times of appropriate particle populations, i.e. τ_{vib}, τ_{el} or τ_{ion}.

In any case, the maximum time is referred to as the total relaxation time of the system. In a thermodynamically closed system, after relaxation, the uniform average temperature for all subsystems

$$T_{\alpha,tr} = T_{\alpha,in} = T$$

is settled out. Then, the gas particles distribution functions on velocities and levels of inner excitation are defined by a Maxwell-Boltzmann distribution (3.22), (3.23) at the equilibrium temperature T, and the equilibrium chemical composition is defined by the ratios of detailed balance.

Following the general definition, if the radiative field is at the equilibrium (i.e. energy distribution of photons is the Planck distribution at temperature T), then the given physical-chemical system is in the state of total thermodynamic equilibrium; otherwise, the system is in the state of local thermodynamic equilibrium (LTE). Rarefied gas of the planetary atmosphere is considered as an open thermodynamic system whose nonequilibrium state is strongly influenced by the solar electromagnetic and corpuscular radiation. It means that a rigorous mathematical description of this system is possible only on the basis of nonequilibrium physical-chemical kinetics (Hirschfelder *et al.*, 1954; Prigogine, 1962; Hochstim (Ed.), 1969; Mitchner and Kruger, 1973; Polak, 1979).

CHAPTER 4

METHODS OF MATHEMATICAL MODELING OF RAREFIED GASES IN PLANETARY AND COMETARY ATMOSPHERES

4.1. Atmospheric Gas: Microscopic Kinetic Approach

The distribution function of atmospheric gas particles by translational and inner degrees of freedom, defining the particle distributions in the phase space $(\mathbf{r}, \mathbf{c})$ and by levels of inner excitation, serves as a basis for the mathematical description of rarefied gas of the upper atmosphere at the microscopic level using methods of nonequilibrium physical-chemical kinetics. These distribution functions result from the solution of Boltzmann-type kinetic equations.

The kinetic description of the rarefied gas of the upper atmosphere was given by Marov and Shematovich (1988), Marov et al. (1990, 1996), Shematovich et al. (1991a). The particles of the upper atmosphere are represented as the set of components

$$\{a\} = \{a^\alpha\}_{(\alpha=0,1,...,M^a)} , \qquad (4.1)$$

where $a^{\alpha=0} = e_a$ is the electron component (thermal electrons), and $a^{\alpha>0} = a^\alpha(\mathbf{i})$ are the neutral and ionized components determined by sets of quantum numbers i. The detailed inner structure of particles $\alpha(\mathbf{i})$ includes the classification of particles by their chemical nature, charge and levels of inner excitation. Thus expression (4.1) corresponds to the representation of the rarefied atmospheric gas as a set of chemically different atoms, molecules and ions, with various levels of rotational, vibrational and electronic excitations.

We first introduce the external factors influencing the upper atmosphere as the set $\{b^\beta\}_{(\beta=1,...,M^b)}$, which includes UV and X-ray solar radiation $(\beta = 1)$, photoelectrons $(\beta = 2)$, precipitating charged particles of magnetospheric origin $(\beta = 3)$, etc. (M^b is the number of different external factors.)

For the sake of convenience of the mathematical treatment, let us formalize the set of aeronomical reactions (3.1) – (3.19). The external energetic influence (photolytic (3.1) – (3.3) or impact (3.4) – (3.6) reactions) on the components of the rarefied atmospheric gas is described by the multichannel process

$$\{s^{\alpha\beta}\} = \{\bigcup_{r=1}^{R^{\alpha\beta}} s_r^{\alpha\beta} \; : \; \alpha + \beta \to \sum_l \alpha_l(r) + \beta'\} \; . \tag{4.2}$$

This process leads to the formation of excited and ionized particles and secondary fluxes of energetic particles $b^{\beta'}$ ($\beta' = 0$ is absorption of particles of primary flux; and $\beta' \neq \beta$ is the formation of the accompanying flux of new particles $b^{\beta'}$). In this expression $R^{\alpha\beta}$ is the number of possible channels of particle interactions. The main channels of process (4.2) are the reactions of excitation, dissociation, and ionization of atmospheric particles by EUV photons, photoelectrons, etc. The detailed structure of process (4.2) channels is determined by the energy E^β of the impacted particle and the reaction energy threshold $\triangle E^{s_r^{\alpha\beta}}$. The probability of implementation of this reaction and possible final states of its products are characterized by the differential scattering function $dI^{s_r^{\alpha\beta}}$ (Hirschfelder et $al.$, 1954; Hochstim (Ed.), 1969; Polak, 1979; Blochintsev, 1961; Sampson, 1965).

Another important process is the collisional redistribution of energy absorbed in reactions (4.2) between gas particles as a multichannel mechanism of relaxation

$$\{s^{\alpha\alpha_1}\} = \{\bigcup_{r=1}^{R^{\alpha\alpha_1}} s_r^{\alpha\alpha_1} \; : \; \alpha + \alpha_1 \to \alpha(r) + \alpha_1(r)\} \; . \tag{4.3}$$

The main channels of process (4.3) are elastic collisions, inelastic collisions of excitation and de-excitation, ion-molecule, chemical and recombination reactions (3.7) – (3.19). These reactions are characterized by their energy output $\triangle E^{s_r^{\alpha\alpha_1}}$. The probability of realization of the elementary act of process (4.3) is defined by the differential scattering function $dI^{s_r^{\alpha\alpha_1}}$.

Thus the set of aeronomical reactions in the upper atmosphere includes processes (4.2), describing the disturbances of the gas state by external energetic influence, and processes (4.3), describing the collisional relaxation of the disturbed gas state. The ratio of kinetic rates of processes (4.2) and (4.3) defines the degree of deviation of the upper atmospheric gas state from LTE. One should also keep in mind that dynamic processes can transport gas state pertubations.

We characterize the state of the rarefied gas of the upper atmosphere at the emicroscopic level by the distribution functions of atmospheric particles (neutral-constituent, ions and thermal electrons) by velocities and states of inner excitation

$$\{F_\alpha(\mathbf{c}, \mathbf{r}, t) = n_\alpha(\mathbf{r}, t) f_\alpha(\mathbf{c}, \mathbf{r}, t)\}_{(\alpha=0,1,...,M^a)} \; .$$

Similarly, for the external fluxes of particles affecting the upper atmosphere we adopt the distribution

$$\{F_\beta(\mathbf{c}^\beta, \mathbf{r}, t) = n_\beta(\mathbf{r}, t) f_\beta(\mathbf{c}^\beta, \mathbf{r}, t)\}_{(\beta=1,\dots,M^b)} \; .$$

Because the external influence is defined by the fluxes of particles injected into 1 steradian (unit of solid angle), we also introduce the parameter which serves as the analogue of the radiation intensity

$$W_\beta(\mathbf{E}^\beta, \Omega) \, d\mathbf{E}^\beta \, d\Omega = c^\beta F(\mathbf{c}_\beta^\beta) \, dc^\beta$$

and represents the quantity of particles with energy from E^β to $E^\beta + dE^\beta$ and velocity direction $\Omega = \mathbf{c}^\beta / c^\beta$ inside the angle $d\Omega$ through per unit area and time. Therefore, for the description of external fluxes of particles the distribution

$$\{W_\beta(E^\beta, \Omega, \mathbf{r}, t)\}_{(\beta=1,\dots,M^a)} \; ,$$

will be used and particle velocities in the processes (4.2) will be defined as $\mathbf{c}^\beta = c(E^\beta)\Omega$. For the electromagnetic solar radiation, the spectral distributions is as follows

$$W_\beta(E^\beta, \Omega, \mathbf{r}, t) = W_\nu(h\nu) \, \delta(\Omega - \Omega_\nu) \; ,$$

where Ω_ν is the flux transfer direction, and $W_\nu(h\nu) \, d(h\nu)$ is the photon flux with energy $(h\nu, h\nu + d(h\nu))$ through 1a unit area in the direction perpendicular to Ω_ν.

In the general case, the evolution of the state of the atmospheric rarefied gas at the microscopic level is evaluated through the solution of Boltzmann-type kinetic equations with a source term (Marov and Shematovich, 1985; Shematovich, 1987; Marov et al., 1990; 1996)

$$\begin{cases} \frac{\partial}{\partial t} F_\alpha + \mathbf{c} \frac{\partial}{\partial \mathbf{r}} F_\alpha + \mathbf{S}_\alpha \frac{\partial}{\partial \mathbf{c}} F_\alpha = \sum_{\alpha'} \sum_\beta Q_\alpha(W_\beta, F_{\alpha'}) + \\ \qquad\qquad\qquad\qquad\qquad + \sum_{\alpha_1} J_{\alpha\alpha_1}(F_\alpha, F_{\alpha_1}) \\ F_\alpha \mid_{t=t_0} = F_\alpha^{(0)}(\mathbf{c}, \mathbf{r}) \; , \quad F_\alpha \mid_{r \in \Gamma(G)} = F_\alpha^{(1)}(\mathbf{c}, t) \; , \\ \qquad\qquad\qquad\qquad\qquad \alpha = 0, 1, \dots, M^a \end{cases} \tag{4.4}$$

jointly with initial and boundary conditions for the atmospheric region $G(\mathbf{r})$ with a boundary surface $\Gamma(G)$. Here, $\mathbf{S}_\alpha = \mathbf{S}_\alpha(\mathbf{c}_\alpha, \mathbf{r})$ is the external force field influencing particles of α component (for radiation $\mathbf{S}_\nu = 0$).

The left-hand part of the kinetic equation (4.4) describes the transport of particles in the phase space $(\mathbf{r}, \mathbf{c})$, while the right-hand part describes the changes of gas state due to the external influence (4.2) and the relaxation collisional processes (4.3).

The first term in the right-hand part of the kinetic equation (4.4) is the source term due to photolytic and impact influence. It defines the production rate of α component particles in the phase space element $(\mathbf{r}, \mathbf{c}, d\mathbf{r}, d\mathbf{c})$ due to the external energetic flux β affecting atmospheric components α'. This rate production term may be written in the form:

$$
\left\{
\begin{aligned}
Q_\alpha(W_\beta, F_{\alpha'}; \mathbf{c}, \mathbf{r}, t) &= \sum_{r=1}^{R^{\alpha'\beta}} \eta_\alpha(s_r^{\alpha'\beta}) \, Q_\alpha^{s_r^{\alpha'\beta}}(\mathbf{c}, \mathbf{r}, t) \\
Q_\alpha^{s_r^{\alpha'\beta}}(\mathbf{c}) &= \int d\mathbf{c}' \, dE^\beta \, dE^{\beta'} \, d\mathbf{\Omega} \, d\mathbf{\Omega}' \, \sigma^{s_r^{\alpha'\beta}}(\mathbf{c}', \mathbf{c}^\beta \to \mathbf{c}, \mathbf{c}^{\beta'}) \times \\
&\qquad\qquad \times W_\beta(E^\beta, \mathbf{\Omega}) F_{\alpha'}(\mathbf{c}') \ .
\end{aligned}
\right.
\tag{4.5}
$$

Here η_α is an indicator of the process $s_r^{\alpha'\beta}$ such that

$$
\eta_\alpha(s_r^{\alpha'\beta} \ : \ \alpha' + \beta \to \sum_l \alpha_l(r) + \beta') =
\left\{
\begin{aligned}
-1 &, \quad \alpha = \alpha' \\
0 &, \quad \alpha \notin \bigcup_l \alpha_l(r), \ \alpha' \\
1 &, \quad \alpha \in \bigcup_l \alpha_l(r) \ .
\end{aligned}
\right.
$$

In this expression $\sigma^{s_r^{\alpha\beta}}$ is the scattering cross-section for dynamic/probabilistic realizations of the elementary act of the interaction (4.2); it is connected with the scattering function for the given process by the following ratio (Sampson, 1965)

$$
|\mathbf{c}^\beta - \mathbf{c}'| \sigma^{s_r^{\alpha'\beta}}(\mathbf{c}^\beta, \mathbf{c}') = \int d\mathbf{c} \, dE^{\beta'} \, d\mathbf{\Omega}' \, I^{s_r^{\alpha'\beta}}(\mathbf{c}', \mathbf{c}^\beta \to \mathbf{c}, \mathbf{c}^{\beta'}) \, E^{\beta'} \, \mathbf{\Omega}' \ .
$$

Let us note that the rates (4.5) for photoprocesses occurring in lines (with a discrete spectrum) can be written through Einstein coefficients (3.29) and (3.30). Consequently, for the processes of spontaneous and induced radiation, as well as resonant line absorption with the average energy $E_{\alpha(i \to i')} = E_\alpha(i) - E_\alpha(i')$, these expressions may be written in the following form (Sampson, 1965):

$$
Q_{\alpha(i)}(\mathbf{c}) = \sum_{i' > i} F_{\alpha(i)}(\mathbf{c}) \left[A_{\alpha(i' \to i)} + \Phi_\beta(E_{\alpha(i' \to i)}) \, B_{\alpha(i' \to i)} \right] +
$$

$$
+ \sum_{i' < i} F_{\alpha(i')} \Phi_\beta(E_{\alpha(i' \to i)}) \, B_{\alpha(i' \to i)} \ ,
$$

where Φ_β is spectral flux of photons $\Phi_\beta(E^\beta) = \int d\mathbf{\Omega} \, W_\beta(E^\beta, \mathbf{\Omega})$.

The second term in the right-hand part of the system (4.4) describes the local changes of the gas state due to collisions of atmospheric particles in the reactions (4.3) and may be represented as the following collision integrals:

$$\begin{cases} J_{\alpha\alpha_1}(F_\alpha, F_{\alpha_1}; \mathbf{c}, \mathbf{r}, t) = \sum_{r=1}^{R^{\alpha\alpha_1}} J_\alpha^{s_r^{\alpha\alpha_1}}(\mathbf{c}) \\ J_\alpha^{s_r^{\alpha\alpha_1}}(\mathbf{c}) = \int d\mathbf{c}_1\, \sigma^{s_r^{\alpha\alpha_1}}(\mathbf{c}, \mathbf{c}_1)\, |\mathbf{c} - \mathbf{c}_1|\, [e^{s_r^{\alpha\alpha_1}} F_{\alpha(r)}(\mathbf{c}') \times \\ \qquad\qquad \times F_{\alpha_1(r)}(\mathbf{c}'_1) - F_\alpha(\mathbf{c}) F_{\alpha_1}(\mathbf{c}_1)] \end{cases} \qquad (4.6)$$

Here $e^{s_r^{\alpha\alpha_1}}$ is the normalized factor of phase volume conservation in chemical reactions. The scattering cross-section for dynamic/probabilistic realization of the elementary act $(\mathbf{c}, \mathbf{c}_1) \to (\mathbf{c}', \mathbf{c}'_1)$ in the reaction (4.3) is connected with the scattering function by the expression

$$|\mathbf{c} - \mathbf{c}_1|\, \sigma^{s_r^{\alpha\alpha_1}}(\mathbf{c}, \mathbf{c}_1) = \int d\mathbf{c}'\, d\mathbf{c}'_1\, I^{s_r^{\alpha\alpha_1}}(\mathbf{c}, \mathbf{c}_1 \to \mathbf{c}', \mathbf{c}'_1) \ .$$

The integrals of collisions (4.6) between charged particles (electrons and ions) can be rewritten in the Landau form (Hochstim (Ed.), 1969; Mitchner and Kruger, 1976):

$$\begin{cases} J^{s_r^{\alpha\alpha_1}}(\mathbf{c}) = -\frac{\partial}{\partial \mathbf{c}} \mathbf{L}_{\alpha\alpha_1}(\mathbf{c}) \\ \mathbf{L}_{\alpha\alpha_1}(\mathbf{c}) = \frac{\chi_{\alpha\alpha_1}}{m_\alpha} \int [\frac{\hat{I}}{|\mathbf{c}-\mathbf{c}_1|} - \frac{(\mathbf{c}-\mathbf{c}_1)(\mathbf{c}-\mathbf{c}_1)}{|\mathbf{c}-\mathbf{c}_1|^3}][\frac{1}{m_{\alpha_1}} F_\alpha(\mathbf{c}) \times \\ \qquad \times \frac{\partial}{\partial \mathbf{c}_1} F_{\alpha_1}(\mathbf{c}_1) - \frac{1}{m_\alpha} F_{\alpha_1}(\mathbf{c}_1) \frac{\partial}{\partial \mathbf{c}} F_\alpha(\mathbf{c})]\, d\mathbf{c}_1 \ , \end{cases}$$

where $\hat{I}$ is the unity tensor; $\chi_{\alpha\alpha} = 2\pi e_\alpha^2 e_\alpha^2 \ln \Lambda_{\alpha\alpha}$; $\ln \Lambda_{\alpha\alpha}$ is the Coulomb logarithm. This modified form of representation is due to the fact that in the case of charge particle interactions there are multiple scatterings of particles under small angles, and consequently, the influence of Coulomb collisions on the distribution functions can be considered as a diffusion in the velocity space.

The system of equations (4.4) – (4.6) is not closed because for its solution the knowledge of external local spectral energy distributions (EUV radiation, photoelectrons, energetic particles of magnetospheric origin, etc.) are required.

In the general case, the spectral distributions of the external sources are defined by the solution of kinetic equations describing their energy transfer in the respective upper atmospheric region

$$\frac{\partial}{\partial t} W_\beta + \mathbf{\Omega} \frac{\partial}{\partial \mathbf{r}} W_\beta + \mathbf{S}_\beta \frac{\partial}{\partial \mathbf{c}^\beta} W_\beta = \sum_\alpha \sum_{\beta'} Q_\beta^{(+)}(W_{\beta'}, F_\alpha) - \\ - \sum_\alpha Q_\beta^{(-)}(W_\beta, F_\alpha) \ . \qquad (4.7)$$

Here $\mathbf{c}^\beta = \mathbf{c}^\beta(E^\beta)\mathbf{\Omega}$, and $\mathbf{S}_\beta$ is the external source. Boundary conditions for the primary external fluxes (solar radiation, precipitating particles, etc.) can be taken on the external boundary of the region considered, while for the secondary fluxes they result from the conditions of local formation of particles in processes (4.2).

In particular, the first term in the right hand part of the equation (4.7) describes the source of secondary energetically active particles formed by absorption ($\beta' \neq \beta$) or by energy scattering of primary particles ($\beta' = \beta$), and is written as follows:

$$
\left\{
\begin{aligned}
& Q_\beta^{(+)}(W_{\beta'}, F_\alpha; E^\beta, \mathbf{\Omega}, \mathbf{r}, t) = \sum_{r=1}^{R^{\alpha\beta'}} Q_\beta^{s_r^{\alpha\beta'}}(E^\beta, \mathbf{\Omega}, \mathbf{r}, t) \\
& Q^{s_r^{\alpha\beta'}}(E^\beta, \mathbf{\Omega}) = \int d\mathbf{c}'\, d\mathbf{c}\, dE^{\beta'}\, d\mathbf{\Omega}' \times \\
& \times \sigma^{s_r^{\alpha\beta'}}(\mathbf{c}, c^\beta, (E^\beta)\mathbf{\Omega} \rightarrow \mathbf{c}', c^{\beta'}(E^{\beta'})\mathbf{\Omega}') W_{\beta'}(E^{\beta'}, \mathbf{\Omega}')\, F_\alpha(\mathbf{c}) \ .
\end{aligned}
\right.
\tag{4.8}
$$

The second term characterizes scattering of particle flux in the process of its interaction with the atmospheric gas and is given by the expression:

$$
Q_\beta^{(-)}(W_\beta, F_\alpha; E^\beta, \mathbf{\Omega}, \mathbf{r}, t) = W_\beta(E^\beta, \mathbf{\Omega}) \sum_{r=1}^{R^{\alpha\beta}} \int d\mathbf{c}\, \sigma^{s_r^{\alpha\beta}}(\mathbf{c}, c^\beta)\, F_\alpha(\mathbf{c}) \ . \tag{4.9}
$$

We may conclude that the set of equations (4.4) - (4.6) and (4.7) – (4.9) strictly defines the state of rarefied atmospheric gas at the microscopic level and its arbitrary pertubations due to the external influence of electromagnetic and corpuscular solar radiation. As a whole, expressions (4.1) – (4.9) underlie the kinetic physical-chemical model of the upper atmosphere.

Let us note that the description of the gas state using the set (4.4) is valid over a wide range of atmospheric conditions and the main constraints imposed on this approach is what is pertinent to the utilization of the kinetic equations (Hirschfelder et al., 1954; Hochstim (Ed.), 1969):

1. collisions of particles are statistically independent (hypothesis of molecular chaos);
2. triple (and higher order) collisions are absent (the gas is rarefied);
3. the collisional time is less than the free path time.

The first condition is performed when the distance between particles is considerably less than the mean free path. It is verified at all altitudes of the upper atmosphere. The second condition is also satisfied in the upper atmosphere because triple collisions begin to play a role only in the mesosphere. Finally, the third condition is valid practically throughout the whole atmosphere, because the time of interaction $\tau_{int} \sim \frac{2\pi\hbar}{kT} \cong 10^{-13} \div 10^{-14}$ s is considerably less than all other characteristic times of the atmospheric gas.

For charged particles, three additional conditions (4-6) must be fulfilled (Klimontovich, 1980):

4. the acceleration of particles in an electric field within a mean free path λ should be considerably less than their thermal energy, $|e_\lambda| E \lambda \ll \bar{E}_{\alpha,tr}$;

5. the magnetic field does not influence collisions of particles, i.e. the Debye radius r_d is less than the Larmor radius, $r_d \ll r_L$;

6. the relaxation time of collective processes in the plasma must be considerably less than the characteristic temporary scales of changes of the distribution functions.

Conditions (4–6) are performed in the ionosphere with a good accuracy.

It is quite evident that from the mathematical viewpoint, the set of kinetic equations has a very complex structure as they are integro-differential and nonlinear by the nature (Bird, 1976; Ferziger and Kaper, 1972; Alekseev, 1982). Therefore, to solve this system, it is necessary to use special approaches which are briefly described in the following sections.

4.2. Regimes of Rarefied Gas Flow in the Upper Atmosphere

Nonequilibrium features of the rarefied atmospheric gas flow emerge from the analysis of the characteristic relaxation time and length scales of the collisional aeronomic processes. Here we shall consider the main processes in the upper atmosphere as related to the nonequilibrium features of the gas and their characteristic parameters.

For the sake of convenience of the qualitative analysis of the characteristic scales of the atmospheric physical and chemical processes, we represent the kinetic equations (4.4) in the following form:

$$
\begin{cases}
\frac{D}{Dt}F_\alpha = Q_\alpha^{(s)} + J_\alpha^{(el)} + J_\alpha^{(r)} \\
F_\alpha \mid_{t=t_0} = F_\alpha^{(0)}(\mathbf{c}, \mathbf{r}) , \quad F_\alpha \mid_{r \in \Gamma(G)} = F_\alpha^{(1)}(\mathbf{c}, t) , \\
\qquad\qquad \alpha = 0, 1, \ldots, M^a
\end{cases}
\tag{4.10}
$$

where

$$
\begin{cases}
\frac{D}{Dt}F_\alpha = \frac{\partial}{\partial t}F_\alpha + \mathbf{c}\frac{\partial}{\partial \mathbf{r}}F_\alpha + \mathbf{S}_\alpha \frac{\partial}{\partial \mathbf{c}}F_\alpha \\
Q_\alpha^{(s)} = \sum_{\alpha'}\sum_{\beta} Q_\alpha(W_\beta, F_{\alpha'}) \\
J_\alpha^{(el)} = \sum_{\alpha_1} J_{\alpha\alpha_1}^{(el)}(F_\alpha, F_{\alpha_1}) \\
J_\alpha^{(r)} = \sum_{\alpha_1} J_{\alpha\alpha_1}^{(r)}(F_\alpha, F_{\alpha_1})
\end{cases}
$$

Here the collisional terms are divided into two subsets:

$$
\sum_{\alpha_1} J_{\alpha\alpha_1}(F_\alpha, F_{\alpha_1}) = \sum_{\alpha_1} J_{\alpha\alpha_1}^{(el)}(F_\alpha, F_{\alpha_1}) + \sum_{\alpha_1} J_{\alpha\alpha_1}^{(r)}(F_\alpha, F_{\alpha_1})
$$

The first term includes only elastic collisions of atmospheric particles, and all other inelastic and chemically reactive collisions are combined in the second term. This separation is convenient, as elastic collisions lead the gaseous medium to local thermal equilibrium, while inelastic collisions, accompanied by energy exchange between internal and translational degrees of freedom and chemical reactions, cause perturbations of the thermal state.

Let us introduce the characteristic microscopic and macroscopic scales of the atmospheric physical and chemical processes based on the molecular level of description. Let T and L be the characteristic macroscopic time and length, within which typical changes of the main macroscopic parameters of gas occur. Further, let v_T be the thermal velocity of gas particles; v_s - the characteristic average velocity of the particles of the external source affecting the atmospheric gas; and n - the characteristic gas density. Let us also introduce some characteristic parameters of the collisional interactions of the gas particles. For the three groups of collisional processes in the right hand part of the equation (4.10), we use the following notation for the characteristic values of cross-sections: σ_{el} - for elastic collisions; σ_r - for inelastic and chemically reactive collisions; σ_s - for external effects of interaction.

To characterize the microscopic scales of the collisional processes, the mean free path length $\lambda \sim (n\sigma)^{-1}$, and the mean free time $\tau \sim \lambda/v_T$ between particle collisions are generally used, each process having its own characteristic values λ and τ defined by its cross-section. For the atmospheric gas, however, the relaxation times for all degrees of freedom have markedly different scales. The following hierarchy of times is usually adopted (Bauer, 1973; Rees, 1989):

$$\tau_{tr} \le \tau_{rot} \le \tau_{vib} \le \tau_{dis} \ll \tau_{ion} \sim \tau_{el} \ ,$$

though in a multicomponent gas an overlap of scales is possible.

To simplify the problem, only the fastest process is usually considered as the key one, and its cross-section is taken as a characteristic value of all processes in a given subset of the collisions. In the rarefied atmospheric gas, an equilibrium on the translational degrees of freedom is most rapidly obtained, after only a few collisions. Therefore, to analyze the flow regimes, the cross-section for the elastic collisions can be adopted as the main characteristic parameter which defines both λ and τ.

Using the characteristic micro- and macroscopic scales, were transform the kinetic equations (4.10) into a dimensionless form (Cercignani, 1988):

$$\frac{D}{Dt}F_\alpha = \frac{1}{\epsilon_s}Q_\alpha^{(s)} + \frac{1}{\epsilon_{el}}J_\alpha^{(el)} + \frac{1}{\epsilon_r}J_\alpha^{(r)} \qquad (4.11)$$

where the following dimensionless parameters we introduced

$$\begin{cases} \epsilon_{el} = \frac{\lambda_{el}}{L}, & \lambda_{el} = (n\sigma_{el})^{-1} \\ \epsilon_r = \frac{\lambda_r}{L} = \epsilon_{el}\frac{\sigma_{el}}{\sigma_r} \\ \epsilon_s = \frac{\lambda_s}{L} = \epsilon_{el}\frac{\sigma_{el}}{\sigma_s}\frac{v_T}{v_s} \end{cases} \qquad (4.12)$$

These parameters characterize the ratio of micro- and macroscopic scales of gas state variations. Depending on their magnitudes, various relaxation properties of the planetary upper atmosphere are possible.

In a simplified qualitative way, these properties can be estimated by the Knudsen number $Kn_{el} = \lambda/L$, where λ is defined by the gas particle density n and the gasdynamical collisional cross-section ($\sigma_{el} \simeq 10^{-15} cm^2$), and L is the characteristic macroscopic scale length change of n. In other words, Kn_{el} is the rate of possible deviation of the gas thermal state from the LTE (Hirschfelder et al., 1954).

It is also known from the kinetic theory of rarefied gases, that when $Kn_{el} \ll 1$ the continuum gas dynamic flow regime occurs. In this case the gas thermal state is close to local equilibrium for the translational degrees of freedom, and is generally determined by the processes of local collisional relaxation (Cercignani, 1988)

$$J_\alpha^{(el)} + \frac{\sigma_s}{\sigma_{el}}\frac{v_s}{v_T}Q_\alpha^{(s)} + \frac{\sigma_r}{\sigma_{el}}J_\alpha^{(r)} = 0 \ . \qquad (4.13)$$

If no external disturbance effects the system, the local equilibrum Maxwellian distribution functions

$$F_\alpha^{(M)} = A_\alpha n_\alpha(t, \mathbf{r})exp(-\frac{m_\alpha(\mathbf{c} - \mathbf{u}_\alpha(t, \mathbf{r}))^2}{2kT_\alpha(t, \mathbf{r})}) \qquad (4.14)$$

emerge as the solution of equations (4.13). Here $\mathbf{u}_\alpha$ is flow velocity of the gaseous α-component and T_α is its local kinetic temperature.

The distinctive feature of the gas dynamic flow regime is that the functions (4.14) have small variations on microscopic scales and are time-dependent only imlicitly, through the macroscopic characteristics. Accordingly, using the methods of perturbation theory in the rarefied gas kinetics (e.g., Hilbert or Chapman-Enskog approaches), it is possible to derive the gas dynamic conservation equations for density, velocity and kinetic temperature, which describe variations of the gas state at the macroscopic scales (T, L) from the kinetic system (4.10). However, one must keep in mind that the gas dynamic approach is valid for the description of the rarefied gas flow only when disturbances caused by external influences or chemical reactions are small, i.e., when $\sigma_r/\sigma_{el} \ll 1$ and $(\sigma_r * v_s/(\sigma_{el} * v_T) \ll 1$.

Clearly, the relaxation properties of the atmospheric gas define the contribution of nonthermal particles in the atmospheric chemistry, dynamics, and energetics. In the case of fast exothermic reactions $\sigma_r \simeq \sigma_{el}$ or high-energy external influences on the atmospheric gas (when $v_s \gg v_T$ and, accordingly, $(\sigma_r * v_s) \simeq (\sigma_{el} * v_T)$), significant formation rates of superthermal and internally excited particles are possible. Then, the additional terms in equations (4.11) or (4.13) become comparable with the elastic collisions term. It results in considerable disturbances of the local distribution functions of gas particles by kinetic energy and in their deviations from the equilibrium functions (4.14). This nonequilibrium regime of atmospheric gas flow even within the limit $Kn_{el} \ll 1$ must be described at the microscopic level, in particular using the theory of linearized Boltzmann equation (Cercignani, 1969, 1988).

The opposite extreme case $Kn_{el} \gg 1$ corresponds to a free molecular (collisionless) regime of atmospheric gas flow. Indeed, in this case, the gas evolution is determined only by the individual dynamic parameters of the particle and the external influences. This regime is described by the collisionless Boltzmann equation with particle sources:

$$\frac{D}{Dt}F_\alpha = Q_\alpha^{(s)} \tag{4.15}$$

In this case, the initial and/or boundary distribution functions evolve to

$$F_\alpha(t, \mathbf{r}, \mathbf{c}) = F_\alpha^{(0)}(\mathbf{r} - \mathbf{r}_\alpha(t) , \mathbf{c} - \mathbf{c}_\alpha(t)) \tag{4.16}$$

along the dynamic trajectories of atmospheric particles

$$\begin{cases} \frac{\partial}{\partial t}\mathbf{r}_\alpha = \mathbf{c}_\alpha \\ \frac{\partial}{\partial t}\mathbf{c}_\alpha = \mathbf{s}_\alpha \end{cases} \tag{4.17}$$

In the free molecular flow regime, the states of rarefied gas are determined only by the initial and boundary conditions and by the external influences. Deviations of the gas state from the equilibrium may be arbitrary large.

A transition regime flow is usually defined as one in which the mean free path is neither very small nor very large in comparison with a characteristic macroscopic scale of the flow (Bird, 1976). From the kinetic theory of rarefied gases it is known that in the case when the disturbances caused by external influences and chemical reactions are small, the gas dynamic approach is sufficiently reliable in the regions where the Knudsen number $Kn_{el} < 0.1$; therefore formally the transition regime is implemented for flows with the Knudsen number greater than 0.1. The upper limit of the transition regime is usually considered as a region where the flow is practically collisionless, i.e. where the Knudsen number is sufficiently greater than unity. This transition flow regime is characterized by distribution functions

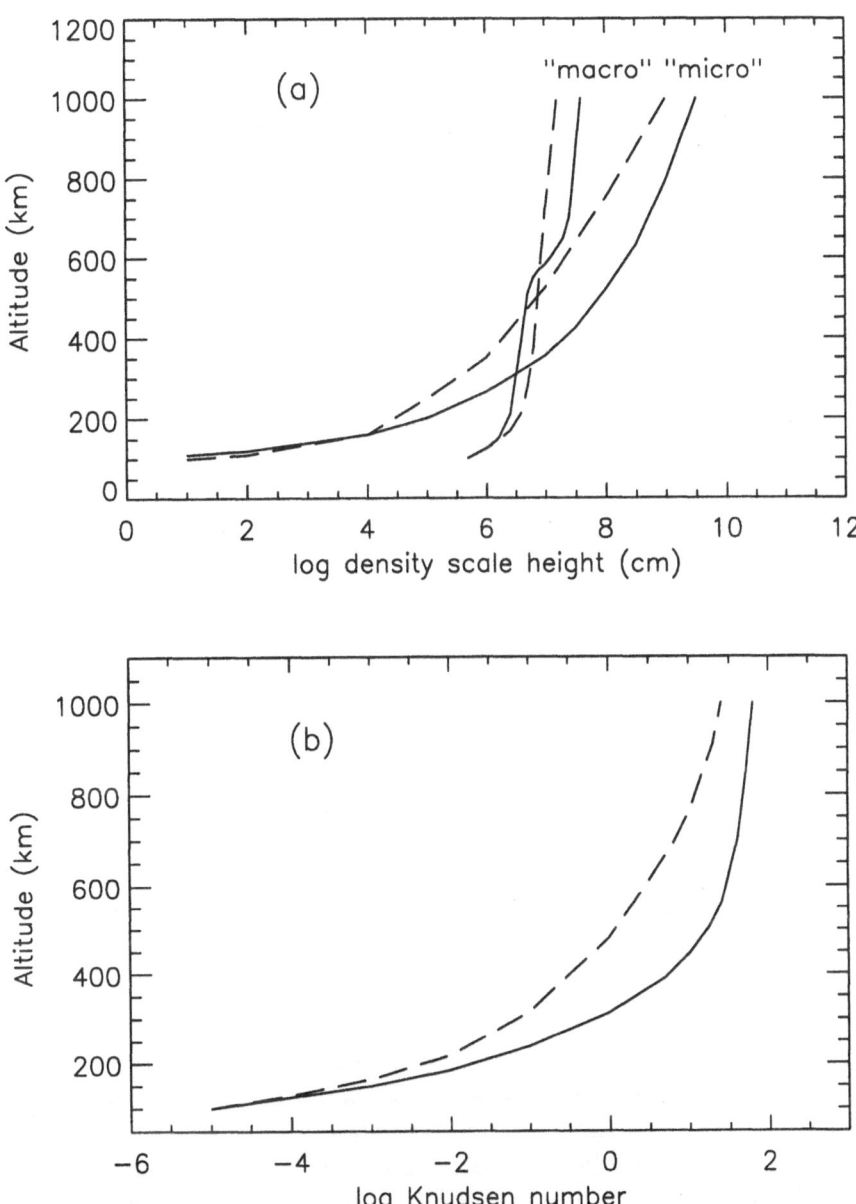

Figure 4.1. The height profiles of micro- and macroscopic scales of density changes in the upper atmosphere for low $F_{10.7}=70$ (solid line) and high $F_{10.7}=230$ (dashed line) solar activity: (a) - mean free path length λ (microscale), and density scale height L (macroscale); (b) - Knudsen number $Kn_{el} = \lambda/L$.

varying both on the microscopic and on macroscopic flow scales, and directly depending on time. The collisional physical and chemical processes in this flow regime may cause considerable disturbances of the distribution functions and therefore one must use the nonlinear collision terms for their description. This means that the gas flows in such transition regimes are only accurately described by the set of Boltzmann type kinetic equations with the source terms (Ivanovskiy et $al.$, 1967; Izakov, 1967; Marov et $al.$, 1990; 1996).

In the planetary upper atmospheres essentially all flow regimes are encountered. Indeed, the gas density varies dramatically with altitude what is characterized by a scale height $H = kT/mg$ (g is acceleration due to gravity; m is average mass of atmospheric gas particles; T is temperature; and k is Boltzmann constant) which can be referred to as the macroscopic characteristic $L = H$. In Fig $4.1a$ the vertical profiles of mean free path λ and H, based on the data of MSIS-86 model for minimum ($F_{10.7} = 70$) and maximum ($F_{10.7} = 230$) solar activity (in standard units 10^{-22} W/m^2Hr) are presented (Marov and Shematovich, 1988). The respective Knudsen number $Kn_{el} = \lambda/H$ characterizing the deviation of the gas state from thermal equilibrium is shown in $Fig.$ $4.1b$. It follows from $Fig.$ 4.1 that at thermospheric heights (90-300 km) the Knudsen number $Kn_{el} \ll 1$ and hence the distribution of gas particles on translational degrees of freedom is close to the local equilibrium. Therefore, at these heights the gas state variations can be described by the gas dynamic approach (Chamberlain, 1978; Marov and Kolesnichenko, 1987; Rees, 1989). In the transition region ($\sim 300 - 500$ km) $Kn_{el} \geq 0.1$. In this region, external energetic sources as well as transport of excited particles from the lower thermosphere may cause significant deviations of the gas state from equilibrium. Finally, at exospheric heights (> 500 km) $Kn_{el} \gg 1$, and thus the atmospheric gas is practically collisionless.

The above treatment of the upper atmosphere stratification is referred to as a simplified approach. In a more detailed consideration, it is necessary to assess more accurately the disturbances caused by solar influence and chemical reactions. This means that significant variations of the gas thermal state may propagate into atmospheric regions where a gasdynamic approach seems plausible for stable conditions. Let us emphasise that external disturbances are proportional to n whereas elastic relaxation rate is proportional to n^2. Therefore, since the atmospheric gas density profile rapidly decreases with height these two opposite processes are divergent and the rate of nonequilibrium state increases with altitude. In other words, the nonequilibrium nature of the gas flow expands over larger atmospheric regions. Consequently, the complete set of nonlinear Boltzmann type kinetic equations with the source terms must be used as the most efficient tool for the evaluation of this extended transition region of the upper atmosphere.

4.3. Basic Approaches to the Kinetic Simulation of Gas Flows in the Transition Regime

In this section we shall briefly discuss the general methods of mathematical simulation of rarefied gas flows, focusing on their application and relevance to the aeronomical problems.

Different approaches were developed to solve the Boltzmann equation, involving both analytical and numerical methods. A detailed analysis of the first category (see e.g. Bird, 1994; Cercignani, 1969, 1988) shows, however, that analytical approaches may require extensive numerical work in order to find a solution for specific flow. In some cases, this effort may exceed what would be required for a direct numerical solution of the problem. Only very few of the 'analytical' methods lead to complete solutions and mostly depend on the numerical procedures adopted. On the other hand, the 'numerical' methods rely on computation from the onset. While analytical methods are almost invariably based on the Boltzmann equation as an accepted mathematical model of the gas flow at the molecular level, a large fraction of the numerical methods is based directly on the simulation of the physics of the flow. According to the Bird's (1994) analysis, these approaches can be used to define two different classes: analytical methods, and simulation methods (i.e. all those using simulating particles or molecules).

To obtain approximate analytical solutions one must include additional assumptions or approximations, because the problem (which involve large disturbances) is unavoidably non-linear. Depending on the type of assumptions, the analytical methods can be divided into two groups: moment methods, where the form of the distribution function is assumed, and methods of Boltzmann equation approximations (model equations). We shall briefly review these methods; the more detailed discussion can be found elsewhere (see, e.g. Bird, 1994; Cercignani, 1969; 1988).

The first approach makes use of the moment equations that are obtained by multiplying the Boltzmann equation by a molecular quantity Q, and then integrating it over the velocity space. Substitution of the various values of Q leads to a series of equations with macroscopic quantities. Utilization of multiplicator $\overline{cQ}$ means that, as Q progresses to successively higher orders of c (particle velocity), each equation involves a moment of progressively higher order. It results in an infinite number of equations and moments. However, the basis of the moment method is that the velocity distribution function f is assumed to conform to some expression that contains a finite set of macroscopic quantities or moments. It allows to eliminate the higher order moments and thus to close the series of equations and form a determinate set.

The history of approximate solutions goes back in time to the first attempts of Maxwell and Boltzmann, Hilbert (1912), Chapman (1916-17) and Enskog (1917). The Chapman-Enskog solution of the Boltzmann equation is the second-order approach based on series expansion of the velocity distribution function f. This series may be written as

$$f = f_0(1 + a_1 Kn + a_2 Kn^2 + ...) ,$$

where coefficients a_n are functions of n (gas density), $\mathbf{u}_0$ (flow velocity), and T (kinetic temperature) only, and Kn is the Knudsen number.

The first-order approach is the local equilibrium or Maxwellian distribution function f, when the gas is fully described by n, $\mathbf{u}_0$, and T. In the equilibrium condition, the viscous stress tensor τ and heat flux vector $\mathbf{q}$ vanish and the conservation equations reduce to the well-known Euler equations of non-viscous gas flow. The second-order approaches were obtained independently by Enskog and Chapman, and these form the basis of the classical work of Chapman and Cowling (1952). The Chapman-Enskog solution leads to the velocity distribution function which also involves only the moments of n, $\mathbf{u}_0$, and T. This distribution function enables τ and $\mathbf{q}$ to be written as products of the coefficients of viscosity and heat conduction with the velocity and temperature gradients, respectively, thus reducing the conservation equations to the Navier-Stokes equations of the continuum gas dynamics (hydrodynamics). The above consideration shows that, from the standpoint of the kinetic theory, both the Euler and Navier-Stokes equations may be regarded as 'five moment' solutions of the Boltzmann equation, the former being valid for the $Kn \to 0$ limit and the latter for $Kn \ll 1$. In the context of continuum gas dynamics, the Euler equations describe reversible adiabatic (isentropic) flows, while the Navier-Stokes equations provide the standard description of viscous flows. This means that the first- and second-order Chapman- Enskog solutions merely reconcile the molecular and continuum approaches for small Knudsen number flows. It is obvious that the expansion used for f must break down as the Knudsen number approaches unity, but there remains the possibility that the use of more than two terms may extend the range of applicability of the continuum approach to higher Knudsen numbers. The third-order approach leads to the Burnett equations. These equations are, however, extremely complex and closed form solutions are not available even for simple flows.

In 1949, H.Grad proposed a systematic method to deal with the solution of the Boltzmann equation, and this is the well known 13-moment method. Grad (1949) put forward an alternative expansion for f as a series of Hermite polynomials with the first-order approach again being the local Maxwellian. The third-order approach leads to the expression for f

as the local Maxwellian multiplied by an expression involving τ and $\mathbf{q}$ as well as n, $\mathbf{u}_0$, and T. The total number of moment equations that are required for a determinate set is equal to the total number of dependent variables in the conservation equations (13). Although there is some 'rationale' in the Grad's approach and his equations give better results than the Navier-Stokes equations for some problems, the common results from their application to specific problems in the transition regime have proved to be rather disappointing.

An absence of known series expansions, resulting in equations with a considerably extended applicability limit in comparison with Navier-Stokes equations shows that, to develop equations suitable for the transition region, new assumptions on distribution function approximations are required. Among the best known new approximations are the Mott-Smith (1951), Liu-Lees (1961), and Ytrehus (1994) solutions. The lesson taught by these solutions is that it is not so useful to look for general methods with the aim of obtaining continuum-like equations, instead, one should rather devise approximate methods for dealing with particular problems. One may conclude that while the results from some applications of the moment method are in remarkably good agreement with experiment, the solutions are not unambiguous because the method is based on several arbitrary assumptions. In particular, these assumptions are associated with the form of the distribution function and with the choice of non-conserved quantities for the additional moment equations. The best selection of some set of parameters for a specific problem is not necessarily a good equivalent for another one.

As it was earlier mentioned, the small-perturbation approach is based on small values of the Knudsen number or its reciprocal in order to obtain approximate solutions for the near continuum and near collisionless regimes, respectively. Some methods may also place constraint on the flow disturbance. In particular, if the disturbance is sufficiently small that the velocity distribution function is only slightly perturbed from the equilibrium, the small-perturbation approach leads to the linearized Boltzmann equation which were treated in detail by Cercignani (1969, 1988).

The model equation approach involves a form of approximation to the Boltzmann equation itself. It is the collision term on the right hand side of the Boltzmann equation that poses the greatest mathematical difficulties, and it is this term that needs to be approximated. The best known model approach is the BGK (Bhatnagar, Gross, and Krook, 1954) equation where the collisional term is presented in the form $n\nu(f_0 - f)$. In these term ν is proportional to n and may also depend on temperature, but it is assumed to be independent on the molecular velocity $\mathbf{c}$. The presence of the local Maxwellian function f_0 means, however, that the equation remains a nonlinear integro-differential equation. This is because f_0 is a function of the stream velocity $\mathbf{u}_0$ and the temperature T which must be obtained as integrals over f. The integration over both f_0 and f must result in identical values of $\mathbf{u}_0$ and T, so that the collision term vanishes when the moments are taken over conserved quantities. This condition ensures that the BGK equation is consistent with the conservation equations.

The BGK equation obviously provides the correct solution for collisionless or free molecular regimes of flow, since it has an immaterial form of the collision term. An exact solution would require to span the complete range of Knudsen numbers in order to be close enough to the correct limits at the extremes of this range. Although approximate collision terms give rise to an indeterminate error in the transition regime, the assumptions inherent to the BGK equation are least likely to introduce distorted effects in the problems that involve small perturbations and the boundary conditions are formulated in terms of equilibrium distributions. Therefore the small-perturbation solutions provide important reference values, especially when they are obtained without resort to arbitrary assumptions about parameters or modifications of the equations. Nonetheless, these solutions are available only for a restricted set of problems with comparatively simple boundary conditions and have limited practical applications.

The kinetic models discussed above have been very useful in obtaining approximate solutions and forming qualitative ideas on the solutions of practical problems. Nevertheless, these methods do not provide us with the detailed and precise analysis of the gas flow in the transition regime. Therefore the implementation of 'numerical' methods is the most appropriate approach available for solving the Boltzmann equation. Various numerical procedures exist which either attempt to solve for f by conventional techniques of numerical analysis or efficiently by-pass the formalism of the integro-differential equation and simulate the physical situation that the equation describes (simulation methods).

Numerical solution of the Boltzmann equation based on finite difference methods meets severe computational difficulties due to a large number of independent variables. In practice, the only approach that has been successfully implemented for inhomogeneous problem solution in more than one space dimension was the Hicks-Yen-Nordsiek technique (Nordsiek, and Hicks, 1967; Yen, Hicks, and Osteen, 1974), which incorporates Monte-Carlo a quadrature method to evaluate the collision term. The method of computational fluid dynamics was extended to two-dimensional steady-state flows by Tcheremissine and Aristov (Aristov and Tcheremissine, 1980; Tcheremissine, 1985). Alternatives of the Monte Carlo procedure for the evaluation of the collision term have been proposed and tested by Tcheremissine (1991) and Tan *et al.* (1989). These methods attempt to reduce the computational load associated with the collision term by concentrating on non-equilibrium portion of the velocity distribution function. The basic disadvantage of these methods is the necessity to specify a bounded grid in velocity space. This poses additional problems which complicate the over-all computational procedure. Besides, chemically reacting and thermally radiating flows (and even simpler flows of polyatomic gases) are hard to evaluate with the same degree of accuracy as what is obtained using the Boltzmann equation for monoatomic nonreacting and nonradiating gases.

The above described approach paved the way to the development of physically based simulation methods. This is computer modeling of the real gas by a large number of simulated molecules, reviewed in detail by Bird (1994).

The first attempt of such method development was undertaken by Alder and Wainwright (1957). The molecular dynamics approach was a determin-istic one. Indeed, while they used probabilistic procedures when setting the initial configuration of molecules, the calculation of the subsequent molec-ular motion, including the collisions and boundary interactions, were de-termined. Collisions occured when particles cross-sections overlaped, and thus the computation time for a straightforward application of the method was proportional to square of the number of simulated molecules. A major difficulty imposed by this method utilization was that for a given molecu-lar size, flow geometry, and gas density, the number of simulated molecules is not a free parameter. Nonetheless, the method of molecular dynamics proved to be valuable for the simulation of dense gases and liquids (Allen and Tildesley 1987), but it was inappropriate for rarefied gases.

The first probabilistic simulation method appeared when the Monte Carlo technique was used (Haviland and Lavin 1962). In this approach a large number of 'test particle' trajectories are computed with the assumed distribution serving as the 'target' gas for the computation of typical inter-molecular collisions. The calculations can be only done if there is already an initial representation of the distribution function over the whole flow field.

As an alternative of the test particle approach, it was suggested to in-troduce a time variable and to follow the trajectories of a very large number of simulated molecules simultaneously. These methods, which started with the pioneer work of Bird on the so called Direct Simulation Monte Carlo (DSMC) method (Bird, 1960) have become a powerful tool for practical calculations. There appeared to be very few limitations to the complexity of the flow fields that this approach can deal with, including chemically reacting and ionized flows.

In the DSMC method, the intermolecular collisions are considered on a probabilistic rather than deterministic basis. The real gas is modeled by a comparatively small number of simulated molecules of the correct physical size. For each of them space coordinates and velocity components (as well as variables describing the internal state, if one deals with polyatomic molecules) are stored in the memory and are modified with time as the molecules are simultaneously getting through representative collisions and boundary interactions in the simulated region of space. The microscopic boundary conditions are specified by the behavior of individual molecules, rather than by distribution function. This facilitates incorporation of complex physical effects, such as chemical reactions, into the simulation. The flow develops from an initial state with time running in physically realistic manner, not stricly coupled with iteration from an initial flow approximation. The steady state solutions are obtained as asymptotic limits of unsteady calculation stages. The simulated physical space is subdivided into cells, which are taken small enough to ensure the solution to be approximately constant through the cell. Time is splitted in discrete steps of size Δt, small with respect to the mean free path time, i.e the time between two subsequent collisions of a molecule. All particles move over distances corresponding to this time step, followed by calculation of a representative set of collisions occurring at this interval. In accordance with the DSMC basic principle (Bird, 1994) the procedure for the probabilistic selection of a representative set of collisions is strongly related with the basic foundation of the kinetic theory. This is why the method has the same constraints as the classical kinetic theory, which is based on the Boltzmann equation. The main limitations are the assumption of molecular chaos and the requirement of a rarefied gas.

The term 'direct simulation Monte Carlo' is a generic term that covers various modifications of the method that have been developed over the years. Simulation procedures are not unique and, given the very rapid increase in applications of direct simulation methods in recent years, many new procedures were introduced. A local consideration of the N-particle gas model, i.e. the analysis of collisional processes in each individually taken cell, is a common feature of different approaches to the DSMC method. Differences mostly concern the procedure of calculation of the collisional relaxation in a cell. The development of the method was basically aimed in devising an effective and mathematically rigorous algorithms of collision frequency. Most eminent break-throughs in the improvements of the traditional schemes were accomplished by Koura (1970), Belotserkovskii and Yanitskii (1975), Deshpande (1978), Nanbu (1980), Yanitskii (1988), Ivanov and Rogasinskii (1988a, 1988b), Babovsky (1989).

The existence of different flow regimes in the upper atmosphere described in a previous section leads to the implementation of various approaches for describing the atmospheric gas flow involving continuum gas dynamic equations, collisionless kinetic equations, and nonlinear Boltzmann equations with source terms. In investigations of the planetary atmospheres all these approaches were used, and a number of specific methods, taking into account peculiar features of the atmospheric gas, were elaborated to solve the appropriate sets of equations.

The most intensive studies were carried out using the gas dynamic approach. The theoretical basis for this approach and its various applications were thoroughly analysed and can be found in particular in the monographs of Banks and Kockarts (1973); Chamberlain and Hunten (1987); Marov and Kolesnichenko (1987); and Rees (1989). However, as we earlier discussed, these methods can be reliably applied only to gas flow regimes for which deviations from the equilibrium state are small. Unfortunately, it is not the case for the planetary atmospheres where various disturbances, caused by external forcing and chemical reactions occur. In some specific cases these disturbances can be especially large; therefore, even in the case of $Kn \ll 1$, a gas dynamical approach must be accompanied by the treatment of specific nonequilibrium processes at the kinetic level (Logan and Mc Elroy, 1977; Solomon, 1983; Shematovich et al., 1991b, 1992). This joint (gas dynamical and kinetic) approach is possible because in a rather dense regions of the atmosphere, linearized kinetic equation can be used. In turn, the collisionless (exospheric) regions of the planetary atmospheres can be succesfully studied using free molecular approach (see, e.g. Chamberlain and Hunten, 1987; Fahr and Shizgal, 1983).

Most difficult for numerical analysis are regions of the upper atmosphere where transition flow regimes occur. This makes it necessary to solve the complete set of nonlinear Boltzmann type kinetic equations with sources. Until recently there were only limited attempts to model gas flow in these atmospheric regions at the kinetic level using either test particle Monte Carlo method (Chamberlain and Smith, 1971; Brinkmann, 1971; Barghouthi et al. 1993; Hodges, 1994), or multistream models (Nagy and Banks, 1970; Torr et al., 1974; Kozyra et al., 1982; Ishimoto et al., 1992). Nevertheless, in these approaches based on linearized kinetic equations, nonlinear effects in the transition regions were not included, although it is known both from theoretical (Rohrbaugh and Nisbet (1973); Shizgal and Lindenfeld, 1980; Marov et al., 1990) and experimental studies (Yee et al., 1980; Hedin, 1989; Cotton et al., 1993) that these effects play an important role in the upper atmosphere. Therefore, for a thorough investigation of these nonlinear and nonequilibrium aeronomic processes, it is necessary to solve the complete set of nonlinear kinetic equations. A significant progress in the kinetic study of rarefied gas transition flows was accomplished only in the latest years. Specifically, this progress is a part of the author's efforts in the development of efficient numerical algorithms for in-depth study of

the aeronomic problems involving of special modifications of the original
DSMC methods. The detailed description of these approaches will be given
in the following Chapters.

CHAPTER 5

KINETIC APPROACH TO THE MODELING OF COLLISIONAL PROCESSES IN A RAREFIED GAS

The kinetics of aeronomic processes in a rarefied atmospheric gas can be formulated as a statistical behavior of an ensemble of atoms, molecules and their ions under the influence of solar electromagnetic and corpuscular radiation. For a given set of elementary processes, the microscopic kinetic description of the atmospheric particle ensemble is characterized by energy distribution functions, populations of quantum levels and energy exchange in the collisional processes. The dynamic evolution of this ensemble is described by transport processes. In general, the system (4.4) must be used (see e.g., Ivanovskiy *et al.*, 1967; Izakov, 1967; Blum *et al.*, 1972; Marov and Shematovich, 1988; Marov *et al.*, 1990, 1996) which can be eficiently evaluated based on a discrete stochastic approximation of the initial microscopic model (Shematovich, 1987; Shematovich *et al.*, 1991a).

In this chapter we shall first begin with a general scheme of stochastic simulation of collisional processes in a rarefied atmospheric gas. Then proceeding from the asymptotic equivalence between dynamic and the stochastic descriptions on the level of one-particle distribution functions, we shall deal with the basic principles of the stochastic model development. The focus will be given to the structural stochastic models as possible modifications of stochastic simulation scheme, which enable to build up numerical models of nonequilibrum relaxation kinetics. The stochastic description of transport processes responsible for the rarefied gas flow will also be considered.

5.1. Stochastic Simulation of a Nonequilibrium Kinetic System

5.1.1. PHYSICAL AND MATHEMATICAL GROUNDS FOR THE METHOD OF STOCHASTIC SIMULATION

A stochastic simulation is based on the relationship between two basic approaches to the description of rarefied gases:

- the consideration of the gas evolution using the dynamic Liouville equation and corresponding kinetic equations (Prigogine, 1962; Ferziger and Kaper, 1972; Montroll and Lebowitz (Eds.), 1983);

- the evaluation of the gas evolution through stochastic laws (random process and its stochastic Kolmogorov equation equivalent) (Leontovich, 1936; Kac, 1963, 1973; Vallander, 1967; Miroshin, 1967; Spohn, 1980).

To establish an asymptotic equivalence between these two approaches, let us consider the gas system as a finite set of particles in the phase space (Γ). Then, using the Liouville equation for a given system of N particles, it is possible to obtain the kinetic equation for the uniform-space evolution of the gas in a Markovian form (Prigogine, 1962)

$$\frac{\partial}{\partial t}\rho_N(t) = K_\Pi \rho_N(t) \; , \tag{5.1}$$

where

$$\rho_N(t) = \bigcup_{i=1}^{N} \mathbf{c}_i(t)$$

is the velocity distribution function, and operator $K_\Pi = \Omega \Psi(0)$ is the operator of collisions. Operator $\Psi(0)$ approximates the collisions in an asymptotic form, i.e. the time spell of collisional interaction is considered as a small value, while the operator Ω takes into account the finite times of the collisional processes. For the operator $\Psi(0)$ the following estimation (Prigogine, 1962) is valid:

$$\Psi(0) \sim \frac{1}{\tau_r} \sim n_0 d_0^2 v_0 \; [1 + \gamma + \gamma^2 + \ldots] \; ,$$

where $\gamma = n_0 d_0^3$, $\tau_r \simeq (n_0 d_0^2 v_0)^{-1}$ (d_0 is the finite radius of interaction potential in particle collisions, v_0 - thermal velocity). The first term of this series expansion is caused by binary collisions, the second - by triple ones, and etc. The operator Ω can be evaluated in the similar manner:

$$\Omega \sim 1 + \gamma + \gamma^2 + \ldots \; .$$

Hence, under the condition of gas dilution ($\gamma \ll 1$) the only binary collisions must be taken into account in the evolution equation. Then $\Psi(0) = \sum_{ij} S_{ij}$, and operators S_{ij} are equal to (Prigogine, 1962; Kac, 1973)

$$S_{ij}\rho_N = V^{-1} \int g_{ij} \, d\sigma_{ij} [\; \rho_N(\mathbf{c}_1, \ldots, \mathbf{c}_i', \ldots, \mathbf{c}_j', \ldots, \mathbf{c}_N) -$$

$$\rho_N(\mathbf{c}_1, \ldots, \mathbf{c}_i, \ldots, \mathbf{c}_j, \ldots, \mathbf{c}_N)] \; , \tag{5.2}$$

where $(\mathbf{c}_i', \mathbf{c}_j')$ are the particle velocities $(\mathbf{c}_i, \mathbf{c}_j)$ after collision, $g_{ij} =| \mathbf{c}_i - \mathbf{c}_j |$, $d\sigma_{ij} = \sigma(g_{ij}, \Omega) \, d\Omega$ is the differential scattering cross-section, V is the gas volume. From this approach it follows that gas evolution can be regarded as a sequence of completed (time of interaction $(t_* \ll t_r)$) and statistically independent scattering acts (collisions).

These basic statements of the kinetic theory open an opportunity to construct a physical-probabilistic analogue of collisional processes in the rarefied gas. To accomplish the goal it is necessary to organize the random process, its master equation being compatible with the kinetic equation of rarefied gas evolution

$$\frac{\partial}{\partial t}\rho_N(t) = \sum_{i,j}^{N} S_{ij}\rho_N(t) \ . \tag{5.3}$$

By definition, this is the Markovian random process which, in the space uniform case, has a jump-like nature. Transitions between the system states are considered to be analogues of collision changes of gas state.

At the second step, using some physical assumptions [hypothesis of molecular chaos (Prigogine, 1962; Kac, 1963, 1973)] and method of asymptotic series expansions (Prigogine, 1962; Gigulev, 1971), the mathematical concept of the initial system evolution in the one-particle phase space (μ - space) both for dynamic and stochastic models must be introduced. In other words, an asymptotic equivalence between the dynamic and stochastic descriptions at the level of one-particle distribution functions, which are the solutions of the Boltzmann kinetic equations, must be found. The mathematical basis for this procedure (construction of the physical-probabilistic analogue) is the conformity of definite-type Markovian processes to nonlinear parabolic equations of the Boltzmann type (Kac, 1963, 1973; McKean, 1966).

The physical assumptions underlying the gas model evaluation such as dilution and finite and fast decreasing radius of interaction potential permit one to specify the class of Markovian processes to satisfy the Boltzmann kinetic equation. For a space-uniform gas, the uniform Markovian process is appropriate.

It also follows from the above mentioned physical assumptions that gas state changes are caused by instant, time-shared collisions, what relates them to the class of uniform jump-like Markovian processes. The structure of differential rates:

$$\begin{cases} dG^{S_r^{\alpha\alpha_1}} = F_\alpha(\mathbf{c})F_{\alpha_1}(\mathbf{c}_1)\, dI^{S_r^{\alpha\alpha_1}}\, d\mathbf{c}\, d\mathbf{c}_1 \\ dI^{S_r^{\alpha\alpha_1}} = \mid \mathbf{c} - \mathbf{c}_1 \mid d\sigma^{S_r^{\alpha\alpha_1}} \end{cases} \tag{5.4}$$

describing the gas state changes in the collisions gives rise to the probabilistic characteristics of the transitions between states of random process.

The conservation laws determine the invariants of transitions and allocate the phase space region where the process state changes occur.

5.1.2. PHYSICAL AND PROBABILISTIC ANALOGUE OF A NONEQUILIBRIUM KINETIC SYSTEM

Using the relationship between the dynamic and probabilistic descriptions of the collisional evolution of a rarefied gas, the physical-probabilistic analogue of the nonequilibrum kinetic model was developed (Zmievskaya $et\ al.$, 1979, 1980, 1983; Pyarnpuu $et\ al.$, 1981; Shematovich, 1980).

The rarefied multicomponent gas $\{a^\alpha\}_{(M^\alpha)}$ is considered as a discrete medium in which the countable set $\{a_i^\alpha\}_{(i=1,\dots,m,\dots)}$ of particles correspond to each a^α component and, consequently, the following relation:

$$\{a^\alpha\}_{(M^\alpha)} \to \bigcup_\alpha^{M^\alpha} \bigcup_i^\infty a_i^\alpha$$

describes the system. In the uniform-space case, each particle of the medium is characterized by its velocity c_i^α and by individual atomic attributes (molecular weight, charge, etc.). Then the state of discrete medium $\{a_i^\alpha\}$ can be described in the Gibbs space $\{\mathbf{C}\} = \cup_{\alpha=0}^{M^\alpha}\{\mathbf{C}^\alpha\}$ by point $\mathbf{C} = \cup_\alpha \cup_{i=1}^\infty c_i^\alpha$, where $\{\mathbf{C}^\alpha\}$ - the velocity space of all particles a_i^α, $(i = 1,\dots,m,\dots)$.

As it was already mentioned, the probabilistic description of the rarefied gas (see, Leontovich, 1936; Vallander, 1967; Miroshin, 1967; Spohn, 1980) is based on the theory of random processes (Kolmogorov, 1938; Feller, 1970; Skorochod, 1983). Using this approach and taking also into account the hypothesis of measure factorization in the Gibbs space $\{\mathbf{C}\}$ (analogue of molecular chaos hypothesis fora rarefied gas), the state of the considered discrete medium can be described by the set $(i = 1,\dots,m,\dots; \alpha = 0, 1,\dots, M)$ of individual probabilistic characteristics:

$$P_i^\alpha(c_i^\alpha(t) \in \Gamma^\alpha) = \int_{\Gamma^\alpha} p_i^\alpha(c_i^\alpha, t)\, dc_i^\alpha \ , \tag{5.5}$$

where $p_i^\alpha(c_i^\alpha, t)$ is the density of probabilistic distribution of particle a_i^α at a given moment of time; the set $\Gamma^\alpha \subset \{\mathbf{C}^\alpha\}$.

The physical assumptions adopted earlier allow to introduce the boundary $(\hat{r}, \hat{t})$ (Klimontovich, 1980), sharing the regions of atomic ($\sim r_* \simeq d_0, \tau_* = d_0/v_0$) and kinetic ($\sim \tau^* \simeq \tau\,(= \lambda/v)$, $r^* = \lambda\,(= (n_0 d_0^2)^{-1})$ space-temporal scales for the appropriate collisional processes. Hence, the probabilistic characteristics (5.5) are smoothed out on scales $(\hat{r}, \hat{t})$, and therefore, these values describe the evolution of gas $\{a^\alpha\}$ in time $\hat{\tau} \ll t \ll \tau^*$ and space $\hat{r} \ll r \ll r^*$ scales (i.e. on the microscopic kinetic level). In a rarefied gas both the volume $\hat{r}^3$ and time interval $\hat{\tau}$ are physical infinitesimals (Klimontovich, 1980), and therefore interactions of particles can be described as instant and local in space transitions (i.e., lifetimes of complexes, forming in the process of interactions of particles, are small as compared to the lifetimes of particles, $\tau_* \ll \tau^*$).

Using the characteristics (5.5), it is possible to determine a random measure in space $\{\mathbf{C}^\alpha\}$:

$$S^\alpha(\Gamma^\alpha, t) = \sum_{i=1}^{\infty} \eta(\mathbf{c}_i^\alpha(t), \Gamma^\alpha); \quad \eta(\mathbf{c}^\alpha, \Gamma^\alpha) = \begin{cases} 1, & \mathbf{c}^\alpha \in \Gamma^\alpha \\ 0, & \mathbf{c}^\alpha \notin \Gamma^\alpha \end{cases} \quad (5.6)$$

This measure determines the number of elements a^α in set Γ^α at time t. At the probabilistic level of description, the functional $S(\Gamma^\alpha, t)$ corresponds to the microscopic phase density (Valander, 1967) and, consequently, the velocity distribution function used in rarefied gas dynamics is connected with probabilistic characteristics (5.5), (5.6) as the moment of measure (5.6). All physical characteristics of the gas can also be considered as moments of measure $S(\Gamma^\alpha, t)$.

The idea on space-temporal locality of the gas particle interactions favors the concept of evolution of the medium as a strict Markovian random process. Respectively, the transitive probability $Q_i^\alpha(\Gamma^\alpha, t' \mid \mathbf{c}_i^\alpha, t)$ is defined as the probability for particle a_i^α, characterized by $\mathbf{c}_i^\alpha$ at time t, to be at time t' with velocity $\mathbf{c}_i^{\alpha\prime}$ in the set $\Gamma^\alpha \subset \{\mathbf{C}^\alpha\}$. This probabilistic characteristic is described by the Chapman-Kolmogorov equation (Kolmogorov, 1938; Feller, 1970):

$$\begin{cases} Q_i^\alpha(\Gamma^\alpha, t' \mid \mathbf{c}_i^\alpha, t) = \int\limits_{\{\mathbf{C}^\alpha\}} Q_i^\alpha(\Gamma^\alpha, t' \mid \mathbf{v}_i^\alpha, t'') \times \\ \qquad\qquad\qquad \times Q_i^\alpha(d\mathbf{C}_{\mathbf{v}_i^\alpha}^\alpha, t'' \mid \mathbf{c}_i^\alpha, t) \\ \mathbf{v}_i^\alpha \in d\mathbf{C}_{\mathbf{v}_i^\alpha}^\alpha \subset \{\mathbf{C}^\alpha\} \,, \quad t < t'' < t' \,. \end{cases} \quad (5.7)$$

In general case, the transitive probability can be written with accuracy up to terms $O(t' - t)$ in the following form:

$$Q_i^\alpha(\Gamma^\alpha, t' \mid t) = \Lambda_{\alpha,i}(\mathbf{c}_i^\alpha \in \Gamma^\alpha; t)(t' - t) + O(t' - t) \,,$$

where

$$\begin{cases} \Lambda_{\alpha,i}(\mathbf{c}_i^\alpha \in \Gamma^\alpha; t) \\ \Lambda_{\alpha,i}(\mathbf{c}_i^\alpha \in \mathbf{C}^\alpha; t) = \lambda_{\alpha,i}(\mathbf{c}_i^\alpha, t) \end{cases}$$

is the modified standard pair of λ-functions for a jump-like Markovian process (Feller, 1970).

It follows from these expressions that the transitive probability for the Markovian process, which describes the sequence of particle collisions in a given discrete medium, can be written in the form

$$Q_i^\alpha(\Gamma^\alpha, t' \mid t) = \sum_{\alpha(r),r} \sum_{i'} \Lambda_{i',i}^{s_r^{\alpha\alpha_1}}(c_i^\alpha \in \Gamma^\alpha; c_{i'}^{\alpha(r)}, t)(t' - t) \ .$$

Here the functions Λ represent the probabilities of particle a_i^α location in the set Γ^α after binary particle interactions in physical-chemical processes $s_r^{\alpha\alpha_1} : \alpha(r) + \alpha_1(r) \to \alpha + \alpha_1$. These functions are determined from the dynamical-probabilistic description of elementary collision acts and are equal to (Zmievskaya et al., 1980):

$$\begin{cases} \Lambda_{i',i}^{s_r^{\alpha\alpha_1}}(\Gamma^\alpha; c_{i'}^{\alpha(r)}, t) = \sum_{\alpha_1(r),r} \sum_{i'_1}^{(i')} \int_{\Gamma^\alpha} dc_i^\alpha \times \\ \times \int_{C^{\alpha_1(r)}} \tilde{I}(\alpha_{F\to I}; c_{i'}^{\alpha(r)}, c_{i'_1}^{\alpha_1(r)} \to c_i^\alpha) \, P_{i'_1}^{\alpha_1(r)} \, dc_{i'_1}^{\alpha_1(r)} \\ \lambda_{i',i}^{s_r^{\alpha\alpha_1}}(c_{i'}^{\alpha(r)}, t) = \sum_{\alpha_1(r),r} \sum_{i'_1}^{(i')} \int_{C^{\alpha_1(r)}} dI(\alpha_{F\to I}) P_{i'_1}^{\alpha_1(r)} \, dc_{i'_1}^{\alpha_1(r)}. \end{cases} \tag{5.8}$$

In the expressions (5.8) the following notations are used: $\sum^{(i')}$, where the summarizing in the case of $\alpha_1(r) = \alpha(r)$ is carried out on $i'_1 \neq i'$. In (5.8) the scattering function

$$\tilde{I}(\alpha_{F\to I}; c_{i'}^{\alpha(r)}, c_{i'_1}^{\alpha_1(r)} \to c_i^\alpha) = \int_{C^{\alpha_1(r)}} dc_{i_1}^{\alpha_1} \times$$
$$\times I^{s_r^{\alpha\alpha_1}}(\alpha_{F\to I}; c_{i'}^{\alpha(r)}, c_{i'_1}^{\alpha_1(r)} \to c_i^\alpha, c_{i_1}^{\alpha_1})$$

is used as well. This function determines the probability of particle formation with velocity c_i^α in collisions of particles with velocities $c^{\alpha(r)}$ and $c^{\alpha_1(r)}$ (process $s_r^{\alpha\alpha_1} : \alpha(r) + \alpha_1(r) \to \alpha + \alpha_1$).

Let us consider the case when the density of transitive probability q_i^α is defined, i.e.

$$Q_i^\alpha(\Gamma^\alpha, t'|v_i^\alpha, t) = \int_{\Gamma^\alpha} dc_i^\alpha q_i^\alpha(c_i^\alpha, t'|v_i^\alpha, t) \ .$$

Then, after substituting expression (5.8) in the Chapman-Kolmogorov equation (5.7) and after taking the derivative on final time t', we obtain the direct Kolmogorov equation:

$$\frac{\partial}{\partial t'} q_i^\alpha(c_i^\alpha, t'|v_i^\alpha, t) = - \sum_{\alpha_1,r,i'}^{(i)} \lambda^{s_r^{\alpha\alpha_1}}(c_i^\alpha, t) q_i^\alpha(c_i^\alpha, t'|v_i^\alpha, t) +$$

$$+ \sum_{\alpha(r),r,i'}^{(i'_1)} \int_{C_r} dc_{i'}^{\alpha(r)} q_{i'}^{\alpha(r)}(c_{i'}^{\alpha(r)}, t'|w_{i'}^{\alpha(r)}, t) \frac{\partial}{\partial V(\Gamma^\alpha)} \Lambda_{i',i|c_i^\alpha}^{s_r^{\alpha(r)\alpha_1(r)}} \quad ,$$

where $\mathbf{C}_r = \mathbf{C}^{\alpha(r)} \setminus \delta_{\alpha(r)\alpha}\mathbf{c}_i^\alpha$, and $V(\Gamma^\alpha) = \int_{\Gamma^\alpha} dc_i^\alpha$.

If the initial particle distributions $\{p_i^\alpha(\mathbf{v}_i^\alpha, t_0)\}$ are known, then, in agreement with the total probability expression, the following relation can be written:

$$p_i^\alpha(\mathbf{c}_i^\alpha, t) = \int_{\mathbf{C}^\alpha} d\mathbf{v}_i^\alpha \ q_i^\alpha(\mathbf{c}_i^\alpha, t'|\mathbf{v}_i^\alpha) \ p_i^\alpha(\mathbf{v}_i^\alpha, t_0) \quad .$$

Using this expression it is also possible to obtain (by multiplying the direct Kolmogorov equation by $\prod_{\alpha,i} p_i^\alpha(\mathbf{v}_i^\alpha, t_0)$ and then integrating it over velocities) the evolutionary equation for a particle of the considered discrete medium:

$$\tfrac{\partial}{\partial t'}p_i^\alpha(\mathbf{c}_i^\alpha, t') = - \sum_{\alpha_1,r,i_1}{}^{(i)} \int_{\mathbf{C}^{\alpha_1}} dc_{i_1}^{\alpha_1} \ dI_{ii_1}^{s_r^{\alpha\alpha_1}} \ p_i^\alpha p_{i_1}^{\alpha_1} +$$

$$+ \sum_{\alpha(r),r,i'}\sum_{\alpha_1(r),r,i_1'}{}^{(i')} \int_{\mathbf{C}^{\alpha(r)}} dc^{\alpha(r)} \int_{\mathbf{C}^{\alpha_1(r)}} dc^{\alpha_1(r)} \tilde{I}_{i'i_1'}^{s_r^{\alpha\alpha_1}} (\alpha_{F\to I})p_{i'}^{\alpha(r)} p_{i_1'}^{\alpha_1(r)} \quad .$$

Finally, using the expression of microscopic reversibility (3.31) and property (3.28) of scattering functions, we can come to the direct Kolmogorov equation for the jump-like Markovian process $\mathbf{C}(t)$:

$$\begin{cases} \tfrac{\partial}{\partial t}p_i^\alpha(\mathbf{c}_i^\alpha, t) = \sum_{\alpha_1,i_1}\sum_r L_{ii_1}^{s_r^{\alpha\alpha_1}} (p_i^\alpha, p_{i_1}^{pha_1}) \\ p_i^\alpha|_{t=0} = {}^{(0)}p_1^\alpha(\mathbf{c}_i^\alpha), \quad \alpha = 0,1,\ldots,M^\alpha; \ i = 1,\ldots,m,\ldots \quad , \end{cases} \qquad (5.9)$$

where ${}^{(0)}p_i^\alpha$ are the given initial densities of probabilistic distributions of particles, and nonlinear operators are defined as

$$L_{ii_1}^{s_r^{\alpha\alpha_1}} (p_i^\alpha, p_{i_1}^{\alpha_1}) = \int_{\{\mathbf{C}^{\alpha_1}\}} [e^{s_r^{\alpha\alpha_1}} \ p_{i'}^{\alpha(r)}(\mathbf{c}_{i'}^{\alpha(r)}) \ p_{i_1'}^{\alpha_1(r)}(\mathbf{c}_{i_1'}^{\alpha_1(r)}) -$$

$$-p_i^\alpha(\mathbf{c}_i^\alpha) \ p_{i_1}^{\alpha_1}(\mathbf{c}_{i_1}^{\alpha_1})] \ dc_{i_1}^{\alpha_1} \ dI_r^{s_r^{\alpha\alpha_1}} (g_{ii_1}^{\alpha\alpha_1}) \quad . \qquad (5.10)$$

These operators correspond to collisional physical and chemical interactions of particles.

Here we emphasize again that direct Kolmogorov equation for the jump-like Markovian process describes the evolution of discrete medium particles at the dynamical-probabilistic level. The equations (5.9) - (5.10) averaged by particle ensemble coincide in the uniform-space case with the Boltzmann equation for elastic collisions, as well as with the Wang-Chang and Uhlenbeck equation in the case of inelastic collisions and with the Mazur equation in the case of chemical reactions (Prigogine, 1962; Uhlenbeck,

1963; Hochstim (Ed.), 1969; Ferziger and Kaper, 1972). The expressions (5.5) - (5.10) determine the probabilistic level of description of collisional processes (4.3) in a rarefied multicomponent gas. These expressions can be also considered as the physical-probabilistic analogue of nonequilibrium kinetic system. This is the basic concept for the study of nonequilibrium collisional kinetics of rarefied multicomponent gas using the well developed direct simulation Monte Carlo methods.

5.1.3. NUMERICAL STOCHASTIC MODEL

The physical-probabilistic analogue of the nonequilibrum kinetic system can be realized in the computer modeling proceeding along the following sequence:

a) approximation of a set of gas particles by the system of finite number of modeling particles (mathematical model);

b) development of a random process the describing evolution of the given numerical system;

c) introduction of the probabilistic description of the numerical model with the involvement of probabilistic approximation of kinetic processes pertinent to the given model;

d) development of algorithm and numerical realization (running) of the model, including evaluation of the efficiency of the algorithm and limits of its applicability.

Based on the methods of direct Monte Carlo simulation (Bird, 1976, 1994; Belotserkovskii and Yanitskii, 1975; Belotserkovskii, 1984) the initial set of particles of a rarefied multicomponent gas is replaced by the mathematical model, i.e. by the numerical system containing a finite number N of modeling particles as follows:

$$
\begin{cases}
\{a^\alpha\}_{(M^\alpha)} \to \{a_i^\alpha\}_{(M^\alpha, N)} \\
N = \sum_{\alpha=0}^{M^\alpha} N^\alpha .
\end{cases}
\tag{5.11}
$$

Particles from set $(1, 2, \ldots, N^\alpha)$ have the velocities $(\mathbf{c}_1^\alpha, \ldots, \mathbf{c}_{\alpha N}^\alpha)$ and correspond to a^α - component of gas. The states of numerical system (5.11) are characterized by discrete representations of kinetic distribution functions $F_\alpha(\mathbf{c}, t) = n_\alpha(t) f_\alpha(\mathbf{c}, t)$, and can be described by the following expression:

$$
\mathbf{C} = \bigcup_{\alpha=0}^{M^\alpha} \mathbf{C}^\alpha , \mathbf{C}^\alpha = \bigcup_{i=1}^{N^\alpha} \mathbf{c}_i^\alpha = \{N^\alpha; \mathbf{c}_1^\alpha, \ldots, \mathbf{c}_{N^\alpha}^\alpha\} .
\tag{5.12}
$$

The evaluation of the uniform jump-like Markovian process $\mathbf{C}(t)$ corresponding to the stochastic nature of changes in the numerical model (5.11) includes the following steps:

a) jump-like transitions $\mathbf{C} \to \mathbf{C}'$ between model states;

b) probabilities of these transitions $w(\mathbf{C} \to \mathbf{C}')$;

c) the probabilistic distribution of time $\tau(\mathbf{C} \to \mathbf{C}')$ between the consequent transitions in the model.

The transition $\mathbf{C} \to \mathbf{C}'$ between model states is realized as an instant change of model particle characteristics in accordance with the dynamical-probabilistic scheme of initial elementary collision acts

$$s_r^{\alpha \alpha_1} \; : \; \mathbf{c}_i^{\alpha}, \mathbf{c}_{i_1}^{\alpha_1} \to \mathbf{c}_{i'}^{\alpha(r)}, \mathbf{c}_{i_1'}^{\alpha_1(r)} \; . \tag{5.13}$$

The new velocity values are obtained using the conservation laws (3.27), and their directions are statistically determined through differential scattering cross-sections $\sigma^{s_r^{\alpha\alpha_1}}(g^{\alpha\alpha_1}, \Omega)$. Therefore, it is possible to write

$$\mathbf{C} \to \mathbf{C}' = \mathbf{C}_{ii_1}^{s_r^{\alpha\alpha_1}} \; , \tag{5.14}$$

where the model state after the transition (5.13) is equal to:

$$\mathbf{C}_{ii_1}^{s_r^{\alpha\alpha_1}} = \begin{cases} \mathbf{C}^{\alpha'}, \quad \alpha' \neq \alpha, \alpha_1, \alpha(r), \alpha_1(r) \\ (-)\mathbf{C}_{i'}^{\alpha} \, (-)\mathbf{C}_{i_1}^{\alpha_1} \\ (+)\mathbf{C}^{\alpha(r)}(\mathbf{c}_{i'}^{\alpha(r)}) \, (+)\mathbf{C}^{\alpha_1(r)}(\mathbf{c}_{i_1'}^{\alpha_1(r)}) \; . \end{cases} \tag{5.15}$$

Here the following set of definitions is used:

$$\mathbf{V}^{\gamma} = \bigcup_{k=1}^{N^{\gamma}} \mathbf{v}_k^{\gamma}, \; (-)\mathbf{V}_i^{\gamma} = \mathbf{V}^{\gamma} \setminus \mathbf{v}_i^{\gamma}, \; (+)\mathbf{V}^{\gamma}(\mathbf{v}_{i'}^{\gamma}) = \mathbf{V}^{\gamma} \cup (\mathbf{v}_{i'}^{\gamma}) \; .$$

The probability of transition (5.14) is the discrete representation of the differential rate (5.4) of the collisional process

$$\begin{cases} w(\mathbf{C} \to \mathbf{C}') = w_{ii_1}^{s_r^{\alpha\alpha_1}} = V^{-1} G_{ii_1}^{s_r^{\alpha\alpha_1}} \\ G_{ii_1}^{s_r^{\alpha\alpha_1}} = |\mathbf{c}_i^{\alpha} - \mathbf{c}_{i_1}^{\alpha_1}| \, \sigma^{s_r^{\alpha\alpha_1}}(\mathbf{c}_i^{\alpha}, \mathbf{c}_{i_1}^{\alpha_1}) \; . \end{cases} \tag{5.16}$$

Here V is the parameter of the model with a dimension of volume (cm^3) which is used for determination of the time scale. The total frequency of transitions in the considered numerical system is equal to:

$$w(\mathbf{C}) = \sum_{\alpha,\alpha_1,r} \sum_{i,i_1} w_{ii_1}^{s_r^{\alpha\alpha_1}} \; . \tag{5.17}$$

Because $\mathbf{C}(t)$ is a strict Markovian jump-like random process (Belotserkovskii and Yanitskii, 1975; Zmievskaya et al., 1980,) the time intervals sharing transitions between model states are defined by the total frequency (5.17), i.e., addressing the whole system state, and are distributed by an exponential law

$$P\{\tau(\mathbf{C} \to \mathbf{C}') \le \tau\} = 1 - \exp(-\omega(\mathbf{C})\tau) \ . \tag{5.18}$$

Finally, expressions (5.11) - (5.18) give an accurate determination of the random process $\mathbf{C}(t)$ describing the evolution of the numerical stochastic model. Referring to the definitions of the random process theory, one can obtain the evolutionary equation of the numerical system (5.11).

Indeed, the transitions (5.14) in the numerical system (5.11) are adequate to the integral transformation (Shematovich, 1980)

$$\hat{f}(\mathbf{X}) = \int f(\mathbf{Y}) \, K(\mathbf{X}, \mathbf{Y}) \, d\mathbf{Y} \ ,$$

where $f(\mathbf{X})$ is the continuous function, and $K(\mathbf{X}, \mathbf{Y})$ is density of stochastic nucleus (Feller, 1970) of the jump-like Markovian process $\mathbf{X}(t)$. For the stochastic model under consideration this transformation takes the following form:

$$\hat{f}(\mathbf{C}) = V^{-1}\omega^{-1}(\mathbf{C}) \sum_{\alpha,\alpha_1,r} \sum_{i,i_1} e^{s_r^{\alpha\alpha_1}} \int f(\mathbf{C}_{ii_1}^{s_r^{\alpha\alpha_1}}) \, g_{ii_1}^{\alpha\alpha_1} \, d\sigma^{s_r^{\alpha\alpha_1}} \ . \tag{5.19}$$

Using (5.19) and the direct Kolmogorov equation for the process $\mathbf{C}(t)$, we obtain the following evolutionary equation of the given stochastic model

$$\frac{\partial}{\partial t}\varphi(\mathbf{C}, t) = \omega(\mathbf{C}) \left[\hat{\varphi}(\mathbf{C}, t) - \varphi(\mathbf{C}, t)\right] \ , \tag{5.20}$$

where $\varphi(\mathbf{C}, t)$ is the density of probabilistic distribution of the model state at time t, and $\hat{\varphi}(\mathbf{C}, t)$ is the integral transformation (5.19) of the given function.

From this analytical consideration it is clear, that the evolution of the stochastic model state (5.11) is defined by the solution of Cauchy problem

$$\begin{cases} \frac{\partial}{\partial t}\varphi(\mathbf{C}, t) = K \ \varphi(\mathbf{C}, t) \\ \varphi|_{t_0} = \varphi_0(\mathbf{C}) \ , \end{cases} \tag{5.21}$$

where the operator K includes all possible collisional transitions (5.14) for the model state $\mathbf{C}$ and is equal to

$$\begin{cases} K \ \varphi = \sum_{\alpha,\alpha_1,r} \sum_{i,i_1} K_{ii_1}^{s_r^{\alpha\alpha_1}} \ \varphi \\ K_{ii_1}^{s_r^{\alpha\alpha_1}} \ \varphi = V^{-1} \int dI_{ii_1}^{s_r^{\alpha\alpha_1}} \ [e^{s_r^{\alpha\alpha_1}}\varphi(\mathbf{C}_{ii_1}^{s_r^{\alpha\alpha_1}}, t) - \varphi(\mathbf{c}, t)] \ , \end{cases} \tag{5.22}$$

$\varphi_0(\mathbf{C})$ being the initial distribution of the numerical model state. Thus, equations (5.21) and (5.22) represent the so-called master equation for the numerical stochastic model and in the case of binary elastic collisions coincide with the well-known Kac-Prigogine equation (5.3) for a N - particle velocity distribution function.

In order to ensure the existence and uniqueness of the solution of equation (5.21) for the stochastic model, one must place constrains on the smoothness of functions included in the structure of the transitive probability densities of process $\mathbf{C}(t)$. To satisfy this condition one must require the continuity of scattering functions

$$dI_{s_r}^{s^{\alpha\alpha1}} = g^{\alpha\alpha1}\sigma_{s_r}^{s^{\alpha\alpha1}}(g^{\alpha\alpha1}, \Omega^{s^{\alpha\alpha1}})$$

$$I_{s_r}^{s^{\alpha\alpha1}} = g^{\alpha\alpha1}\sigma^{s^{\alpha\alpha1}}(g^{\alpha\alpha1})$$

for all physical and chemical processes with collision parameters in the following intervals

$$\{0 < g^{\alpha\alpha1} \leq g_{max}; \ 0 < \Omega^{s^{\alpha\alpha1}} \leq 4\pi\} \ ,$$

where $g_{max} < \infty$ is the finite parameter, determined by the given initial state of the numerical model.

In this case the following statement is valid: if the distribution of the initial state of the numerical stochastic model has a continuous density $\varphi_0(\mathbf{C})$, and if the scattering functions are continuous as well, then the density $\varphi(\mathbf{C}, t)$ is the unique solution of the Cauchy problem (5.21) on the class of continuous density functions (Shematovich, 1980).

It should be noted that equation (5.21) is linear and can be effectively solved using the analogue algorithms of the Monte-Carlo method (Shematovich, 1979). Suitable algorithms developed for solving the master equations for stochastic models of specific physical-chemical systems are discussed below.

An approximation of the probabilistic description of a space-uniform gas $\{a_i^\alpha\}_{(M^\alpha, \infty)}$ by the numerical stochastic model $\{a_i^\alpha\}_{(M^\alpha, N)}$ can be obtained using the relationship between one-particle evolutionary equations of a discrete medium and numerical system.

We introduce the following functions:

$$\varphi^{(\alpha,\dots)}(\mathbf{C}^{(\alpha,\dots)}, t) = \int \varphi(\mathbf{C}, t) \prod_\gamma {}^{(\alpha,\dots)} d\mathbf{C}^\gamma \ , \tag{5.23}$$

where

$$\begin{cases} d\mathbf{C}^\gamma = \prod_{i=1}^{N^\gamma} dc_i^\gamma; \quad \mathbf{C}^{(\alpha,\dots)} = (\mathbf{C}^\alpha, \dots) \\[2mm] \prod_\gamma^{M^\alpha} {}^{(\alpha,\dots)} d\mathbf{C}^\gamma = \dots \times d\mathbf{C}^{\alpha-1} \times d\mathbf{C}^{\alpha+1} \times \dots \end{cases} \tag{5.24}$$

They can be considered as densities of the common distribution of states of a^α - type particle subsystems of the model at time t. Let us note that these density functions are symmetric relative to the particle subsystems rearrangement

$$\varphi^{(\alpha,\alpha_1,\dots)}(\mathbf{C}^{(\alpha,\alpha_1,\dots)},t) = \varphi^{(\alpha_1,\alpha,\dots)}(\mathbf{C}^{(\alpha_1,\alpha,\dots)},t) \ .$$

As a first step, equation (5.21) is integrated on the velocities of all modeling particles, except for the velocities of the particles of given subsystem, in yielding the set of evolutionary equations for the subsystems of the initial numerical model:

$$\begin{cases} \frac{\partial}{\partial t}\varphi^{(\alpha)}(\mathbf{C}^\alpha,t) = \sum\limits_{\alpha_1,r}\sum\limits_{i,i_1}(i)K_{\alpha,ii_1}^{s_r^{\alpha\alpha_1}}(\varphi^{(\alpha,\alpha_1,\alpha(r),\alpha_1(r))}) \\ \varphi^{(\alpha)}(\mathbf{C}^\alpha)|_{t_0} = {}^{(0)}\varphi^{(\alpha)}(\mathbf{C}^\alpha), \quad \alpha = 0,1,\dots,M^\alpha \ , \end{cases} \qquad (5.25)$$

where operators K are equal

$$\begin{cases} K_{\alpha,ii_1}^{s_r^{\alpha\alpha_1}}\ \varphi^{(\alpha,\alpha_1,\alpha(r),\alpha_1(r))} = V^{-1}\int dI_{ii_1}^{s_r^{\alpha\alpha_1}} \times [e^{s_r^{\alpha\alpha_1}}\times \\ \varphi^{(\dots)}(\mathbf{C}_{ii_1}^{s_r^{\alpha\alpha_1}},t) - \varphi^{(\dots)}(\mathbf{C}^{(0)},t)]\,d\mathbf{C}^{\alpha_1}\,d\mathbf{C}^{\alpha(r)}\,d\mathbf{C}^{\alpha_1(r)} \ . \end{cases} \qquad (5.26)$$

In this expression $\mathbf{C}_{ii_1}^{s^{\alpha\alpha_1}}$ is the $(\alpha,\alpha_1,\alpha(r),\alpha_1(r))$ state of the subsystem of the stochastic model after the transition (5.14).

We notice that the Cauchy problem (5.25) is equivalent to problem (5.21); however, this set of equations is not closed for functions $\varphi^{()}$. To solve the problem (5.21), one can use the standard procedure of approximate expansion methods. This approach is analogous to the method of asymptotic decomposition, which is usually used in the kinetic theory, in particular, for solving the BBGKI (Bogolubov-Born-Green-Kirkwood-Irwine) hierarchy equations (Prigogine, 1962; Gigulev, 1971).

To apply this approach, we introduce the following expressions for functions

$$\varphi_{(s,\dots)}^{(\alpha,\dots)}(\mathbf{C}_{(s,\dots)}^{(\alpha,\dots)},t) = J_{(s,\dots)}^{(\alpha,\dots)}\int \varphi^{(\alpha,\dots)}(\mathbf{C}^{(\alpha,\dots)},t)\prod_\gamma^\alpha\prod_{i=s+1}^{N^\gamma} dc_i^\gamma \ , \qquad (5.27)$$

where

$$J_{(s,\dots)}^{(\alpha,\dots)} = \prod_\gamma^\alpha (N^\gamma)! \ / \ (N^\gamma - s_\gamma)!$$

$$\mathbf{C}_{(s,\dots)}^{(\alpha,\dots)} = \{\mathbf{C}_{(s)}^\alpha,\dots\}, \quad \mathbf{C}_{(s)}^\alpha = \bigcup_{i=1}^s \mathbf{c}_i^\alpha \ .$$

Functions (5.27) may be called (n, S) - particle densities of the probabilistic distribution of $s, \ldots$ particles from $\alpha, \ldots$ subsystems of the model; $n = \sum_\gamma \sum_\nu \delta_{\gamma\nu}$; $S = \sum_\gamma^{\alpha, \ldots} s_\gamma$. The normalization condition for functions (5.27) has the form

$$\int \varphi_{(s, \ldots)}^{(\alpha, \ldots)} (\mathbf{C}_{(s, \ldots)}^{(\alpha, \ldots)}, t) \prod_\gamma^{\alpha, \ldots} dC_{(s_\gamma)}^\gamma = J_{(s, \ldots)}^{(\alpha, \ldots)} . \tag{5.28}$$

Functions (5.27) are symmetric relative to the rearrangement of the arguments within the limits of the given subsystem particles and relative to the rearrangement of groups of the model subsystem particles.

At the next step, integrating the set of equations (5.25) by $\prod_\gamma^{\alpha, \ldots} \prod_{i=s+1}^{N^\gamma} dc_i^\gamma$, we obtain the following system

$$\frac{\partial}{\partial t} \varphi_{(s)}^{(\alpha)} (\mathbf{C}_{(s)}^{(\alpha)}, t) = D_\alpha + \tilde{D}_\alpha , \tag{5.29}$$

where the operators D_α and $\tilde{D}_\alpha$ are equal to:

$$D_\alpha = \sum_{\alpha_1, r} \sum_i \sum_{i_1}^{s_1} D_{ii_1}^{s_r \alpha \alpha_1} (\varphi_{(s, s_1, s(r), s_1(r))}^{(\alpha, \alpha_1, \alpha(r), \alpha_1(r))}) = \int K_\alpha^{s_r \alpha \alpha_1} (\varphi^{()}) \prod_\gamma^{\alpha, \alpha_1, \ldots} \prod_{j=s_\gamma+1}^{N^\gamma} dc_i^\gamma$$

$$\tilde{D}_\alpha = \sum_{\alpha_1, r} \sum_{i=1}^s (N^{\alpha_1} - s_1) \int D_{is_1+1}^{s_r \alpha \alpha_1} dc_{s_1+1}^{\alpha_1} ,$$

respectively. In order to derive equation (5.29), the symmetry of collisions and the property of functions $\varphi_{()}^{()}$ earlier described were used. To clarify this set of equations we introduce the normalized density

$$f_{(s, \ldots)}^{(\alpha, \ldots)} = \theta_{(s, \ldots)}^{(\alpha, \ldots)} \varphi_{(s, \ldots)}^{(\alpha, \ldots)} , \tag{5.30}$$

where

$$\theta_{(s, \ldots)}^{(\alpha, \ldots)} = \prod_\gamma^{\alpha, \ldots} \theta_{s_\gamma}$$

$$\theta_{s_\gamma} = V^{-s}\gamma [N^\gamma (N^\gamma - 1) \times \ldots \times (N^\gamma - s_\gamma + 1)] .$$

After substituting (5.30) into the system (5.29) the latter takes the following form

$$\frac{\partial}{\partial t} f_{(s)}^{(\alpha)} = \tilde{D}_\alpha + \delta D_\alpha , \tag{5.31}$$

where $\delta = n/N = 1/V$; $n = \sum_\gamma^{M^\alpha} n_\gamma$, $n_\gamma = N^\gamma/V$.

To find out the asymptotic solution of the chain of equations (5.31), we consider the following limits $(N^\alpha \to \infty, \ V \to \infty, \ N^\alpha/V \le B < \infty)$, under which the solution can be presented as a series expansion in the form

$$
\begin{cases}
f_{(s,...)}^{(\alpha,...)} = \sum_{r=0}^{\infty} {}^{(r)}f_{(s,...)}^{(\alpha,...)}\delta^r \\
{}^{(r)}f_{(s,...)}^{(\alpha,...)} = 0, \quad r < 0 \ .
\end{cases}
\tag{5.32}
$$

Note that due to the linearity of operators D_α and $\tilde{D}_\alpha$, the following expression may be used.

$$
{}^{(r)}D_\alpha = \sum_{\alpha_1,r} \sum_i^{s} \sum_{i_1}^{s_1} D_{\alpha,ii_1}^{s_r^{\alpha\alpha_1}} \left({}^{(r)}f_{(s,...)}^{(\alpha,...)}\right) \ .
$$

Substituting (5.32) into (5.31), with account for the comparison by degrees of δ, we obtain the coupled system of equations

$$
\frac{\partial}{\partial t}{}^{(r)}f_{(s,...)}^{(\alpha,...)} = {}^{(r-1)}D_\alpha + {}^{(r)}\tilde{D}_\alpha \ .
\tag{5.33}
$$

Using expansion (5.32), the uniform equation for the zero degree of δ approximation takes the form:

$$
\frac{\partial}{\partial t}{}^{(0)}f_{(s,...)}^{(\alpha,...)} - {}^{(0)}\tilde{D}_\alpha = 0 \ .
\tag{5.34}
$$

Assume that solution of the evolution equation (5.21) exists and is unique, and the particles forming initial condition $C|_{t=0}$ are statistically independent, i.e.

$$
\varphi_0(C) = \prod_\alpha \prod_i {}^{(0)}\varphi_i^\alpha(c_i^\alpha) \ .
\tag{5.35}
$$

Then, at all $t > 0$ and $s \ge 1$ the following statements are valid:

a) the limits of distribution functions under the following conditions N^α, and $V \to \infty$, and $N^\alpha/V < \infty$ exist

$$
\lim \varphi_{(s,...)}^{(\alpha,...)}(C_{(s,...)}^{(\alpha,...)}, t) = P_{(s,...)}^{(\alpha,...)}(C_{(s,...)}^{(\alpha,...)}, t) \ ;
\tag{5.36}
$$

b) the limited functions $P_{(s,...)}^{(\alpha,...)}$ satisfy the Boltzmann constraint (molecular chaos hypothesis)

$$
P_{(s,...)}^{(\alpha,...)}(C_{(s,...)}^{(\alpha,...)}, t) = \prod_\gamma^{\alpha,...} \prod_{i=1}^{s_\gamma} p_i^\gamma(c_i^\gamma, t) \ ,
\tag{5.37}
$$

where functions $p_i^\alpha(\mathbf{c}_i^\alpha, t)$ are the solutions of the system of Boltzmann type equations

$$\begin{cases} \frac{\partial}{\partial t} p_i^\alpha(\mathbf{c}_i^\alpha, t) = \sum_{\alpha_1, r} \sum_{i_1} {}^{(i)} L_{i i_1}^{s_r^{\alpha \alpha_1}} (p_i^\alpha, p_{i_1}^{\alpha_1}) \\ p_i^\alpha \big|_{t=0} = {}^{(0)} \varphi_i^\alpha(\mathbf{c}_i^\alpha), \quad \alpha = 0, 1, \ldots, M^\alpha; \; i = 1, \ldots, N^\alpha \end{cases} \tag{5.38}$$

and operators L are defined by the expressions (5.10).

Using condition (5.35) and normalization ratio (5.30) it is possible to write

$$f_{(s,\ldots)}^{(\alpha,\ldots)} \big|_{t=0} = \{ \prod_\gamma^{\alpha,\ldots} n_\gamma^{s_\gamma} \} \times \{ \prod_\gamma^{\alpha,\ldots} \prod_{i=1}^{s_\gamma} {}^{(0)} \varphi_i^\gamma(\mathbf{c}_i^\gamma) \} \times \{ \prod_\gamma^{\alpha,\ldots} (1 + \sum^{s_\gamma - 1} z^{(s_\gamma)} \delta^r) \} \; ,$$

where $z_r^{(s_\gamma)} = n_\gamma^{-r} d(s_\gamma, r)$, and d-Stirling number of 1-st kind: $d(s_\gamma, 0) = 1$, $d(s_\gamma, 1) = \frac{1}{2} s_\gamma (s_\gamma - 1)$, etc. Hence the initial conditions for density ${}^{(0)}f$ may be written in the form

$$^{(r)} f_{(s,\ldots)}^{(\alpha,\ldots)} \big|_{t=0} = \{ \prod_\gamma^{\alpha,\ldots} (1 + \sum_{r=1}^{s_\gamma - 1} z_r^{(s_\gamma)} \delta^r) - 1 \} \times \prod_\gamma^{\alpha,\ldots} \prod_{i=1}^{s_\gamma} f_i^\gamma(\mathbf{c}_i^\gamma) \big|_{t=0} \; , \tag{5.39}$$

where the following definition for the initial one-particle density

$$f_i^\gamma \big|_{t=0} = f_{(1)}^{(\gamma)} \big|_{t=0}$$

is adopted. Then at $r = 0$ we have the uniform system of equations:

$$\begin{cases} \frac{\partial}{\partial t} {}^{(0)} f_{(s,\ldots)}^{(\alpha,\ldots)} - {}^{(0)} \tilde{D}_\alpha = 0 \\ {}^{(0)} f_{(s,\ldots)}^{(\alpha,\ldots)} \big|_{t=0} = \prod_\gamma^{\alpha,\ldots} \prod_{i=1}^{s_\gamma} f_i^\gamma(\mathbf{c}_i^\gamma) \end{cases} \tag{5.40}$$

From the theorem proofing the molecular chaos propagation (Grunbaum, 1971) it follows that for the chain of equations (5.40) with factorized initial conditions, the Boltzmann property is valid (Prigogine, 1962):

$$^{(0)} f_{(s,\ldots)}^{(\alpha,\ldots)} (\mathbf{C}_{(s,\ldots)}^{(\alpha,\ldots)}, t) = \prod_\gamma^{\alpha,\ldots} \prod_{i=1}^{s_\gamma} f_i^\gamma(\mathbf{c}_i^\gamma) \; . \tag{5.41}$$

It brings an evidence, as it follows from (5.40), (5.41), and (5.29), that the distribution function f_i^γ is the solution of the Cauchy problem (5.38).

This is the zero order solution in the series expansion (5.32). To obtain the solution of higher order $r \geq 1$, one must evaluate the chains (5.33) of non-uniform equations with initial conditions (5.39), which gives rise to all terms of expansion (5.32) under $r \geq 1$.

Using expressions (5.30) and (5.32), it is also possible to get the following asymptotic expansions:

$$\varphi_{(s,\ldots)}^{(\alpha,\ldots)}(t) = \sum_{r=0}^{\infty} {}^{(r)}\varphi_{(s,\ldots)}^{(\alpha,\ldots)}(t)\delta^r \ ,$$

and

$$^{(0)}\varphi_{(s,\ldots)}^{(\alpha,\ldots)}(t) = \{\prod_{\gamma}^{\alpha,\ldots} n_\gamma^{s_\gamma}\}^{-1} \ {}^{(0)}f_{(s,\ldots)}^{(\alpha,\ldots)}(t) \ .$$

They can be simplified if the following definition is introduced

$$P_{(s,\ldots)}^{(\alpha,\ldots)}(\mathbf{C}_{(s,\ldots)}^{(\alpha,\ldots)},t) = {}^{(0)}\varphi_{(s,\ldots)}^{(\alpha,\ldots)}(\mathbf{C}_{(s,\ldots)}^{(\alpha,\ldots)},t)$$

to finally obtain the following asymptotic form

$$\varphi_{(s,\ldots)}^{(\alpha,\ldots)}(\mathbf{C}_{(s,\ldots)}^{(\alpha,\ldots)},t) = P_{(s,\ldots)}^{(\alpha,\ldots)}(\mathbf{C}_{(s,\ldots)}^{(\alpha,\ldots)},t) \times \{1 + O(N^{-1})\} \ .$$

These asymptotic estimates ensure a mathematical basis for the development of a numerical stochastic model consisting of a finite number of modeling particles. As it was shown above, this model is asymptotically equivalent to the initial physical-probabilistic analogue, and the evolution of this numerical model is described by the linear master equation (5.21). The linearity of the master equation is a significant advantage of this stochastic approach, which allows to considerably simplify the procedure of numerical simulation.

5.1.4. BASIC ALGORITHMS

We have seen that an asymptotic equivalence between the dynamic and stochastic approaches at the level of one-particle distribution functions serves as a basic concept to replace the common kinetic description of a rarefied gas evolution by its physical-probabilistic analogue. This concept makes it possible to use the direct simulation of collisional processes instead of the solution of set of Boltzmann type kinetic equations, based on the stochastic interpretation of the collisional evolution of a rarefied gas. To realize this approach it is necessary to develop an appropriate set of algorithms. Only the basic procedure for the algorithm construction will be considered in this section. Algorithms developed for specific aeronomic problems are described in more detail in the following sections together with their applications for the solution of these problems.

The random process $\mathbf{C}(t)$ describing the evolution of the numerical stochastic model (5.11) induces the uniform Markovian chain $^{(m(t))}\mathbf{C}$, where $m(t)$ is the number of transitions (5.13) during time t. It provides a baseline for the algorithm realization of the model accordingly, because $^{(m(t))}\mathbf{C}$ follows explicitly from (5.13) - (5.18) and can be treated as the analogue algorithm of the Monte-Carlo method for the solution of master equation (5.21) (Yanitskii, 1975, 1991; Shematovich, 1979).

The realization of the Markovian chain $^{(m(t))}\mathbf{C}$ on a discrete time grid $t_\delta = \delta\Delta t$, $\delta = 0, 1, \ldots$ includes the following steps:

a) it is accepted that at time $t_{\delta-1}$

$$m = 0, \ T^{(m)} = 0, \ ^{(m)}\mathbf{C} = \mathbf{C}(t_{\delta-1}) \ ;$$

b) the frequency $\omega^{(m)} = \omega(^{(m)}\mathbf{C})$ is calculated and through distribution (5.18) the waiting time $\tau^{(m)}$ at the next transition is defined using

$$P\{\tau^{(m)} \leq \tau\} = 1 - \exp(-\omega^{(m)}\tau) \ , \tag{5.42}$$

and then the overall transition time is accumulated as

$$T^{(m)} = T^{(m-1)} + \tau^{(m)} \ ;$$

c) the condition for termination of the numerical realization is checked out as follows

$$T^{(m-1)} \leq \Delta t < T^{(m)} \ ;$$

d) if this condition is not satisfied, the next transition (5.13) comes from step (b) in accordance with probabilities (5.16) and (5.17), and in response to the following assumed relation

$$^{(m)}\mathbf{C} = {}^{(m-1)}\mathbf{C}_{ii_1}^{s_r^{\alpha\alpha_1}} \ ;$$

e) when the calculations are completed and running on a time interval Δt is stopped, the following assumed relation is applied:

$$\mathbf{C}(t_\delta) = {}^{(m-1)}\mathbf{C} \ .$$

Algorithm steps (a) - (e) are analogous to the Monte-Carlo procedure for solving the evolutionary model equation. The procedure involves approximation of the expression (4.9) at algorithm step Δt in the following form

$$P\{\tau^{(m)} \leq \tau\} = 1 - \exp[-\omega(\mathbf{C}(t_{\delta-1}))\tau] \ . \tag{5.43}$$

This means that waiting time of the next transition is defined by the total frequency, calculated at time $t_{\delta-1}$ (on the previous time step) for all transitions during Δt. This algorithm in the case $\Delta t \ll 1$ is a good approximation of the exact process earlier described (Yanitskii, 1991).

It is worth noting that the direct numerical realization of the sequence (a)-(e) is very complicated since after each transition (5.13) it is necessary to calculate the collision frequency. This procedure is not economical as well, as the number of operations performed is proportional to the square of total number of modeling particles N^2. To simplify the process and to reduce the computational cost it is necessary to develop the procedure linearly depending on the number of particles. Various effective ($\sim N$) numerical schemes for solving the evolutionary equation were suggested (see, e.g. Ivanov et $al.$, 1987; Ivanov and Rogazinskii, 1988, 1991; Bird, 1994).

It was shown that the algorithm under consideration utilizes the Markovian chain $\{\mathbf{C}(t)\}$, which on the grid $\{t_\delta\}$ approximates the state $\mathbf{C}(t)$ of the numerical model. On the other hand, the probabilistic distribution $\varphi(\mathbf{C}, t_\delta)$ is the solution of the model evolutionary equation, i.e.

$$\varphi(\mathbf{C}, t_\delta) = [1 - \Delta t \omega(\mathbf{C}(t_{\delta-1}))]\varphi(\mathbf{C}, t_{\delta-1}) + \Delta t K \varphi(\mathbf{C}, t_{\delta-1}) . \qquad (5.44)$$

Hence, it is confirmed that the algorithm (5.43) corresponds to the Monte-Carlo analogue method of solution on the finite-differential scheme (5.44) of the evolutionary equation of the numerical stochastic model.

Due to the linearity of the master equation (5.21) [and, consequently, of the analog algorithms of the Monte-Carlo method for its solution (Ivanov et $al.$, 1987; Yanitskii, 1991)] the macroparameters of the gas may be calculated as a superposition of the random process $\mathbf{C}(t)$ trajectories:

$$D_\alpha(t) = < d_\alpha(\mathbf{c})\varphi(\mathbf{C}, t) >_\mathbf{c} = \frac{1}{L} < d_\alpha(\mathbf{c}; \{^{(l)}\mathbf{C}(t)\}_{(L)}) >_L ,$$

where the (l)- realization of $\mathbf{C}(t)$ process is defined as $^{(l)}\mathbf{C}(t)$ $(l = 1, \ldots, L;$ L- is the number of realizations);

$$\{^{(l)}\mathbf{C}(t)\}_{(L)} = \bigcup_{l=1}^{L} {}^{(l)}\mathbf{C}(t) ;$$

and definition $< \ldots >_L$ corresponds to the values of the $\mathbf{C}(t)$ trajectory realizations averaged along L.

An accuracy of the macroparameter calculations may be evaluated as $O(\Delta t + (N * L)^{-\frac{1}{2}} + N^{-1})$ (Shematovich, 1979), where L is the number of realizations. This relation allows us to optimize the parameters $\Delta t, N, L$ of the algorithmic realization of the stochastic model.

5.2. Structural Stochastic Model for Kinetic Relaxation Systems

For a rarefied atmospheric gas involving kinetics of aeronomic relaxation processes (4.3) the system of Boltzmann type kinetic equations can be written in the form

$$\begin{cases} \frac{\partial}{\partial t} F_\alpha(\mathbf{c}, t) = \sum_{\alpha_1} \sum_r J^{s_r^{\alpha\alpha_1}} (F_\alpha, F_{\alpha_1}) \\ F_\alpha \mid_{t=0} = F_\alpha^{(0)}(\mathbf{c}) \ . \end{cases} \tag{5.45}$$

This system is defined by the dynamic and probabilistic characteristics (scattering functions) of particle interactions in elementary acts of processes (4.3), as well as by the energy distribution functions of colliding particles by translational and inner degrees of freedom. In applications to the aeronomy problems, rarefied multicomponent gases are usually characterized by very different scales of interaction parameters of collisions for gas species, such as cross-sections of elastic and inelastic interaction, and by substantial deviations of the energy distribution functions from equilibrium (Chamberlain, 1978; Rees, 1989; Marov et al., 1990). Accordingly, the multichannel processes of relaxation have a complex structure of kinetic scales as well, characterizing the rates of energy exchange between translational and inner degrees of freedom.

In this section we describe the modified method of stochastic simulation which enables us to develop numerical models of the nonequilibrum relaxation kinetics of the chemical reactions. In these models, detailed structures of the kinetic scales of the aeronomic processes are taken into account, and they are applied to investigations of the multicomponent gas of the upper atmosphere (Pyarnpuu and Shematovich, 1985, 1987; Pyarnpuu et al., 1986). Based on this approach a set of improved numerical procedures, called structural stochastic models, is developed.

5.2.1. CHARACTERISTIC TIME SCALES OF RELAXATION

Let us consider the characteristic time scales for a local kinetics (5.45) of collisional physical-chemical processes in a rarefied multicomponent gas. The average time $\tau_\alpha^{s_r^{\alpha\alpha_1}}$ between collisions of a particle α with an arbitrary particle of type α_1 in collisional process $s_r^{\alpha\alpha_1}$ is defined by the expression

$$\tau_\alpha^{s_r^{\alpha\alpha_1}} = \{n_{\alpha_1} q^{s_r^{\alpha\alpha_1}}\}^{-1} \ , \tag{5.46}$$

where the reaction rate $q^{s_r^{\alpha\alpha_1}}$ is equal to

$$q^{s_r^{\alpha\alpha_1}} = \int d\mathbf{c} \, d\mathbf{c}_1 \, dI^{s_r^{\alpha\alpha_1}} f_\alpha(\mathbf{c}) f_{\alpha_1}(\mathbf{c}_1) \ .$$

Consequently, the average time between the consecutive collisions with participation of α type particles is equal to

$$\frac{1}{\tau_\alpha} = \sum_{\alpha_1, r} (\tau_\alpha^{s_r^{\alpha\alpha_1}})^{-1} \ . \tag{5.47}$$

The kinetic scale (5.47) defines the characteristic time interval on which a change of velocity distribution functions $f_\alpha(\mathbf{c})$ due to collisional processes involving α-type particles occurs. From (5.47) it follows that the scale τ_α is defined by the dynamical-probabilistic characteristics $I^{s_r^{\alpha\alpha_1}}$ of collisional processes, averaged by the thermal state of the colliding particles.

Let T will be the characteristic macroscopic time of gas state change. Then parameter $\varepsilon_\alpha = \tau_\alpha/T$ defines the nonequilibrium degree of thermal state of a given gas component. Three characteristic cases are possible:

First, a weak deviation of the state of the α component from local equilibrium by translational degrees of freedom (Struminskii, 1982), when $\varepsilon_\alpha \ll 1$. In this case at $t' = t/T$ the kinetic equation for the given gas component takes the form

$$\frac{\partial}{\partial t'} F_\alpha(\mathbf{c}, t') = \sum_{\{\hat{s}_r^{\alpha\alpha_1}\}} \frac{1}{\varepsilon^{\hat{s}_r^{\alpha\alpha_1}}} J^{\hat{s}_r^{\alpha\alpha_1}} + \sum_{\{\tilde{s}_r^{\alpha\alpha_1}\}} \frac{1}{\varepsilon^{\tilde{s}_r^{\alpha\alpha_1}}} J^{\tilde{s}_r^{\alpha\alpha_1}} , \qquad (5.48)$$

where the fast and slow (in comparison with the adopted time scale) collision processes are defined as follows

$$\{\hat{s}_r^{\alpha\alpha_1}\} = \{s_r^{\alpha\alpha_1} : \tau_\alpha^{s_r^{\alpha\alpha_1}} \geq \tau_\alpha\}$$

$$\{\tilde{s}_r^{\alpha\alpha_1}\} = \{s_r^{\alpha\alpha_1}\} / \{\hat{s}_r^{\alpha\alpha_1}\} ,$$

and the collision integrals are represented in dimensionless forms. The solution at zero order by parameter ε_α for kinetic equation (5.48) is defined from the condition

$$\sum_{\{\tilde{s}_r^{\alpha\alpha_1}\}} J^{\tilde{s}_r^{\alpha\alpha_1}} = 0 .$$

This condition is satisfied by local Maxwellian distributions (3.22) (Struminskii, 1982)

$$f_\alpha(t') = f_\alpha^{(M)}(\mathbf{c}^\alpha; T_\alpha(t')) , \qquad (5.49)$$

where T_α is local kinetic temperature of α-component particles. Hence, the distributions (5.49) implicitly depend on time t'.

Second, a strongly nonequilibrium thermal state of gas characterized by condition $\varepsilon_\alpha \sim 1$, at which the characteristic temporary scales of micro- and macro- parameter changes are nearly the same, and the thermal state of the gas is defined by the solution of the initial kinetic system (5.45). In this case the distribution functions depend directly on time t':

$$f_\alpha = f_\alpha(\mathbf{c}, t') .$$

Third, possible thermal steady state of gas component α when collisional processes weakly influence thermal state of particles or thermal energy inflow from external sources, i.e. thermostat component domains. This particular case corresponds to the condition $\varepsilon_\alpha \to \infty$, and the solution of equation (5.45) for thermal state of thermostat component corresponds to the given initial distribution

$$f_\alpha(t') = f_\alpha^{(0)}(\mathbf{c}) \ . \tag{5.50}$$

The general structure of the collisional processes in a rarefied multicomponent gas can be formally determined by the set of parameters $\{\varepsilon_\alpha = \tau_\alpha/T\}_{(M^\alpha)}$. In accordance with expressions (5.47) - (5.50), the initial set of gas particles is represented as:

$$\begin{cases} \{a^\alpha\}_{(M^\alpha)} = \{a^{\hat{\alpha}}\}_{(M^\alpha)} \cup \{a^{\tilde{\alpha}}\}_{(M^\alpha)} \cup \{a^{\bar{\alpha}}\}_{(M^\alpha)} \\ M^\alpha = \hat{M}^\alpha + \tilde{M}^\alpha + \bar{M}^\alpha \ , \end{cases} \tag{5.51}$$

where

$$\begin{cases} \{a^{\tilde{\alpha}}\}_{(\tilde{M}^\alpha)} = \{a^\alpha \ : \ \varepsilon_\alpha \ll 1\}_{(M^\alpha)} \\ \{a^{\hat{\alpha}}\}_{(\hat{M}^\alpha)} = \{a^\alpha \ : \ \varepsilon_\alpha \sim 1\}_{(M^\alpha)} \\ \{a^{\bar{\alpha}}\}_{(\bar{M}^\alpha)} = \{a^\alpha \ : \ \varepsilon_\alpha \gg 1\}_{(M^\alpha)} \ . \end{cases} \tag{5.52}$$

From a mathematical view point, expressions (5.52) characterize various patterns of distribution $f_\alpha(\mathbf{c})$ on the macroscopic time scale T; in particular, for component $\tilde{\alpha}$, it reflects possible small deviations from the local equilibrium distributions $f_{\tilde{\alpha}}(\mathbf{c}, t') = f_{\tilde{\alpha}}^{(M)}(\mathbf{c}; \ T_{\tilde{\alpha}}(t'))$ and for component $\hat{\alpha}$ - possible significant changes $f_{\hat{\alpha}} = f_{\hat{\alpha}}(\mathbf{c}, t')$, while for component $\bar{\alpha}$ the distributions do not vary on scales $\sim T$ at all, i.e. $f_{\bar{\alpha}} = f_{\bar{\alpha}}^{(0)}(\mathbf{c})$. In turn, from a physical view point, condition (5.52) can be partially or completely satisfied in a rarefied multicomponent gas if either the background gas is impacted by stationary or periodic (of finite time duration) fluxes of energetic particles, or the background gas consists of particles variing by mass, density, and interaction cross-sections (which is, for example, the case for the processes of formation and/or relaxation of energetic admixture). In every particular case of (5.51), the form of (5.52) is defined by the structure of physical and chemical model of the respective kinetic system, i.e. an absence of each subsystem (5.52) is admitted.

If there is a structure with a given hierarchy of kinetic scales such as (5.52), then it is possible to use a separate kinetic description of the different subsystems. In this case, in accordance with (5.48) - (5.50), thermostat components are defined by the initial steady-state distributions $f_{\bar{\alpha}}^{(0)}$, and components $\tilde{\alpha}$ are described by balance equations for density and temperature, obtained from (5.45) by averaging through Maxwellian distributions (5.49). Then the volume kinetics of collisional processes with a detailed structure of kinetic scales (5.52) is defined by the following set of equations:

$$\begin{cases} \frac{\partial}{\partial t'} F_{\hat{\alpha}}(\mathbf{c}, t') = \sum_{\alpha_1, r} J_{\hat{\alpha}}^{s_r^{\hat{\alpha}\alpha_1}}(F_{\hat{\alpha}}, F_{\alpha_1}) \\ \frac{d}{dt'}\{n_{\tilde{\alpha}}(t')\, D_{\tilde{\alpha}}(t')\} = \sum_{\alpha_1, r} V_{\tilde{\alpha}}^{s_r^{\tilde{\alpha}\alpha_1}}(d_{\tilde{\alpha}}) \\ \frac{d}{dt'} n_{\tilde{\alpha}}(t') = \sum_{\alpha_1, r} V_{\tilde{\alpha}}^{s_r^{\tilde{\alpha}\alpha_1}} \\ F_{\tilde{\alpha}}(t') = n_{\tilde{\alpha}}(t') f_{\tilde{\alpha}}^{(0)}(\mathbf{c}) \ ; \quad F_{\hat{\alpha}}\big|_{t'=0} = F_{\hat{\alpha}}^{(0)}(\mathbf{c}) \\ n_{\tilde{\alpha}} D_{\tilde{\alpha}}\big|_{t'=0} = \int d\mathbf{c}\ d_{\tilde{\alpha}}(\mathbf{c}) F_{\tilde{\alpha}}^{(0)}(\mathbf{c}) \ , \end{cases} \quad (5.53)$$

where $d_{\tilde{\alpha}}(\mathbf{c}) = \{1, \frac{1}{2} m_{\tilde{\alpha}} c^2\}$ are the molecular entities corresponding to the local density and kinetic temperature of the component $\tilde{\alpha}$. The collision terms are specified as follows

$$J^{s_r^{\hat{\alpha}\alpha_1}} = \begin{cases} J^{s_r^{\hat{\alpha}\hat{\alpha}_1}}(F_{\hat{\alpha}}, F_{\hat{\alpha}_1}) \\ J^{s_r^{\hat{\alpha}\tilde{\alpha}_1}}(F_{\hat{\alpha}}, n_{\tilde{\alpha}_1} f_{\tilde{\alpha}_1}^{(M)}) \\ J^{s_r^{\hat{\alpha}\tilde{\alpha}_1}}(F_{\hat{\alpha}}, n_{\tilde{\alpha}_1} f_{\tilde{\alpha}_1}^{(0)}) \ , \end{cases} \quad (5.54)$$

and terms $J^{s_r^{\hat{\alpha}\tilde{\alpha}_1}}$ and $J^{s_r^{\hat{\alpha}\tilde{\alpha}_1}}$ are linear, while $J^{s_r^{\hat{\alpha}\hat{\alpha}_1}}$ is the nonlinear one. The corresponding moments emerging from collision terms

$$V_{\tilde{\alpha}}^{s_r^{\tilde{\alpha}\alpha_1}}(d_{\tilde{\alpha}}) = \int d\mathbf{c}\ d_{\tilde{\alpha}}(\mathbf{c}) J_{\tilde{\alpha}}^{s_r^{\tilde{\alpha}\alpha_1}}(n_{\tilde{\alpha}} f_{\tilde{\alpha}}^{(M)}, F_{\alpha_1}) \quad (5.55)$$

may be cast in the form:

$$V^{s_r^{\tilde{\alpha}\alpha_1}} = \sum_{\tilde{\alpha}_1} V^{s_r^{\tilde{\alpha}\tilde{\alpha}_1}} + \sum_{\hat{\alpha}_1} V^{s_r^{\tilde{\alpha}\hat{\alpha}_1}} + \sum_{\tilde{\alpha}_1} V^{s_r^{\tilde{\alpha}\tilde{\alpha}_1}} \ . \quad (5.56)$$

The moments $V^{s_r^{\tilde{\alpha}\tilde{\alpha}_1}}$, $V^{s_r^{\tilde{\alpha}\hat{\alpha}_1}}$ and $V^{s_r^{\tilde{\alpha}\tilde{\alpha}_1}}$ can be expressed through kinetic rates of respective collisional processes, for example

$$V^{s_r^{\tilde{\alpha}\tilde{\alpha}_1}}(d_{\tilde{\alpha}} = 1) = n_{\tilde{\alpha}(r)} n_{\alpha_1(r)} q^{s^{\tilde{\alpha}(r)\tilde{\alpha}_1(r)}} - n_{\tilde{\alpha}} n_{\tilde{\alpha}_1} q^{s_r^{\tilde{\alpha}\tilde{\alpha}_1}} \ ,$$

etc.

The kinetic system (5.53) describes the process of relaxation of a rarefied gas characterized by the structure (5.52) of collision scales. It should be noted that in the cases $\varepsilon \ll 1$, or $\varepsilon \gg 1$, i.e when it is possible to exclude an implicit dependence on time, the solution of the initial system can be easily obtained. In particular, it is well known that:

- for the thermodynamically closed kinetic systems ($M^{\bar{\alpha}} = 0$) gas relaxation leads to an equilibrium state characterized by a common kinetic temperature

$$f_\alpha(\mathbf{c}, t') \to f_\alpha^{(M)}(\mathbf{c}; T_\alpha(t')) \stackrel{t' \to \infty}{\Rightarrow} f_\alpha^{(M)}(\mathbf{c}; T) \ ;$$

- for thermodynamically open kinetic systems, the steady state

$$f_\alpha(\mathbf{c}, t') \stackrel{t' \to \infty}{\Rightarrow} f_{\bar{\alpha}}^{(0)}(\mathbf{c})$$

which adopts the external conditions is obtained.

In the kinetic theory of rarefied gases there are a set of analytical and numerical methods which can be applied for the investigation of extreme cases of gas states (near thermal equilibrium when $\varepsilon \ll 1$, and close to the steady state when $\varepsilon \to \infty$) (Ferziger and Kaper, 1973). Analysis of strongly nonequilibrium systems ($\varepsilon \sim 1$) is especially complicated due to mathematical difficulties, and to advance the solution of the problem, new simulation methods have to be applied. One of the most suitable approaches for these considerations is the earlier mentioned DSMC method. Its modification (Pyarnpuu and Shematovich, 1985, 1987) based on structural models ensures the solution of kinetic system (5.53) using the most effective algorithms.

5.2.2. STRUCTURAL STOCHASTIC MODELS

In response to the general approach to the development of physical–probabilistic analogue, the initial system of particles (5.51) of rarefied multicomponent gas is replaced by the set of finite number of modeling particles:

$$\{a^{\tilde{\alpha}}, a^{\hat{\alpha}}, a^{\bar{\alpha}}\}_{(\tilde{M}^\alpha, \hat{M}^\alpha, \bar{M}^\alpha)} \to \{a_i^{\tilde{\alpha}}, a_i^{\hat{\alpha}}, a_i^{\bar{\alpha}}\}_{(M^\alpha, \tilde{N}, \hat{N}, \bar{N})} =$$

$$\{a_i^{\tilde{\alpha}}\}_{(\tilde{M}^\alpha, \tilde{N})} \cup \{a_i^{\hat{\alpha}}\}_{(\hat{M}^\alpha, \hat{N})} \cup \{a_i^{\bar{\alpha}}\}_{(\bar{M}^\alpha, \bar{N})} , \tag{5.57}$$

where

$$N = \tilde{N} + \hat{N} + \bar{N};$$

$$\tilde{N} = \sum_{\tilde{\alpha}} N^{\tilde{\alpha}}, \quad \hat{N} = \sum_{\hat{\alpha}} N^{\hat{\alpha}}, \quad \bar{N} = \sum_{\bar{\alpha}} N^{\bar{\alpha}} \ .$$

The state of each numerical model subsystem (5.57) is characterized by a discrete form of the kinetic distribution functions $F_\alpha = n_\alpha(t) f_\alpha(\mathbf{c}, t)$, in accord with the detailed structure (5.52). Then, for the component $\hat{\alpha}$ characterized by a nonequilibrum state on translational degrees of freedom, the corresponding subsystem of model particles will be described by the following state vector

$$\mathbf{C}^{\hat{\alpha}} = \bigcup_{i=1}^{N^{\hat{\alpha}}} \mathbf{c}_i^{\hat{\alpha}} \ , \tag{5.58}$$

where particle velocities are constrained by the distribution $f_{\hat{\alpha}}(\mathbf{c}, t)$. For the component $\tilde{\alpha}$, described by local Maxwellian distribution function (5.49), the following characteristic is used:

$$\begin{cases} \mathbf{C}^{\tilde{\alpha}} = \sum_{i=1} \mathbf{d}_{\tilde{\alpha}}(\mathbf{c}_i^{\tilde{\alpha}}) \\ \mathbf{c}_i^{\tilde{\alpha}} = \mathbf{c}(f_{\tilde{\alpha}}^{(M)}(\mathbf{c}; \mathbf{D}_\alpha(t))) \ . \end{cases} \tag{5.59}$$

In (5.59) the collisional changes of molecular entities $\mathbf{d}_{\tilde{\alpha}}(\mathbf{c}) = \{1, \frac{1}{2}m_\alpha c_\alpha^2\}$ are taken into account. These collision characteristics determine the dependence of the distributions $f_{\tilde{\alpha}}^{(M)}$ on time. Finally, for the $\bar{\alpha}$-type component, only the concentration of these particles is important

$$\mathbf{C}^{\bar{\alpha}} = \{N^{\bar{\alpha}}; \ \mathbf{c}_i^{\bar{\alpha}} = \mathbf{c}(f_{\bar{\alpha}}^{(0)}(\mathbf{c}))\} \ , \tag{5.60}$$

in other words, for the collisional processes involving $\bar{\alpha}$-component particles, the particle velocities are defined from the given distribution $f_{\bar{\alpha}}^{(0)}$. Then, state of the system (5.57) can be described as follows:

$$\mathbf{C} = (\cup_{\tilde{\alpha}} \mathbf{C}^{\tilde{\alpha}}) \cup (\cup_{\hat{\alpha}} \mathbf{C}^{\hat{\alpha}}) \cup (\cup_{\bar{\alpha}} \mathbf{C}^{\bar{\alpha}}) \ . \tag{5.61}$$

From the probabilistic description of collisional processes in the rarefied multicomponent gas, it is evident that the evolution of state (5.61) of the numerical system (5.57) is represented by the random process $\mathbf{C}(t)$, relating to the class of uniform jump-like Markovian processes. Then the transitions between states of the numerical system are realized as the instant changes of modeling particle characteristics in collisions

$$\{\mathbf{c}_i^\alpha, \mathbf{c}_{i_1}^{\alpha_1}\} \rightarrow \{\mathbf{c}_{i'}^{\alpha(r)}, \mathbf{c}_{i_1'}^{\alpha_1(r)}\} \ , \tag{5.62}$$

modeling particle velocities being constrained by conservation laws of mass, momentum, and energy. In turn, the scattering direction is defined by the differential function $dI^{\alpha\alpha_1}_{s_r}$.

The relationship between numerical system states (5.61) due to the realization of transition (5.62) is defined by the following expressions (Pyarnpuu and Shematovich, 1985):

- for interacting particles

$$\begin{cases} N^\alpha = N^\alpha - 1 \\ \mathbf{C}^\alpha = \begin{cases} \mathbf{C}^{\hat\alpha} \setminus \mathbf{c}_i^{\hat\alpha} \\ \mathbf{C}^{\tilde\alpha} - \mathbf{d}_{\tilde\alpha}(\mathbf{c}_i^{\tilde\alpha}) \,, \quad \alpha = \alpha, \alpha_1 \\ \mathbf{C}^{\tilde\alpha} \end{cases} \end{cases} \quad (5.63)$$

where $\mathbf{c}_i^{\tilde\alpha} = \mathbf{c}(f_{\tilde\alpha}^{(M)}(D_{\tilde\alpha}))$ is the thermal velocity of $\tilde\alpha$-component parti- cle, deduced from the Maxwellian distribution (5.49); while for thermostat component only the modeling particle population is changed;

- for particles formed in transition (5.62)

$$\begin{cases} N^{\alpha'} = N^{\alpha'} + 1 \\ \mathbf{C}^{\alpha'} = \begin{cases} \mathbf{C}^{\hat\alpha'} \cup \mathbf{c}_i^{\hat\alpha'} \\ \mathbf{C}^{\tilde\alpha'} + \mathbf{d}_{\tilde\alpha'}(\mathbf{c}_{i'}^{\tilde\alpha'}) \,, \quad \alpha' = \alpha(r), \alpha_1(r) \\ \mathbf{C}^{\tilde\alpha'} \end{cases} \end{cases} \quad (5.64)$$

Thus, in the numerical system analysis one must take into account the following key-features: for $\hat\alpha$-components - the detailed changes of the mod- eling particle velocities (because their velocity distribution directly depends on time); for $\tilde\alpha$-components - the changes of invariants $\mathbf{d}_{\tilde\alpha}$ in transitions (because these invariants determine the dependence of distributions (5.49) on time). In this latter case the steady-state velocity distribution function (5.50) is used.

The probabilities of transitions (5.62) represent kinetic rates of the re- spective collisional processes in the finite system of modeling particles

$$\omega_{ii_1}^{s_r^{\alpha\alpha_1}} = V^{-1} q_{ii_1}^{s_r^{\alpha\alpha_1}} = |\mathbf{c}_i^\alpha - \mathbf{c}_{i_1}^{\alpha_1}| \sigma^{s_r^{\alpha\alpha_1}}(\mathbf{c}_i^\alpha, \mathbf{c}_{i_1}^{\alpha_1}) \,. \quad (5.65)$$

For $\alpha, \alpha_1 \in \{\tilde\alpha, \bar\alpha\}$ components, the velocities of these modeling particles are defined from distributions $f_{\tilde\alpha}^{(M)}$ and $f_{\tilde\alpha}^{(0)}$. Because random transitions belong to the class of jump-like Markovian processes, the time intervals between interactions (5.62) are characterized by the following distribution

$$\begin{cases} P\{\tau(\mathbf{C} \to \mathbf{C}^{s_r^{\alpha\alpha_1}}) \leq \tau\} = 1 - \exp\{-\omega(\mathbf{C})\tau\} \\ \omega(\mathbf{C}) = \sum\limits_{\alpha,\alpha_1,r} \sum\limits_{i,i_1} \omega_{ii_1}^{s_r^{\alpha\alpha_1}} \,. \end{cases} \quad (5.66)$$

Expressions (5.57) - (5.66) defining the random process $\mathbf{C}(t)$, describe how the state of the numerical stochastic model evolves and therefore charac- terizes the collisional relaxation (5.53) in a rarefied multicomponent gas.

In support of the theory of random processes and the procedure discussed in Sections 5.1.1-5.1.2, the above consideration reveals the intrinsic relationship between the probabilistic description of the structural stochastic model (5.57) - (5.66) and the kinetic description of collisional relaxation (5.53). This argues again for the stochastic model to serve as an approximation of the physical-probabilistic analogue of collisional relaxation in rarefied gas by the numerical system with a finite number of modeling particles (Pyarnpuu and Shematovich, 1985).

5.2.3. FEATURES OF THE ALGORITHMIC REALIZATION

As it was shown in Section 5.1.4 the structural stochastic model can be numericaly evaluated following the standard scheme of random process $C(t)$ simulation through analogous algorithms of the Monte-Carlo technique (Shematovich, 1979, 1982). Here we shall consider possible algorithmic modifications of this scheme, taking into account some specifics of the model structures.

Because the random process $C(t)$ of the stochastic model is the strictly Markovian, in order to determine time sequences of its state transitions it is necessary to calculate all transition probabilities for each successive state of the model. These calculations use enormous computer time because they require $\sim N^2$ operations. Utilization of structural stochastic models allows to relief the situation and to develop more economical algorithms. Indeed, for multiscale kinetic systems, the velocity distribution functions may be verified depending on how nonequilibrium the model states are. This means that in the procedure of evaluation of collisional processes it is possible to use collision terms averaged through the local equilibrium distributions (5.49) or steady-state distributions (5.50). We now discuss this approach in more detail.

We have seen that a state of the numerical model is defined by expression (5.61). Transitions between states occur due to interactions between model particles of various (kinetic, gas dynamic, and equilibrium) subsystems, such that

$$C \to C' \; = \; C_{ii_1}^{s_r^{\alpha\alpha_1}} \; , \quad \alpha\alpha_1 = \hat{\alpha}\tilde{\alpha}, \; \tilde{\alpha}\bar{\alpha}, \; \hat{\alpha}\bar{\alpha} \; , \qquad (5.67)$$

or due to particle interactions within the subsystem, such that

$$C \to C' \; = \; C_{ii_1}^{s_r^{\alpha\alpha_1}} \; , \quad \alpha\alpha_1 = \hat{\alpha}\hat{\alpha}, \; \tilde{\alpha}\tilde{\alpha}, \; \bar{\alpha}\bar{\alpha} \; . \qquad (5.68)$$

The dynamic realization of transitions (5.67) follows the schemes:

$$\begin{cases} \mathbf{c}_i^{\hat{\alpha}} + \begin{cases} \mathbf{c}_{i_1}^{\tilde{\alpha}_1}(f_{\tilde{\alpha}_1}^{(M)}) \\ \mathbf{c}_{i_1}^{\tilde{\alpha}_1}(f_{\tilde{\alpha}_1}^{(0)}) \end{cases} \rightarrow \{\mathbf{c}_{i'}^{\alpha(r)}, \, \mathbf{c}_{i_1'}^{\alpha_1(r)}\} \\ \mathbf{c}_i^{\tilde{\alpha}}(f_{\tilde{\alpha}}^{(M)}), \, \mathbf{c}_{i_1}^{\tilde{\alpha}_1}(f_{\tilde{\alpha}_1}^{(0)}) \rightarrow \{\mathbf{c}_{i'}^{\alpha(r)}, \, \mathbf{c}_{i_1'}^{\alpha_1(r)}\} \ . \end{cases} \quad (5.69)$$

In turn, probabilities of the transitions (5.69) are described by the expressions:

$$\omega_{ii_1}^{s_r \, \alpha\alpha_1} = \begin{cases} V^{-1} \mid < c^{\tilde{\alpha}_1} > \mathbf{e} - \mathbf{c}_i^{\hat{\alpha}} \mid \sigma^{s_r \, \hat{\alpha}\tilde{\alpha}_1} \\ V^{-1} \mid < c^{\tilde{\alpha}_1} > \mathbf{e} - \mathbf{c}_i^{\hat{\alpha}} \mid \sigma^{s_r \, \hat{\alpha}\tilde{\alpha}_1} \\ V^{-1} \mid < c^{\tilde{\alpha}_1} > \mathbf{e}_1 - < c^{\tilde{\alpha}} > \mathbf{e} \mid \sigma^{s_r \, \tilde{\alpha}\tilde{\alpha}_1} = \\ \qquad V^{-1} < I^{s_r \, \tilde{\alpha}\tilde{\alpha}_1} >_{(f_{\tilde{\alpha}}^{(M)}, f_{\tilde{\alpha}}^{(0)})} \ , \end{cases} \quad (5.70)$$

where the following mean velocities for gas dynamical and steady-state subsystems are used:

$$\begin{cases} < c^{\tilde{\alpha}} > = \int d\mathbf{c}^{\tilde{\alpha}} |\mathbf{c}^{\tilde{\alpha}}| f_{\tilde{\alpha}}^{(M)}(\mathbf{c}; \mathbf{D}_{\tilde{\alpha}}(t)) \\ < c^{\tilde{\alpha}} > = \int d\mathbf{c}^{\tilde{\alpha}} |\mathbf{c}^{\tilde{\alpha}}| f_{\tilde{\alpha}}^{(0)}(\mathbf{c}) \ , \end{cases} \quad (5.71)$$

and $\mathbf{e}$ is direction vector normalized to unity.

Thus, only for $\hat{\alpha}$ - component interactions, the probabilities are determined by individual characteristics of the modeling particles. In this case, the probabilities of transitions (5.67), (5.68) are defined as follows:

$$\omega^{s_r \, \alpha\alpha_1} = \begin{cases} \sum_i \{\omega_i^{s_r \, \tilde{\alpha}\tilde{\alpha}_1} = N^{\tilde{\alpha}_1} \mid < c^{\tilde{\alpha}_1} > -c_i^{\hat{\alpha}} \mid \sigma^{s_r \, \tilde{\alpha}\tilde{\alpha}_1}\} \\ \sum_i \{\omega_i^{s_r \, \tilde{\alpha}\tilde{\alpha}_1} = N^{\tilde{\alpha}_1} \mid < c^{\tilde{\alpha}_1} > -c_i^{\hat{\alpha}} \mid \sigma^{s_r \, \tilde{\alpha}\tilde{\alpha}_1}\} \\ N^{\tilde{\alpha}} N^{\tilde{\alpha}_1} < I^{s_r \, \tilde{\alpha}\tilde{\alpha}_1} >_{(f_{\tilde{\alpha}}^{(M)}, f_{\tilde{\alpha}}^{(0)})} \ . \end{cases} \quad (5.72)$$

It should be also noted that for simulation step (5.67), the dynamical realization is conducted by a scheme similar to (5.69), taking into account the real velocities of interacting modeling particles. However, for transitions $\tilde{\alpha}\alpha_1$ and $\tilde{\alpha}\tilde{\alpha}_1$ the appropriate collision probabilities can be determined by analytical expressions or by the following numerical approximation:

$$\omega^{s_r \, \alpha\alpha_1} = \begin{cases} V^{-1} N^{\tilde{\alpha}} N^{\tilde{\alpha}_1} < I^{s_r \, \tilde{\alpha}\tilde{\alpha}_1} >_{(f_{\tilde{\alpha}}^{(M)}, f_{\tilde{\alpha}_1}^{(M)})} \\ V^{-1} N^{\tilde{\alpha}} N^{\tilde{\alpha}_1} < I^{s_r \, \tilde{\alpha}\tilde{\alpha}_1} >_{(f_{\tilde{\alpha}}^{(0)}, f_{\tilde{\alpha}_1}^{(0)})} \ . \end{cases} \quad (5.73)$$

This means that, in order to calculate the probabilities (5.73) the rate coefficients (5.4) of the respective physical and chemical processes must be used.

Analytical expressions or numerical estimates of transition frequencies (5.72) and (5.73) allow to simplify the calculations of the probabilistic characteristics of the $C(t)$ process because these expressions estimates are proportional to the number of modeling particles N (instead of N^2). This results in a significant acceleration of the computer realization of the algorithms (Pyarnpuu et al., 1986).

5.3. Stochastic Modeling of Rarefied Gas Flows

In order to analyze space-temporary variations of rarefied gas states caused by transport processes, one must know the solution of system (4.4) in the collisionless approximation, i.e. the solution of the following problem

$$\begin{cases} \frac{\partial}{\partial t}F_\alpha + \mathbf{c}\frac{\partial}{\partial \mathbf{r}}F_\alpha + \mathbf{S}_\alpha\frac{\partial}{\partial \mathbf{c}}F_\alpha = 0 \\ F_\alpha \mid_{t=t_0} = F_\alpha^{(0)}(\mathbf{c},\mathbf{r}) \\ F_\alpha \mid_{\mathbf{r}\in\Gamma} = F_\alpha^{(b)}(\mathbf{c},t) \end{cases} \tag{5.74}$$

using the appropriate initial and boundary conditions. The solution of this problem is commonly defined by the transfer of initial and boundary distribution functions along dynamic trajectories of atmospheric particles $\mathbf{r}_\alpha(t)$, $\mathbf{c}_\alpha(t)$ in external field $\mathbf{S}_\alpha = \mathbf{S}_\alpha(\mathbf{r},\mathbf{c})$ (Bird, 1976, 1994)

$$F_\alpha(\mathbf{c},\mathbf{r},t) = F_\alpha^{0,b}(\mathbf{c}-\mathbf{c}_\alpha(t),\mathbf{r}-\mathbf{r}_\alpha(t)) \ . \tag{5.75}$$

However, for numerical realizations, it is usually more convenient to use the evident statement about the invariance of the distribution functions along particle trajectories rather than analytical representation (5.75).

Let us consider a space-nonuniform modification of the stochastic simulation method for the microscopic kinetic model (4.4). This modification is based on the probabilistic concept of the collisionless movement and collisions of gas particles at rest. The stochastic description of space-nonuniform physical-probabilistic analogue of the microscopic kinetic model (4.4) is defined from the solution of the direct Kolmogorov equation together with the given initial and boundary conditions:

$$
\begin{cases}
\frac{\partial}{\partial t}p_i^\alpha + c_i^\alpha \frac{\partial}{\partial r}p_i^\alpha + S_\alpha \frac{\partial}{\partial c}p_i^\alpha = \sum_{\alpha'i'}\sum_{\beta,r,j} L_{i'j}^{s_r^{\alpha'\beta}}(p_{i'}^{\alpha'}) + \\
\qquad\qquad\qquad\qquad + \sum_{\alpha_1 i_1;r} L_{ii_1}^{s_r^{\alpha\alpha_1}}(p_i^\alpha, p_{i_1}^{\alpha_1}) \\
p_i^\alpha \big|_{t_0} = {}^{(0)}p_i^\alpha(c,r) \qquad \alpha = 1,\ldots,M \\
p_i^\alpha \big|_{r\in\Gamma} = {}^{(b)}p_i^\alpha(c,t) \qquad i = 1,\ldots,n,\ldots,
\end{cases}
\tag{5.76}
$$

where the rarefied gas is considered as a discrete stochastic medium (Zmievskaya et al., 1980; Pyarnpuu et al., 1981). Here $p_i^\alpha(c,r,t)$, as commonly adopted, is the density of probabilistic distribution of particle a_i^α in phase space (c,r). Equation (5.76) describes a strictly Markovian process of mixed type, because of the space-temporal locality of gas particle interactions (Zmievskaya et al., 1980; Skorochod, 1983). This process describes continuous changes of gas state due to the movement in the external field and jump-like changes of state due to instant interactions (collisions) between particles. Equations (5.76), averaged on an ensemble of particles, coincide with the initial system (4.4) of the kinetic equations.

In accordance with the theory of stochastic processes (Feller, 1970; Skorochod, 1983) the direct Kolmogorov, equation (5.76) can be replaced by an equivalent set of stochastic differential equations:

$$
\begin{cases}
dr_i^\alpha(t) = c_i^\alpha(t)\, dt \\
dc_i^\alpha(t) = S_\alpha(c_i^\alpha, r_i^\alpha)\, dt + \\
\quad + \sum_{\alpha'i'}\sum_{\beta,j,r}\int_\theta \mathbf{v}_\alpha^{(1)}(\theta; r_{i'}^{\alpha'}, r_j^\beta, c_{i'}^{\alpha'}, c_i^\beta) w_{i'j}^{s_r^{\alpha'\beta}}(d\theta\cdot dt) + \\
\quad + \sum_{\alpha_1 i_1;r}\int_\theta \mathbf{v}_\alpha^{(2)}(\theta; r_i^\alpha, r_{i_1}^{\alpha_1}, c_i^\alpha, c_{i_1}^{\alpha_1}) w_{ii_1}^{s_r^{\alpha\alpha_1}}(d\theta\cdot dt) \\
r_i^\alpha \big|_{t_0} = {}^{(0)}r_i^\alpha, \quad c_i^\alpha \big|_{t_0} = {}^{(0)}c_i^\alpha \\
c_i^\alpha \big|_{r\in\Gamma} = {}^{(b)}c_i^\alpha(t)\ .
\end{cases}
\tag{5.77}
$$

Here $w_{i'j}^{s_r^{\alpha'\beta}}(d\theta\cdot dt)$ and $w_{ii_1}^{s_r^{\alpha\alpha_1}}(d\theta\cdot dt)$ are the Poisson measures (Skorochod, 1983), determining the probabilities of particle interactions in the elementary acts of aeronomical processes (4.2), (4.3), and functions $\mathbf{v}_\alpha^{(1)}(\theta)$, $\mathbf{v}_\alpha^{(2)}(\theta)$ characterize jumps of particle velocities in these interactions. Accordingly, equations (5.76), (5.77) determine the microscopic probabilistic model of the rarefied gas flow. These equations are analogues of Hamilton equations for the gas dynamic description.

Proceeding further from stochastic simulation approach, the numerical stochastic model with a finite number of modeling particles is taken as the physical-probabilistic analogue of rarefied gas flow. As it was shown earlier, this model is a discrete approximation of the kinetic system (4.4), and it is defined by the following features:

a) The space-nonuniform physical-probabilistic analogue (5.77) is approximated on a selected grid configuration of the gas flow region by a numerical system with a finite number of modeling particles:

$$\left\{ \begin{array}{l} G(\mathbf{r};\Gamma) = \bigcup\limits_{L=1}^{\bar{L}} \bigcup\limits_{l=1}^{\bar{l}} G_{L,l}(\mathbf{r};\Gamma_L) \\ \{a_i^\alpha\}_{(M^\alpha,\infty)} \to \{a_i^\alpha\}_{(M^\alpha,N^\alpha)} = \bigcup\bigcup \{a_i\}_{(M^\alpha,N^\alpha(L,l))}^\alpha \\ N^\alpha = \sum\limits_{L,l} N^\alpha(L,l) \quad . \end{array} \right. \qquad (5.78)$$

This means that gas medium taken in each grid cell is replaced by a system of a finite number of modeling particles in response to the given physical-chemical structure of the multicomponent gas and following the initial and boundary conditions. The modeling particle is associated with a point in the phase space

$$\mathbf{x}_i^\alpha = (\mathbf{c}_i^\alpha, \mathbf{r}_i^\alpha) \quad , \qquad (5.79)$$

and consequently, the state of the overall numerical system is referred to as an element of the whole phase space;

b) The evolution of the numerical system has stochastic nature and is described by equations (5.77). The method of splitting by the physical processes is incorporated, and hence the system's evolution on a time step Δt obeys two procedures in each cell:

- collisionless movement of the modeling particles in the external field $\mathbf{S}_\alpha$ in the region considered:

$$\left\{ \begin{array}{ll} d\mathbf{r}_i^\alpha(t) = \mathbf{c}_i^\alpha(t)\, dt & l = 1, \ldots, \bar{l}_L \\ d\mathbf{c}_i^\alpha(t) = \mathbf{S}_\alpha(\mathbf{c}_i^\alpha, \mathbf{r}_i^\alpha)\, dt & i = 1, \ldots, N^\alpha(L,l) \ ; \end{array} \right. \qquad (5.80)$$

- jump-like changes of modeling particle states in collisional interactions in each of the cells [defining the local kinetics of aeronomical processes (4.2) and (4.3)]:

$$\left\{ \begin{array}{l} d\mathbf{r}_i^\alpha(t) = 0 \\ \\ d\mathbf{c}_i^\alpha(t) = \sum\limits_{\alpha' i'} \sum\limits_{\beta,j,r} \int_\theta \mathbf{v}_\alpha^{(1)}(\theta; \mathbf{x}_{i'}^{\alpha'}, \mathbf{x}_j^\beta) w_{i'j}^{s_r^{\alpha'\beta}} \, (d\theta \cdot dt) + \\ \\ \quad + \sum\limits_{\alpha_1 i_1; r} \int_\theta \mathbf{v}_\alpha^{(2)}(\theta; \mathbf{x}_i^\alpha, \mathbf{x}_{i_1}^{\alpha_1}) w_{ii_1}^{s_r^{\alpha\alpha_1}} \, (d\theta \cdot dt) \ ; \end{array} \right. \qquad (5.81)$$

Let us note that equations (5.80) characterize the numerical system with a variable number of modeling particles, since particles can leave the flow region. The particle flows through the boundary of the region are defined by an iterative process or from the boundary conditions (Bisikalo and Shematovich, 1988).

c) Trajectories of modeling particles on time interval Δt for the procedure (5.80) result from the solution of a proper dynamic problem for all particles, using finite-differentional schemes or direct simulation of particle movement (Bird, 1976, 1994; Belotserkovskii and Yanitskii, 1975; Bisikalo and Shematovich, 1988). For the procedure (5.81),

the local kinetics of collisional processes is realized with the use of basic algorithms of numerical stochastic models.

The space-nonuniform stochastic model (5.80) - (5.81) involving the numerical consideration of modeling particle trajectories and their collisions gives rise to the statistical evaluation of the particle distribution functions in the region under consideration and consequently, allows to obtain the solution of the initial kinetic system (4.4). The required accuracy is accomplished by the accumulation of statistics.

We may conclude that a stochastic system with a finite number of modeling particles which parallels space-nonuniform stochastic model (5.78)-(5.81) may serve as a good approximation to the probabilistic description (5.77) of the rarefied atmospheric gas flow including the microscopic kinetic description (4.4).

CHAPTER 6

NUMERICAL KINETIC MODELS
FOR AERONOMY APPLICATIONS

We shall now proceed from the general scheme of stochastic simulation of collisional processes in the rarefied gas to its application to atmospheric studies taking into account specific features of the atmospheric processes. As it follows from the structure of the Boltzmann-type kinetic equation (4.4), these features must be introduced in the structure of equations through the transport and particle source terms stipulated by external influences. The transport processes of atmospheric dynamics, being defined by pressure gradients, are also affected by gravitational and magnetic fields. These planetary fields are particularly responsible for complex trajectories of the modeling particles in the numerical models. In turn, the source terms are caused by the electromagnetic and corpuscular solar radiation, as well as by chemical reactions occurring in the upper atmosphere. The presence of particle sources is a typical feature of planetary aeronomy. In particular, these source terms are responsible for the formation of nonthermal particles in the rarefied atmospheric gas. Their contribution can be correctly evaluated only on the basis of numerical kinetic models which requires, however, some modification of the general scheme of the stochastic simulation. We shall consider these modifications in the framework of the general evaluation and development of numerical kinetic models for aeronomical applications.

6.1. Kinetic Models of Atmospheric Sources of Nonthermal Particles

The microscopic kinetic system (4.4) describing the behavior of the atmospheric gas differs from the systems traditionally used in rarefied gas dynamics by the presence of particle source terms. The characteristic features of the particles, supplied by solar radiation and chemical reactions, are their nonequlibrium distribution in translational and internal degrees of freedom. These so-called superthermal particles play an important role in the atmospheric chemistry, as well as in the processes of planetary evolution through atmospheric escape.

One of the main sources of superthermal and excited particles in atmospheric gas are the exothermic chemical reactions. Particles formed in these reactions have a broad kinetic energy distribution with a maximum at the reaction energy release (Whipple, 1974). These chemical sources are included into the standard procedure of stochastic simulation of nonequilibrium collisional processes in the rarefied gas (see Chapter 5), since in the model realization all possible (elastic, inelastic, and chemical) collisional processes are taken into account. We note that these chemical sources are especially important in those atmospheric regions where the direct influence of the external fluxes is vanishingly small. This situation exists in lower regions of the upper atmosphere, while in higher, more rarefied, regions the external fluxes may play a dominant role. Because these sources are not included into the general simulation model, a special procedure of their description and contribution must be developed.

The external fluxes such as EUV solar photons, photoelectrons, and ions of magnetospheric origin result in the formation of excited and ionized particles in photolytic and high-energy particle impact processes (4.2). The local kinetics of these interactions is described by the following system of equations

$$\begin{cases} \frac{\partial}{\partial t}F_\alpha(\mathbf{c},t) = \sum_\beta \sum_{\alpha'} Q_\alpha(F_{\alpha'}, W_\beta) \\ \frac{\partial}{\partial t}W_\beta(E^\beta, \mathbf{\Omega}, t) = \sum_\alpha \{\sum_{\beta'} Q_\beta^{(+)}(F_\alpha, W_{\beta'}) - Q_\beta^{(-)}(F_\alpha, W_\beta)\} \\ F_\alpha \mid_{t_0} = F_\alpha^{(0)}(\mathbf{c}) , \qquad\qquad \alpha = 0, 1, \ldots, M^a \\ W_\beta \mid_{t_0} = W_\beta^{(0)}(E^\beta, \mathbf{\Omega}) , \qquad \beta = 1, \ldots, M^b . \end{cases} \tag{6.1}$$

In addition to the sources of nonthermal atmospheric particles, this kinetic system also describes the local depletion of the incident external fluxes.

To solve the system (6.1), an efficient numerical model was developed (Shematovich, 1982, 1987; Marov et al., 1990, 1996). The physical–probabilistic analogue of the initial kinetic system is realized as the numerical system $N = N^a + N^b$ in which the subsystem of N^a modeling particles corresponds to the rarefied atmospheric gas

$$\{a^\alpha\}_{(M^a)} \to \{a_i^\alpha\}_{(M^a, N^a)}, \quad N^a = \sum_\alpha N^\alpha , \tag{6.2}$$

and another subsystem of N^b modeling particles corresponds to the external fluxes of particles

$$\{b^\beta\}_{(M^b)} \to \{b_j^\beta\}_{(M^b, N^b)}, \quad N^b = \sum_\beta N^\beta . \tag{6.3}$$

The state of subsystem (6.2) can be described by the vector

$$\begin{cases} \mathbf{C}_a = \overset{M^a}{\underset{\alpha=1}{\cup}} \mathbf{C}^\alpha \\ \mathbf{C}^\alpha = \overset{N^\alpha}{\underset{i=1}{\cup}} \mathbf{c}_i^\alpha \leftarrow F_\alpha(\mathbf{c}) \ . \end{cases} \qquad (6.4)$$

This vector approximates the distribution functions F_α by the system of a finite number of particles. The state of the subsystem (6.3) can be described by the vector

$$\begin{cases} \mathbf{C}_b = \overset{M^b}{\underset{\beta=1}{\cup}} \mathbf{C}^\beta \\ \mathbf{C}^\beta = \overset{N^\beta}{\underset{j=1}{\cup}} \mathbf{c}_j^\beta = \overset{N^\beta}{\underset{j=1}{\cup}} \overset{N^{\beta_j}}{\underset{k=1}{\cup}} c^\beta(E_j^\beta)\Omega_k^\beta \leftarrow W_\beta(E_j^\beta, \Omega) \ , \end{cases} \qquad (6.5)$$

where $c^\beta(E^\beta)$ and Ω_k^β are the discrete forms of particle fluxes having an energy E^β and a direction of flux propagation Ω^β; $N^\beta = \sum_j N_j^\beta$. In the general case, the state of the numerical system (6.1)-(6.2) is characterized by the point $\mathbf{C}_{ab} = \mathbf{C}_a \cup \mathbf{C}_b$ in the $3N$-dimensional phase space of the model.

The evolution of the given numerical system is described by the uniform Markovian process $\mathbf{C}_{ab}(t)$ involving instant interactions between particles of subsystems (6.1) and (6.3) which correspond to elementary acts of photolytic and impact processes (4.2). The jump-like transitions between states of the system have the following dynamic-probabilistic scheme

$$\mathbf{C}_{ab} \to \mathbf{C}'_{ab} \ : \ \mathbf{c}_i^\alpha, c^\beta(E_j^\beta)\Omega_k^\beta \to \underset{l}{\cup} c_{i_l}^{\alpha_l(r)}, c^{\beta'}(E^{\beta'}, \Omega^{\beta'}) \ . \qquad (6.6)$$

Transitions (6.6) occur as instant changes of the parameters of the interacting modeling particles and obey the constraints placed by the conservation laws of mass, momentum, and energy. They result in the following change of numerical system state

$$\mathbf{C}_{ab} \to \mathbf{C}'_{ab} = \mathbf{C}_{ab,ij_k}^{s_r^{\alpha\beta}} = \mathbf{C}_{a,i}^{s_r^{\alpha\beta}} \cdot \cup \mathbf{C}_{b,j_k}^{s_r^{\alpha\beta}} \ , \qquad (6.7)$$

where the states of each of the subsystems are equal to

$$\mathbf{C}_{a,b}^{s_r^{\alpha\beta}}, \ \mathbf{C}_{b,j_k}^{s_r^{\alpha\beta}} = \begin{cases} \mathbf{C}^{\alpha_1}; \ \mathbf{C}^{\beta_1}, \quad \alpha_1 \neq \alpha, \alpha_l(r), Vl, \ \beta_1 \neq \beta, \beta' \\ (-)\mathbf{C}_i^\alpha; \ (-)\mathbf{C}_{j_k}^\beta \\ (+)\mathbf{C}^{\alpha_l(r)}(c_{i_l}^{\alpha_r(l)}), Vl; \ (+)\mathbf{C}^{\beta'}(c_{j'}^{\beta'}) \ . \end{cases} \qquad (6.8)$$

The transition probabilities (6.6) are the discrete representations of the kinetic rates of photolytic and impact processes (4.2), collecting all final states of the reaction products, and may be written in the following form

$$\omega(\mathbf{C}_{ab} \to \mathbf{C}'_{ab}) = \omega^{s_r^{\alpha\beta}}_{ab,ij_k} = V^{-1} g_{ij}^{\alpha\beta} \, \sigma^{s_r^{\alpha\beta}} (\mathbf{c}_i^{\alpha}, c^{\beta}(E_j^{\beta})\, \mathbf{\Omega}_k^{\beta}) \ . \tag{6.9}$$

The total frequency of collisions in the numerical system is equal to

$$\omega(\mathbf{C}_{ab}) = \sum_{\alpha,\beta,r} \sum_{i,j_k} \omega^{s_r^{\alpha\beta}}_{ab,ij_k} \ , \tag{6.10}$$

and the time between consequent transitions is uniformely distributed as follows

$$\tau(\mathbf{C}_{ab} \to \mathbf{C}'_{ab}) = \omega^{-1}(\mathbf{C}_{ab}) \ . \tag{6.11}$$

Therefore, the expressions (6.2) to (6.11) define constructively the random process $\mathbf{C}_{ab}(t)$, describing evolution of the stochastic model under consideration (Shematovich, 1987; Marov and Shematovich, 1987). Based on this approach the transition probability for a given jump-like Markovian process may be written in the form:

$$\left\{ \begin{array}{l} Q_t(\mathbf{C}_{ab}, \mathbf{C}'_{ab}) = \sum_{\alpha,\beta,r} \sum_{i,j} \delta(\mathbf{C}'_{ab} - \mathbf{C}^{s_r^{\alpha\beta}}_{ab,ij}) \, Q_t(\mathbf{C}_{ab} \to \mathbf{C}^{s_r^{\alpha\beta}}_{ab,ij}) \\[2mm] Q_t(\mathbf{C}_{ab} \to \mathbf{C}^{s_r^{\alpha\beta}}_{ab,ij}) = A^{s_r^{\alpha\beta}} \dfrac{dI^{s_r^{\alpha\beta}}(\mathbf{c}_i^{\alpha}, \mathbf{c}_j^{\beta})}{I^{s_r^{\alpha\beta}}(\mathbf{c}_i^{\alpha}, \mathbf{c}_j^{\beta})} \times \\[4mm] \qquad\qquad \times \delta(\mathbf{d}^{s_r^{\alpha\beta}}(\mathbf{C}_{ab}) - \mathbf{d}^{s_r^{\alpha\beta}}(\mathbf{C}^{s_r^{\alpha\beta}}_{ab,ij})) \ , \end{array} \right. \tag{6.12}$$

where $\mathbf{d}^{s_r^{\alpha\beta}}(\mathbf{C})$ are the invariants (mass, momentum, and total energy) of collisional process $s^{\alpha\beta}$, and $A^{s_r^{\alpha\beta}}$ are the normalization constants. Using expression (6.12) and the direct Kolmogorov equation for jump-like process $\mathbf{C}_{ab}(t)$ it can be shown that the density

$$\varphi(\mathbf{C}_{ab}, t) = \int \varphi(\mathbf{C}'_{ab}) \, Q_t(\mathbf{C}'_{ab}, \mathbf{C}_{ab}) \, d\mathbf{C}'_{ab}$$

of the probabilistic distribution of the model state $\mathbf{C}_{ab}$ at time t is described by the following equation:

$$\left\{ \begin{array}{l} \frac{\partial}{\partial t} \varphi(\mathbf{C}_{ab}, t) = K_{ab}\, \varphi(\mathbf{C}_{ab}, t) \\[2mm] \varphi|_{t=0} = \varphi_0(\mathbf{C}_{ab}) \ . \end{array} \right. \tag{6.13}$$

The operator K_{ab} comprises all possible collisional transitions (6.6) for the model state $\mathbf{C}_{ab}$ and is equal to:

$$\left\{ \begin{array}{l} K_{ab}\varphi = \sum_{\alpha,\beta,r} \sum_{i,j} K^{s_r^{\alpha\beta}}_{ab,ij}\varphi \\[2mm] K^{s_r^{\alpha\beta}}_{ab,ij}\varphi = V^{-1} \int dI^{s_r^{\alpha\beta}}_{ij} \, [e^{s_r^{\alpha\beta}} \varphi(\mathbf{C}^{s_r^{\alpha\beta}}_{ab,ij}) - \varphi(\mathbf{C}_{ab})] \ . \end{array} \right. \tag{6.14}$$

In turn, the Cauchy problem (6.13)-(6.14) characterizes the probabilistic description of the numerical stochastic model (6.2) to (6.11) and is an approximation of the initial kinetic system (6.1) of source terms by the numerical system with a finite number of particles (Shematovich, 1987; Marov and Shematovich, 1987). Similarly to the general procedure, in the numerical experiment the probabilistic distribution of the stochastic model state is determined by statistics, accumulated in a computer simulation of random process $\mathbf{C}_{ab}(t)$. This procedure is conducted using the analoguous algorithms of the Monte-Carlo method (Shematovich, 1987).

The jump-like Markovian process $\mathbf{C}_{ab}(t)$ is simulated by the following set of operations:

a) the possible number of photolytic and impact transitions with the involvement of the probabilistic scheme (6.6) for the time interval Δt is determined by the expression

$$m(\Delta t) = \omega(\mathbf{C}_{ab}(t)) \, \Delta t \quad ; \tag{6.15}$$

b) the changes of the model states in transitions (6.6) in the event (6.15) form the Markovian chain

$$^{(m)}\mathbf{C}_{ab} = {}^{(m-1)}\mathbf{C}_{ab,ij}^{s_r^{\alpha\beta}} \quad m = 1,\ldots,m(\Delta t) \quad ; \tag{6.16}$$

c) the parameters of actual (m') transition between these Markovian chain states are defined through the hierarchy of conditional probabilities

$$m' = S_{r'}^{\alpha'\beta'}(a_{i'}^{\alpha'}, b_{j'}^{\beta'}) \rightarrow \begin{cases} W^\alpha = \frac{\omega^\alpha}{\omega_{ab}} \xrightarrow{\zeta_1} \alpha' \\[2mm] W^\beta = \frac{\omega^{\alpha'\beta}}{\omega^{\alpha'}} \xrightarrow{\zeta_2} \beta' \\[2mm] W^r = \frac{\omega_r^{s\,\alpha'\beta'}}{\omega^{\alpha'\beta'}} \xrightarrow{\zeta_3} r' \\[2mm] W^i = \frac{\omega_{i'i}^{s\,\alpha'\beta'}}{\omega_{i'}^{s\,\alpha'\beta'}} \xrightarrow{\zeta_4} i' \\[2mm] W^j = \frac{\omega_{i'j}^{s\,\alpha'\beta'}}{\omega_{i'}^{s\,\alpha'\beta'}} \xrightarrow{\zeta_5} j' \quad , \end{cases} \tag{6.17}$$

where ζ_i are random values from interval $[0,1]$, and W are conditional probabilities, for example

$$W^\alpha = \sum_{\beta,r}\sum_{i,j} \omega_{ij}^{s_r^{\alpha\beta}} / \omega_{ab}(\mathbf{C}_{ab}) \quad ;$$

d) the interrelation of the modeling particle characteristics at the transition (6.17) is defined by expression (6.8);

e) the seeked state of the model is equal to

$$\mathbf{C}(t + \triangle t) = {}^{(m(\triangle t))}\mathbf{C}_{ab} \ . \tag{6.18}$$

The algorithm defined by the expression (6.15) through (6.18) gives the representative statistical sample of random process $\mathbf{C}_{ab}(t)$ and allows to calculate the local kinetic characteristics of source terms of superthermal and excited particles in the upper atmosphere. The final energy distribution function of the atmospheric species results from the balance of the processes of particle production in all possible sources and their relaxation.

6.2. General Scheme of the Numerical Kinetic Model for Atmospheric Gas Flows

The advantage of a stochastic simulation and a numerical realization of its physical-probabilistic analogue for the kinetic collisional processes involved is that, instead of a direct solution of stochastic equations, the repetitive stochastic process based on a Monte-Carlo technique is used. All the necessary physical characteristics are derived from the computer simulation of this stochastic process. The main subsequent elements of the numerical evaluation of kinetic system (4.4) can be summarized as follows:

- splitting up of the system (4.4) between source reactions producing superthermal particles, collisional relaxation of these particles in physical and chemical reactions, and collisionless movement of the atmospheric gas in the planetary gravity/magnetic fields;

- stochastic simulation of the local kinetics of collisional physical and chemical processes using the analogous Monte-Carlo algorithms;

- calculation of the atmospheric particle trajectories in the collisionless approach using finite-difference schemes.

In order to investigate space-temporal variations of the rarefied atmospheric gas at the kinetic level and to develop an effective numerical model the following peculiarities of the atmospheric gas must be taken into account:

- the gas parameters change strongly in the considered regions of the upper atmosphere, i.e. the atmospheric gas flows are basically gradient flows;

- the significant background difference in concentrations of gas components is complicated by the minor admixtures of photochemical origin formed in the atmosphere.

These circumstances make it necessary to introduce some substantial modifications of the numerical stochastic models to be pertinent for the kinetic and dynamic investigations of the upper atmospheric gas.

The effective approach to the study of gradient gas flow is closely related to the development of an appropriate space grid. Because the gas density is the most rapidly varying parameter of the upper atmosphere, exponentially decreasing with height, the characteristic macroscopic scale of the gas flow is defined by the density gradients. Therefore, to simplify the simulation procedure, it is reasonable to divide the region considered $G(\mathbf{r}; \Gamma)$ of the atmosphere into sequence of segments

$$G(\mathbf{r}; \Gamma) = \bigcup_{L=1}^{\bar{L}} G_L(\mathbf{r}; \Gamma_L) \ . \tag{6.19}$$

These segments are defined in such a manner that the characteristic macro-parameters (especially density) of the atmospheric gas in the segment G_L change within one order of magnitude. Thereupon each segment G_L is divided into a sequence of cells

$$G_L(\mathbf{r}; \Gamma) = \bigcup_{l=1}^{\bar{l}_L} G_{L,l}(\mathbf{r}; \Gamma_L) \ . \tag{6.20}$$

The cell sizes $\Delta r_{L,l} = d(G_{L,l}(\mathbf{r}; \Gamma_L))$ are defined by the local (at this altitude) microscopic scale, i.e. by the local mean free path length - $\Delta r_{L,l} \leq \lambda_L$.

Thus, the characteristic size of cells prove to be in accord with the local gas parameters. This means that changes of the distribution functions in each cell are small. Respectively, in every segment it is possible to use with sufficient efficiency the method of splitting the gas flow evolution between collisionless movement and collision kinetics.

The formal algorithm of the solution of kinetic equations system (4.4) on the discrete time grid $t_\delta = \delta \Delta t$, $\delta = 0, 1, \ldots$ may be implemented as follows (Marov and Shematovich, 1988). Let the solution at time $t_{\delta-1}$ be known

$$\begin{cases} R(t_{\delta-1}) = \{R(t_{\delta-1}, L, l)\}_{(\bar{L}, \bar{l})} \\ R(t_{\delta-1}, L, l) = \{F_\alpha(c, r_{L,l}, t_{\delta-1}), W_\beta(E^\beta, \Omega, r_{L,l}, t_{\delta-1})\}_{(M^a, M^b)} \end{cases} \tag{6.21}$$

i.e. on space grid (6.19)-(6.20) of the region considered $G(\mathbf{r})$ the microscopic gas state and external fluxes are given at time $t_{\delta-1}$. Following the splitting procedure the solution of problem (4.4) for the given set of cells on a time step $\Delta t \leq \tau_L$ (τ_L - local free path time) is obtained through the algorithmic steps which include:

- collisionless dynamics of atmospheric particles in the segment considered G_L under specified initial and boundary conditions

$$\begin{cases} \frac{\partial}{\partial t}\tilde{F}_\alpha + \mathbf{c}\frac{\partial}{\partial \mathbf{r}}\tilde{F}_\alpha + \mathbf{S}_\alpha\frac{\partial}{\partial \mathbf{c}}\tilde{F}_\alpha = 0 \\ \tilde{F}_\alpha(t_{\delta-1}) = F_\alpha(\mathbf{c},\mathbf{r}_L,t_{\delta-1}) \\ \tilde{F}_\alpha \mid_{\mathbf{r}\in\Gamma_L(G)} = F_\alpha^{(b)}(\mathbf{c},t) \ , \end{cases} \tag{6.22}$$

boundary conditions for inner segments (segments not contiguous to the borders of the region considered G) being determined by an iterative procedure of particle flow calculations in the whole set of segments (Bisikalo and Shematovich, 1988);

- local kinetics of collisional aeronomic processes (4.2)-(4.3) in each of the cells (6.20)

$$\begin{cases} \frac{\partial}{\partial t}\hat{F}_\alpha = \sum_\beta\sum_{\alpha'} Q_\alpha(\hat{F}_{\alpha'},W_\beta) + \sum_{\alpha_1} J_{\alpha\alpha_1}(\hat{F}_\alpha,\hat{F}_{\alpha_1}) \\ \hat{F}_\alpha(t_{\delta-1}) = \tilde{F}_\alpha(\mathbf{c},\mathbf{r}_{L,l},t_\delta) \ . \end{cases} \tag{6.23}$$

The initial conditions for step (6.23) are the solutions obtained from step (6.22).

In this approach the common solution of system (4.4) at time t_δ may be written as $F_\alpha(\mathbf{c},\mathbf{r},t_\delta) = \hat{F}_\alpha(\mathbf{c},\mathbf{r},t_\delta)$. As a result, the evolutionary algorithm of the numerical realization of the microscopic kinetic model (4.4) based on the method of splitting by the physical processes is implemented as the set of solutions of the above mentioned steps (6.22)-(6.23), i.e. as follows

$$R(t_\delta) = \hat{L}\bar{L}R(t_{\delta-1}) \ , \tag{6.24}$$

where operators $\bar{L}$ and $\hat{L}$ define the change of gas state due to collisionless movement of particles and due to collisional particle interactions, accordingly.

It should be noted that to solve the problem (6.23) it is also necessary to use the splitting procedure by the following physical-chemical processes:

- formation of the superthermal and excited atmospheric particles due to the external influences (4.2) on the rarefied atmospheric gas

$$\begin{cases} \frac{\partial}{\partial t}\hat{F}_\alpha^{(1)} = \sum_\beta\sum_{\alpha'} Q_{\alpha'}(\hat{F}_{\alpha'}^{(1)},W_\beta) \\ \hat{F}_\alpha^{(1)}(t_{\delta-1}) = \tilde{F}_\alpha(t_{\delta-1}) \ ; \end{cases} \tag{6.25}$$

- collisional relaxation of atmospheric gas by the set of physical and chemical chemical reactions (4.3)

$$\begin{cases} \frac{\partial}{\partial t}\hat{F}_\alpha^{(2)} = \sum_{\alpha'} J_{\alpha\alpha_1}(\hat{F}_\alpha^{(2)},\hat{F}_{\alpha_1}^{(2)}) \\ \hat{F}_\alpha^{(2)}(t_{\delta-1}) = \hat{F}_\alpha^{(1)}(t_\delta) \ , \end{cases} \tag{6.26}$$

and hence operator $\hat{L} = \hat{L}_{ab} \cdot \hat{L}_{aa}$.

Finally, for the numerical simulation of gas state on microscopic level, the algorithm (6.24) is realized in the following manner:

- on step (6.22) - by exact solution of particle dynamics in gravitational and magnetic planetary fields, using finite-differential or Monte-Carlo analogue schemes for solving collisionless kinetic equations (Bird, 1976; Belotserkovskii and Yanitskii, 1975; Bisikalo and Shematovich, 1988);

- on step (6.23) - by consideration of collisional processes, using the method of stochastic simulation (Bird, 1976, 1994; Marov et al., 1990, 1996).

Generally, the realization of algorithms to solve the problem (6.22) does not meet serious mathematical difficulties (Bird, 1976, 1994), though the problem (6.23) and especially kinetic relaxation (6.26) from the mathematical point of view are more complex because the collision terms (4.6) are nonlinear and multi-dimensional.

A significant difference in concentrations of gas components causes an additional complication of the algorithm realization. To improve the numerical efficiency of the stochastic simulation method in the case of multicomponent gas the schemes with statistical particle weights are used (Bird, 1976, 1994; Korolev and Yanitskii, 1983; Bisikalo and Shematovich, 1988). In this case, the numbers of modeling particles for different species characterized by their own concentrations in the numerical system remain the same but their weights (essentially the number of real particles represented by one modeling particle) are different. In other words, if for some gas components the condition $n_{\alpha_1}(L, l) \ll n_{\alpha_2}(L, l)$ is satisfied, then in numerical scheme these components are represented by the same number of modeling particles $N^{\alpha_1}(L, l) \cong N^{\alpha_2}(L, l)$, and their phase characteristics are corrected by the parameters $W^\alpha(L, l)$

$$\mathbf{z}_i^\alpha = (\mathbf{c}_i^\alpha, \mathbf{r}_i^\alpha, W^\alpha(L, l)) \ , \qquad (6.27)$$

where the statistical weight of the modeling particle of a given component is equal to $W^\alpha(L, l) = n_\alpha(L, l) \ / \ N^\alpha(L, l)$.

The general scheme of algorithm realization of the numerical stochastic model including the weights (6.27) remains basically the same. Two points must be, however, articulated: first, in the local kinetics evaluation, statistical weights of modeling particles are to be taken into account to properly calculate a hierarchy of collision probabilities and also collision patterns. Second, in the collision simulations it is possible to use both nonconservative schemes, when only the averaged collision invariants are preserved (Bird, 1976, 1994), and conservative schemes, when collision invariants are preserved in each collision (Korolev and Yanitskii, 1983, 1985; Bisikalo and Shematovich, 1988).

As an example of the conservative scheme for simulated collision, the velocity of a particle having a weight W^{α_1} (in the case $W^{\alpha_1} < W^{\alpha_2}$) is changed, and the velocity of a particle having larger weight W^{α_2} is equal to the sum of the previous velocity (before collision), multiplied by $(1 - W^{\alpha_1}/W^{\alpha_2})$, and the velocity after collision, multiplied by $W^{\alpha_1}/W^{\alpha_2}$. A similar approach is also applied to the collision modeling of superthermal particles of photochemical origin with the main components of the upper atmosphere, each collision having probabilities of both elastic and inelastic interactions proportional to the partial cross-sections and statistical weights of particles (Shematovich et $al.$, 1991a). This modified general algorithm gives rise to an improved efficiency of the numerical stochastic simulation that is especially important in the complex cases of the study of nonequilibrum multicomponent gas of the upper atmosphere (Marov et $al.$, 1990, 1996).

6.3. Some Additional Remarks on the Numerical Models

In the upper planetary atmosphere all possible flow regimes (from continuum gas dynamic to collisionless) usually occur. Numerical kinetic models we dealt with are generally adjustable for different flow regimes. Nevertheless, since the mathematical structure of these numerical models is very complicated, their utilization is only justified for the atmospheric regions where one deals with a strongly nonequilibrium gas flow, specifically in the transition (between gas dynamic and collisionless) region where the Knudsen number $Kn \sim 1$. It is well known that in this transition region gas, the flow regime is characterized by distribution functions varying both at the microscopic and macroscopic flow scales, and is directly depending on time. The collisional physical and chemical processes may cause considerable disturbances of the distribution functions, and therefore their description requires the use of the nonlinear collisional terms. The simulation of these collisional terms gives rise to the main dificulties of the respective kinetic model realizations. There are, however, some approaches where these nonlinear collisional terms for both gas dynamic and collisionless flow regimes can be significantly simplified.

For the atmospheric layers located above the transition region where the flow becomes essentially collisionless (i.e. where $Kn \gg 1$) only the external flux influence and transport of the modeling particles can be retained in the numerical model. Removing the collisional terms from the model allows to use a rather simple and very effective numerical algorithms. In particular, these regions of the atmosphere can be effectively modeled using the Monte Carlo probe particle methods.

In contrast, for the atmospheric layers located below the transition region the gas dynamic approach is sufficiently reliable when disturbances caused by external influences and chemical reactions are small. In this case the frequency of elastic collisions in the kinetic system is high enough (i.e., there is a Maxwellian velocity distribution of gas particles), and these relaxation systems are described by a set of gas dynamic equations

$$\frac{\partial}{\partial t}\{n_{\tilde{\alpha}}(t)D_{\tilde{\alpha}}(t)\} = \sum_{\tilde{\alpha}_1,r} n_{\tilde{\alpha}}(t)n_{\tilde{\alpha}_1}(t)V^{s_r^{\tilde{\alpha}\tilde{\alpha}_1}}(d_{\tilde{\alpha}}; f_{\tilde{\alpha}}^{(M)}, f_{\tilde{\alpha}_1}^{(M)}) \ , \qquad (6.28)$$

where the kinetic coefficients relating to the molecular properties transport are defined similarly to (5.55), i.e.

$$V^{s_r^{\tilde{\alpha}\tilde{\alpha}_1}}(d_{\tilde{\alpha}}) = \int dc\, dc_1\, d_{\tilde{\alpha}}(\mathbf{c}) J^{s_r^{\tilde{\alpha}\tilde{\alpha}_1}}(f_{\tilde{\alpha}}^{(M)}, f_{\tilde{\alpha}_1}^{(M)}) \ . \qquad (6.29)$$

Because the system (6.28) describes the collisional relaxation of a multi-component gas in the case of small disturbances of the gas thermal state by physical and chemical processes it is pertinent to models of the local kinetics of photochemical reactions in the mesosphere and lower thermosphere (see e.g., McEwan and Phillips, 1975; Chamberlain, 1978; Marov and Kolesnichenko, 1987; Rees, 1989).

However, it is important to emphasize that the numerical stochastic model for gas dynamic systems is a version of the previously described structural stochastic model ($\hat{M}^\alpha = 0$, $\bar{M}^\alpha = 0$). Indeed, the state of the gasdynamic stochastic model is characterized by the following value

$$\mathbf{C} = \bigcup_{\tilde{\alpha}} \mathbf{C}^{\tilde{\alpha}}, \quad \mathbf{C}^{\tilde{\alpha}} = \sum_{i=1}^{N^{\tilde{\alpha}}} d_{\tilde{\alpha}}(\mathbf{c}_i^{\tilde{\alpha}}) \ , \qquad (6.30)$$

where $d_{\tilde{\alpha}}(\mathbf{c}_i^\alpha)$ are the microscopic attributes for the set $\mathbf{D}_{\tilde{\alpha}}$ of gas macroparameters (particle number density, kinetic temperature, mean mass velocity, etc.), and the velocities of the modeling particles correspond to distribution $f_{\tilde{\alpha}}^{(M)}(\mathbf{c}; \mathbf{D}(t))$. This means that implementation of the collision transition $s_r^{\tilde{\alpha}\tilde{\alpha}_1}$ conforms the discrete change of populations $N^{\tilde{\alpha}}$ and balances values $\mathbf{d}_{\tilde{\alpha}}$ for the dynamic scheme of collision $(\mathbf{c}_i^{\tilde{\alpha}}, \mathbf{c}_{i_1}^{\tilde{\alpha}_1}) \rightarrow (\mathbf{c}_{i'}^{\tilde{\alpha}(r)}, \mathbf{c}_{i'_1}^{\tilde{\alpha}_1(r)})$ as well. In turn, the transition probabilities depend on time through the gas macrocharacteristics

$$\omega_{ii_1}^{s_r^{\tilde{\alpha}\tilde{\alpha}_1}} = V^{-1} < I_{ii_1}^{s_r^{\tilde{\alpha}\tilde{\alpha}_1}} >_{(f_{\tilde{\alpha}}^{(M)}, f_{\tilde{\alpha}_1}^{(M)})} = V^{-1} q^{s_r^{\tilde{\alpha}\tilde{\alpha}_1}}(\mathbf{D}_{\tilde{\alpha}}(t), \mathbf{D}_{\tilde{\alpha}_1}(t)) \ . \qquad (6.31)$$

Expressions (6.30)-(6.31) define the modification of the general algorithm. Essentially, in the construction of the Markovian chain $^{(m)}\mathbf{C}$ one must recalculate only probabilities (6.31) at each algorithm step, taking into account the weak changes of macrocharacteristics $\mathbf{D}_{\tilde{\alpha}}$ in time. This modification allows to study the atmospheric gas flow in a gas dynamic approximation when deviations from thermal equilibrium are small.

Nevertheless, when the source terms responsible for the formation of significant amount of superthermal and internally excited particles are involved, a microscopic description of these particles must be obtained. In this case, the main atmospheric gas species can be also described by the system of gas dynamic equations, though the contribution of small admixtures formed by external flux impacts and chemical reactions must be evaluated incorporating the linearized Boltzmann kinetic equation. This combined approach is the most convenient procedure for the gas flow evaluation in this region, and it is efficiently implemented by the structural stochastic models. A set of effective algorithms developed in response to the peculiarities of a numerical models ensures a thorough study of the different regimes of atmospheric gas flows.

CHAPTER 7

PRODUCTION OF NONTHERMAL PARTICLES BY ELECTROMAGNETIC AND CORPUSCULAR SOLAR RADIATION

We already pointed out that, although the abundances of energetically active (superthermal, internally excited, and ionized) particles formed due to electromagnetic and corpuscular solar radiation are small compared to the dominant neutral atmospheric species, they play an important role in the physical and chemical processes of the upper atmosphere. Their nonequilibrium properties caused by excess kinetic and inner energy can be thoroughly studied only on a microscopic kinetic level and are intrinsically connected with self-consistent consideration of the following major problems (Marov et al., 1990, 1996):

- numerical evaluation of the production rates of energetic particles;

- influence of active particles on dynamics, energy balance, and other properties of the upper atmosphere, including their contribution to the problems of atmospheric photochemistry and formation of hot planetary coronas.

This chapter is dedicated to the problem of nonthermal particle formation by solar radiation. The question of their influence on the background atmospheric gas will be discussed in the following chapters.

7.1. Elementary Processes of Nonthermal Particles Formation Induced by the Solar EUV Flux

In order to correctly evaluate the absorption of the solar radiation in a planetary atmosphere, the transport equation for photons must be solved (see e.g., Sampson, 1965; Marov and Kolesnichenko 1987). However, in the upper atmosphere, where the intrinsic atmospheric radiation is negligible, a simple expressions for the absorption of the solar UV radiation may be used:

$$W_\beta(E^\beta, h) = W_{\beta,\infty}(E^\beta) \exp(-\tau_\beta(E^\beta, h)) \ , \qquad (7.1)$$

where $W_{\beta,\infty}$ is the unattenuated UV photon flux outside the atmosphere, and h is the height. In (7.1) the vertical optical depth of an atmospheric layer (h, ∞) can be written in the following form:

$$\tau_\beta(E^\beta, h) = \int_h^\infty \sum_\alpha \sum_{r=1}^{R^{\alpha\beta}} Ch(\theta_\odot, X_\alpha) \, n_\alpha(h) \, \sigma^{s_r^{\alpha\beta}}(E^\beta) \, dz \ . \tag{7.2}$$

In this expression, the spherical shape of the atmosphere is taken into account through the Chapman function $Ch(\theta_\odot, X_\alpha)$. This function useful for the solution of aeronomical problems is usually tabulated (see e.g. Banks and Kockarts, 1973), and can be also approximated by the following expression:

$$\begin{cases} \sec\theta_\odot \ , \quad \theta_\odot < 75^\circ \\[2mm] (\tfrac{1}{2}\pi X_\alpha \sin\theta_\odot)^{\frac{1}{2}} \exp(\tfrac{1}{2}X_\alpha \cos^2\theta_\odot)(1 \pm erf(\tfrac{1}{2}X_\alpha \cos^2\theta_\odot)) \ , \\[2mm] \qquad \theta_\odot > 75^\circ, \quad (\pm) \to \theta_\odot \gtrless 90^\circ \ . \end{cases} \tag{7.3}$$

Here $\theta_\odot$ is the solar zenith angle and $X_\alpha = \frac{R_p+h}{H_\alpha}$, where $H_\alpha = \frac{kT}{m_\alpha g}$ is the scale height for component α, and R_p is the planet's radius.

To describe the elementary processes induced by solar EUV flux let us consider the structure of photolytical processes (3.1) - (3.3). These processes can be represented by the following scheme:

$$\alpha + h\nu \to \begin{cases} \alpha_1^{(0)}(i_1^{(0)}) + \alpha_2^{(0)}(i_2^{(0)}) \\[2mm] \alpha^{(+)}(i^{(+)}) + e_\nu \\[2mm] \alpha_1^{(0)}(i_1^{(0)}) + \alpha_2^{(+)}(i_2^{(+)}) + e_\nu \ , \end{cases} \tag{7.4}$$

where $\alpha^{()}(i^{()})$ are chemically active radicals $\alpha^{(0)}$ of photolytic origin or ions $\alpha^{(+)}$ in electronically excited states $i^{()}$. In this case, the kinetic characteristics of the photolysis process (7.4) are determined following expression (4.5).

The kinetic rate of photolytic formation of neutral or charged products (7.4) in the upper atmosphere at height h and at zenith angle $\theta_\odot$ is equal to:

$$q_\alpha(\mathbf{r}) = \sum_{\alpha'} \int dc' Q_\alpha(W_\nu, F_{\alpha'} \, ; \, \mathbf{c}, \mathbf{r}) \ .$$

Taking into account the weak dependence of cross-sections of the photolysis process (7.4) on the atmospheric particle velocity and integrating over the velocity space, we obtain the standard expressions for photodissociation rates usually used in aeronomy:

$$\begin{cases} q_{\alpha^{()}(i^{()})}(h, \theta_\odot) = \sum_{\alpha',r} q_{r,\alpha \to \alpha^{()}(i^{()})}^{()}(h, \theta_\odot) \\[3mm] q_{r,\alpha \to \alpha^{()}(i^{()})}^{()} = \int \sigma_{r,\alpha \to \alpha^{()}(i^{()})}^{()}(E^\nu) \, n_\alpha(h, \theta_\odot) \, W_\nu(E^\nu; h, \theta_\odot) \, dE^\nu \ , \end{cases} \tag{7.5}$$

where $\sigma^{()}_{r,\alpha \to \alpha^{()}(i^{()})}$ is the partial cross section of process (7.4) corresponding to the formation of particles $\alpha^{()}(i^{()})$. The attenuated value $W_\nu(E^\nu; h, \theta_\odot)$ of the initial flux during its propagation in the atmosphere to height h at zenith angle $\theta_\odot$ is defined here by the expressions (7.1) - (7.3).

Dissociative collisions in process (7.4) lead to the excitation of electronic levels of molecule $a^\alpha(E^*)$ characterized by energies greater than the dissociation threshold $E_\alpha^* > E_\alpha^{(0)}$. The decay of these excited states of the molecule results in the formation of dissociation products in various internally excited states, and their thermal state is characterized by the energy excess

$$\Delta E_{tr}^\alpha = E^\nu - (E_\alpha^{(0)} + E_{\alpha_1}(i_1^{(0)}) + E_{\alpha_2}(i_2^{(0)})) \ . \tag{7.6}$$

This energy excess transforms into kinetic energy of dissociation products in proportion to their masses

$$\begin{cases} \Delta E_{tr}^{\alpha_1^{(0)}} = \Delta E_{tr}^\alpha \dfrac{m_{\alpha_2^{(0)}}}{(m_{\alpha_1^{(0)}} + m_{\alpha_2^{(0)}})} \\[3mm] \Delta E_{tr}^{\alpha_2^{(0)}} = \Delta E_{tr}^\alpha - \Delta E_{tr}^{\alpha_1^{(0)}} \ . \end{cases} \tag{7.7}$$

Then, in agreement with expression (4.5), the kinetic energy spectrum of the dissociation products is defined by the following expression:

$$\begin{cases} q_{\alpha_l^{(0)}}(E^{\alpha_l^{(0)}}, h) = \sum_{\alpha,r} \int dE^\nu \, \sigma^{(0)}_{r,\alpha \to \alpha_l^{(0)}}(E^\nu) \, W_\nu(E^\nu, h, \theta_\odot) \times \\ \hspace{6cm} \times n_\alpha(h, \theta_\odot) \\[2mm] E^\nu = \sum_r E_{tr}^{\alpha_l^{(0)}(r)} + E_\alpha^{(0)} \ ; \quad \alpha_l^{(0)} \in \bigcup_l \alpha_l^{(0)}(r) \ . \end{cases} \tag{7.8}$$

In order to calculate the characteristics of the photodissociation processes, a modification of the stochastic model described in Chapter 6 must be used. This modification is based on the following assumptions:

- the thermal velocity distribution functions used for the determination of the state vector (6.4) are at local equilibrium;

- calculations of the dynamical-probabilistic characteristics of the transitions (6.6) in the numerical model are constrained by the condition (Sampson, 1965)

$$|E^\nu \, \Omega^\nu / c| \ll |m_\alpha c^\alpha| \ .$$

This condition means that solar UV photons are characterized by a small momentum and permits to simplify the calculation of the formation rate of the photodissociation products.

Photoionization collisions in the process (7.4) are accompanied by the formation of fast electrons. These electrons have energies equal to the excess of energy of the absorbed photon over the ionization threshold, i.e.

$E = E^\nu - E_\alpha^{(+)}(i)$. Thus the energy spectrum of electrons formed in the photoionization process is equal to:

$$\begin{cases} q_e(E, h, \theta_\odot) = \sum_{\alpha, r} \int \sigma^{(+)}_{r, \alpha \to \alpha^{(+)}(i^{(+)})}(E^\nu)\, n_\alpha(h, \theta_\odot) \times \\ \qquad\qquad\qquad\qquad \times W_\nu(E^\nu; h, \theta_\odot)\, dE^\nu \qquad\qquad (7.9) \\ E^\nu = E + E_\alpha^{(+)}(i^{(+)}) \ ; \end{cases}$$

These primary electrons are energetically active particles and may cause additional ionization and dissociation of the atmospheric constituents. To determine the local characteristics of the impact effects and the subsequent thermalization of the fluxes of photoelectrons formed by photoionization, a modification of the general stochastic algorithm of external influences described in Chapter 6 must be used. It is worthwhile to note that in the general case the given stochastic model may be considered as the probabilistic analogue of the multistream method (Strickland et al., 1976; Stamnes, 1980), which has been used to study photoelectron thermalization and transport.

Let us consider in more detail the modification of the initial stochastic model for studies of the electron impact effect and photoelectron thermalization. Usually, in the analysis of impact effects, only the superthermal electrons responsible for the excitations of electronic states of atmospheric particles, dissociation, and ionization are considered (i.e., electrons with energy $\geq 2-4$ eV). Electrons with lower energies (suprathermal electrons) are responsible for the excitation of rotational and vibrational levels and have some specific features (Krinberg, 1978). The numerical study of this type of electrons can be successfully carried out using algorithms of structural stochastic simulation.

It is possible to use the following simplifications of the initial kinetic equation (4.7) to study superthermal electrons:

- collisions between superthermal electrons can be ignored because their density is small. The elastic interaction of these electrons with thermal electrons $\alpha = 0$ is described by the linear operator (Krinberg, 1978; Rees, 1989)

$$n_{\alpha=0} \frac{\partial}{\partial E}[L_e(E) W_{e_\nu}(E)]$$

and in the lower thermosphere, where the density of thermal electrons is also small, can be omitted;

- elastic collisions of superthermal electrons with heavy particles can be ignored because the energy loss in elastic collisions $2m_e/m_{\alpha>0}\, E$ is less than that in inelastic collisions $\Delta E^{s_r^{\alpha e}}$. In this approach only angular scattering of the initial fluxes by elastic collisions must be retained;

- thermalization of superthermal electrons can be considered as a steady-state process because the relaxation time for their energy τ_E is less than the characteristic time scales τ of density $n_\alpha(t)$ and source $Q_e(t)$ changes.

Using these assumptions, the kinetic equation (4.7) for photoelectron flux, averaged over angles Ω (i.e., the equation for isotropic energy flux distribution), may be written in the following form:

$$Q_e^{(\nu)}(E) + Q_e^{(e\nu)}(E) - W_e(E)L^{(-)}(E) = 0 \ , \tag{7.10}$$

where

$$\begin{cases} Q_e^{(\nu)}(E) = \sum_{\alpha,r} n_\alpha \int_{E_\alpha^{(+)}}^{\infty} dE^\nu \, d\sigma_{r,\alpha\to\alpha(+)}^{(+)}(E^\nu)W_\nu(E^\nu) \ , \\ \qquad\qquad\qquad\qquad\qquad E = E^\nu - E_\alpha^{(+)} \\ Q_e^{(e\nu)}(W_e,E) = \sum_{\alpha,r} n_\alpha \int_{E_1'}^{\infty} dE' \, d\sigma^{s_r^{\alpha e}}(E')W_e(E') \ , \\ \qquad\qquad E = E' - \Delta E^{s_r^{\alpha e}} \ ; \quad E_1' = E + \Delta E^{s_1^{\alpha e}} \\ L^{(-)}(E) = \sum_{\alpha,r} n_\alpha \sigma^{s_r^{\alpha e}}(E) \ . \end{cases} \tag{7.11}$$

Here $Q_e^{(\nu)}$ is the source function of photoelectrons with energy E formed by photoionization and $Q_e^{(e\nu)}$ is the source function of electrons with energy E formed by degradation of more energetic electrons with $E' > E$. It is interesting to note that the thermalization process (process of energy loss) has a discrete nature defined by thresholds $\Delta E^{s_r^{\alpha e}}$ of excitation, dissociation and ionization reactions (3.4)-(3.6).

To simplify the expression for $Q^{(e\nu)}$ it is necessary to take into account specific features of impact ionization $s_r^{\alpha e}(+) = {}^{(+)}s_r^{\alpha e} : \alpha + e_\nu \to \alpha^{(+)}(r) + e_\nu' + e'$, accompanying the formation of secondary electrons with energies $0 \le E_{e'} \le (E_e - E_\alpha^{(+)})/2$. This means that in (7.11) the term of impact ionization can be rewritten in the form

$$Q_e^{(e\nu)}(W_e,E) = {}^{(+)}Q_e^{(1)}(W_e,E) + {}^{(+)}Q_e^{(2)}(W_e,E) \ , \tag{7.12}$$

where the functions $Q^{(1)}$ and $Q^{(2)}$ correspond to the scattering of the primary flux and to the formation of secondary electrons due to impact ionization, respectively:

$$
\begin{cases}
{}^{(+)}Q_e^{(1)}(E) = \sum_{\alpha,r} n_\alpha \int_0^E dE' \, d\sigma_{r,\alpha\to\alpha(+)}^{(e)}(E'', E') W_e(E'') , \\
\qquad\qquad\qquad\qquad E'' = E_\alpha^{(+)} + E + E' \\
{}^{(+)}Q_e^{(2)}(E) = \sum_{\alpha,r} n_\alpha \int_{E_2}^\infty dE'' \, d\sigma_{r,\alpha\to\alpha(+)}^{(e)}(E'', E) W_e(E'') , \\
\qquad\qquad\qquad\qquad E_2 = E_\alpha^{(+)} + 2E .
\end{cases}
\tag{7.13}
$$

Here E'' is the energy of the impact electron, E' is the energy of the secondary electron, and $d\sigma_{r,\alpha\to\alpha(+)}^{(e)}$ are the partial cross-sections of impact ionization depending on the energy of the electrons (Krinberg, 1978; Rees, 1989).

In order to determine the steady state distribution of photoelectron flux under given solar and geophysical conditions, the solution of equation (7.10) must be obtained, using the iterative procedure in the form of the following expansion

$$
W_e(E) = \sum_{s=0}^M L_1^{(s)}(W_e^{(0)}) ,
\tag{7.14}
$$

where

$$
\begin{cases}
W_e^{(0)}(E) = Q_e^{(\nu)}(E) \,/\, L^{(-)}(E) \\
L_1(W_e) = Q_e^{(e_\nu)}(W_e, E) \,/\, L^{(-)}(E) ,
\end{cases}
\tag{7.15}
$$

and operator $L_1^{(s)}$ is s-times recurrent realization of operator L_1.

Expression (7.14) is the exact solution of the initial stationary equation (7.10) which allows to determine the energetic spectrum $W_e(E)$. Let us note that, because the source $Q_e^{(\nu)}(E) = 0$ at $E > E_{max}$ (E_{max} is the maximum energy of electrons formed by ionization by EUV solar radiation), the quantity $M(E)$ of expansion (7.14) terms for a given electron energy is equal to the integer part of the value $(E_{max} - E)/E_{min}$, where $E_{min} = min(E_r^{s_r^{\alpha e}})$ is the minimum value of the excitation energy. It means that the electrons under consideration reduce their energy to this value as a result of finite number of inelastic collisions. At each collision the electron loses a discrete portion of energy, i.e. a cascade of discrete energy losses occurs.

The stochastic model for calculations of the steady state distribution (7.14) of the photoelectron flux and accordingly, for calculations of the kinetic characteristics of the impact effect, is built up on the basis of the algorithm described in Chapter 6 through the calculation of all possible photoelectron collisions with the ambient gas. To perform these calculations one has to define the energy interval $(E, E + dE)$ for the photoelectron flux and statistically evaluate the cascade of the flux $W_e(E)\,dE$ impact effects.

This procedure of photoelectron collision calculation must be repeated several times and stopped when the electron energy becomes less than E_{min}. Formally, this procedure is equivalent to to $M(E)$ times realization of operator L_1. The steady state distribution of photoelectron flux is determined by the recurrent procedure for the whole energy spectrum, i.e. by calculations of all possible cascades (7.14) - (7.15).

For the calculation of the kinetic rates formation of energetically active particle by photodissociation and electron impact of the main components of the upper atmosphere, statistical algorithms (Shematovich, 1982, 1987) were developed. In these algorithms, all possible photodissociation and electron impact collisions occurring during a time unit at a given height under fixed solar and geophysical conditions are statistically evaluated. The probabilistic realization for various channels of photolytic and electron impact interactions is supported by the data on partial and total cross-sections and on the photon energy. Accumulation of the representative statistics gives rise to the evaluation of the kinetic rates with sufficient accuracy.

7.2. Geophysical Input Parameters

In order to describe properly the elementary processes of nonthermal particles formation induced by the solar EUV radiation it is necessary first of all to specify the chemical composition of the atmosphere. We shall consider these processes with primary application to the Earth's upper atmosphere (Shematovich, 1987; Marov and Shematovich, 1986; Marov et al., 1990; Gérard et al., 1991), though their principal features with the account for different composition are relevant to the atmospheres of other planets and hence may be extensively extended.

The main channels of absorption of the incident solar short-wave (UV, EUV, and soft X-rays) radiation in the upper atmosphere are the reactions of photodissociation and direct and dissociative ionization (3.1) - (3.3) of the main atmospheric components. In these reactions the absorbed energy of UV and EUV photons is shared between the kinetic energy of dissociation products and photoelectrons as well as the inner energy of the produced radicals and ions. The kinetics of these energetically active atmospheric particles is defined by the following solar- and geophysical parameters:

- spectral flux of UV and EUV solar photons;

- detailed differential cross-sections of photoionization, photoabsorption, and electron impacts for the main atmospheric species;

- composition and number densities of the neutral upper atmosphere.

In the studies of the upper atmosphere photochemistry the following wavelengths of UV and EUV radiation are the most important (see e.g., Banks and Kockarts, 1973):

- solar UV radiation in the Schumann-Runge continuum $1250 \div 1750$ Å that is responsible for the heating of the lower thermosphere and is the main source of atomic oxygen formation due to O_2 photodissociation;

- solar radiation in line H $Ly\alpha$ 1216 Å that is the main source of ions in the D region of ionosphere due to ionization of NO;

- solar EUV radiation in the wavelength $150 \div 1030$ Å that is the primary source of thermospheric heating due to its effective absorption by the main atmospheric components, and is also responsible for the formation of the E and F regions of the ionosphere;

- solar radiation in the wavelength $10 \div 150$ Å (soft X-rays) that contributes substantially to the photochemistry and ionization of the lower thermosphere, especially during high solar activity.

In the last decades series of the international space programs involving such satellites as *Atmosphere Explorer - C, D, E*, and *Solar Mesosphere Explorer* have been undertaken, which provided quite reliable measurements of solar UV radiation fluxes $F_\infty(\lambda)$ throughout the whole 11-year cycle of the solar activity (Hinteregger *et al.*, 1981; Lean, 1987). The results of these measurements and respective data bases became the ground for aeronomical applications (Hinteregger *et al.*, 1981; Torr and Torr, 1985; Tobiska and Barth, 1990; Tobiska, 1991; Richards *et al.*, 1994). For the sake of convenience the fluxes of EUV radiation are presented in 39 intervals, subdividing the initial wavelength range

$$\{F_\infty(\lambda)\} = \{F_\infty(\lambda_j) \ , \quad j = 1, \ldots, J = 39\} \ .$$

This format includes two intervals $j = 1$ ($18 \div 30$ Å) and $j = 2$ ($31 \div 50$ Å) for the soft X-ray radiation; 20 intervals, averaging by 50 Å for the longer wavelength of EUV; and 17 main spectral lines (HeI, HeII, CIII, OII, etc.) and atomic continua (HeI $\sim$, HI $\sim$, etc.). A similar form of representation is used for UV radiation in the Schumann-Runge continuum ($1250 \div 1750$ Å) (Torr *et al.*, 1980).

We used in the computer modeling the improved model of solar EUV radiation flux SERF2 (Tobiska, 1991). This model contains the EUV fluxes for a given period in the wavelength $18 \div 1050$ Å, as well as chromospheric emissions based on the measurements in the H $Ly\alpha$ and HeI 10830 Å lines, correlated with emissions from solar coronal and transition regions in the radio wavelength 10.7 cm.

Vast experimental and theoretical data on photoabsorption and photoionization cross-sections for the main atmospheric components were compiled, and they were coordinated with the models of UV and EUV fluxes

(Torr et al., 1979, 1980; Kirby et al., 1979; Conway, 1988; Fennelly and Torr, 1992). The followed improvements, especially for cross-sections of atomic oxygen photoionization in the wavelength shorter than 250 Å, were reviewed by (Conway, 1988; Fennelly and Torr, 1992). These data including photoabsorption cross-sections, partial and total cross-sections of ionization and dissociation can be considered to as quite reliable for the modeling of aeronomic processes.

Experimental and theoretical data on the cross-sections of atmospheric components for impact energetic electrons based on the measurements of Stamnes and Rees (1983), Richards and Torr (1988), Rees (1989) were compiled by Green and Stolarski (1972), Peterson et al. (1973), Jackman et al. (1977). In our calculations we mostly used data on the respective cross-sections deduced from the numerical approximations (Peterson et al., 1973; Jackman et al., 1977) as well as from the experimental data available (Zipf and McLaughlin, 1978; Richards and Torr, 1988; Rees, 1989).

The composition and density of the planetary upper atmospheres are used to be taken from empirical models based on experimental data available, usually rather limited for atmospheres of planets other than the Earth. But even for the Earth's upper atmosphere, while vast experimental data about physical and chemical parameters and their variations depending on solar and geophysical conditions were accumulated, these data poorly represent every specific point of the atmosphere. The reason is that empirical models compile local dispersed measurements carried out only along satellite trajectories or from a separate radar station and limited radar soundings. To compensate for this incompleteness and to accommodate the prognosis purposes, different phenomenological approaches are used, which involves revealing of some empirical dependencies from detailed statistical analysis of experimental data. Usually, these dependencies characterize the most pronounced variations of main atmospheric parameters depending on solar and geophysical data, such as coordinates, season, local solar time, solar and geomagnetic activity.

Reviewing the existing semi-empirical models of the Earth's upper atmosphere, one may assert that each of them has definite advantages and disadvantages. Every model is based on different (but always limited) sets of experimental data and, consequently, cannot represent with the same degree of accuracy all atmospheric parameters of interest at specific sites and times for different solar and geomagnetic conditions. Therefore, systematic distinctions between models are common in terms of approximation the real situation in the thermosphere (examples of the models comparison are given by Barlier and Berger, 1983; Marov and Kolesnichenko, 1987; Hedin, 1988).

We can briefly summarize the results of these models analysis as follows. The widely used Jacchia models and, especially, its upgrated version (Jacchia, 1977) based on the mass-spectrometric and satellite's drag data describes correctly enough season-latitudinal and diurnal variations of atmospheric parameters in the various phases of the solar cycle, as well as the midlatitude thermospheric response to geomagnetic perturbations. More advantageous, however, turned out the models based on the principle of decomposition of number density data in spherical harmonics (Hedin, 1987; 1991). They allowed to reveal some additional features in the variations of thermospheric parameters obscured in the models $J70$, $J71$, such as the He diurnal variations, the ratio of O to N_2 concentrations, the shift of temperature global maximum position to higher latitudes, etc. It may be pointed out that, at the present time, the $MSIS - 83, -86, -90$ models principally based on this approach quite satisfactorily describe the structure and variations of the thermosphere in the height range 140-600 km for a large range of solar and geomagnetic conditions. This is why the height profiles of concentrations of the main neutral components of the upper atmosphere $n_\alpha(h)$, $\alpha = O_2$, N_2, O were taken from models $MSIS - 86$ and $MSIS - 90$ (Hedin, 1987; 1991) in our numerical simulations.

Many important aeronomy problems currently address the physical and chemical processes in the lower thermosphere, i.e. at heights of 80 to 150 km. This interest is specifically focused on the transition region between the mesosphere and the thermosphere which essentially influences the overall thermospheric-exospheric behavior. Indeed, the processes of energy absorption of UV radiation and precipitated energetic particles reach their maximum rates in the lower thermosphere and thus define the thermal state and physical and chemical composition of this and upper regions of the atmosphere. Hence, a major part of our study was focused on the lower thermosphere for characteristic cases of both low ($F_{10.7} = 70$) and high ($F_{10.7} = 245$) solar activity, though mainly under quiet geomagnetic conditions at equatorial latitude, and these results are discussed in the following sections.

7.3. Photoionization and Kinetics of Photoelectrons

As it was earlier mentioned, the processes of direct and dissociative ionization of the main atmospheric components by solar EUV radiation are accompanied by the formation of energetically active particles - atomic and molecular ions in various excited states and primary ('fresh') photoelectrons. The results of calculations of photoionization kinetics for the Earth's upper atmosphere are shown in $Figs.$ $7.1a$, b and $7.2a$, b, where the total $^{(\nu)}q^{(+)} = \Sigma^{(\nu)}q_\alpha^{(+)}$ and partial $^{(\nu)}q_\alpha^{(+)}$ photoionization rates depending on the solar zenith angle are presented for low and high levels of solar activity.

These results show that, at the heights ≥ 250 km, attenuation of the EUV radiation flux is not significant, and consequently, the height profile of photoionization is defined by the density of the dominant neutral component - atomic oxygen, i.e. $^{(\nu)}q^{(+)} \cong q_0^{(+)}(h) \sim n_0(h)$. At the heights ≤ 250 km, the EUV radiation is absorbed by N_2 and O_2 ionization continua, and the photoionization rates reach their maximum values $\sim 1 \div 4 \times 10^3$ cm^{-3} s^{-1} depending on the solar and geophysical conditions. In general, the photoionization rates at these heights are defined by the ratio of two competitive processes: as height decreases both density of the neutral component and absorption rate of the EUV radiation increase. The radiation ionizing O_2 ($\lambda \leq 1026$ Å) and soft X-ray ($\lambda \leq 100$ Å) penetrate to the heights of $\sim 90 \div 120$ km. In this region the photoionization is mainly defined by the dissociative ionization of O_2 and N_2 due to X-rays (soft X rays below the 50 Å, XUV below 250 Å). Dissociative ionization leads to the formation of high energy photo- and Auger electrons, as well as to the formation of dissociative products characterized by an excess of kinetic energy (Gardner and Samson, 1975; Locht et al., 1992; Van Zyl and Stephen, 1994). Total and partial photoionization rates strongly depend on the solar zenith angle (local solar time).

In our stochastic model of photodissociation processes (3.1)-(3.3), we used the complete set of photolytic collisions what allowed us to accumulate sufficient statistical data and, consequently, to evaluate the production rates and to obtain the energy spectrum of electrons produced due to photoionization. In Figs. 7.3a1, a2, 7.4b1, b2 and 7.5a, b the results of the calculations of the energy spectrum (7.9) of primary photoelectrons at 100 and 200 km are presented for low and high solar activity, and at 160 km - for different local solar times.

In general, the shape of the spectrum is characterized by a smooth decrease with energy, but in the energy range 20 to 30 eV a system of sharp peaks due to the absorption of the HeII line by ionization continua of N_2, O_2, and O occur. The electron distributions formed by photoionization reflect specific features of the solar EUV radiation spectrum and include the following characteristic spectral intervals:

- photoelectrons with small (≤ 1 eV) energies, formed by near-threshold ionization and causing thermal heating and excitation of rotational levels of the atmospheric components;

- suprathermal photoelectrons with energies of 1 to 10 eV, formed by absorption of radiation in lines HeI, OV, etc., resulting in the Coulomb heating of the thermal electrons and excitation of vibrational and electronic states;

- superthermal electrons with energies ≥ 10 eV formed essentially by ionization in the intensive HeII spectral line and causing secondary impact

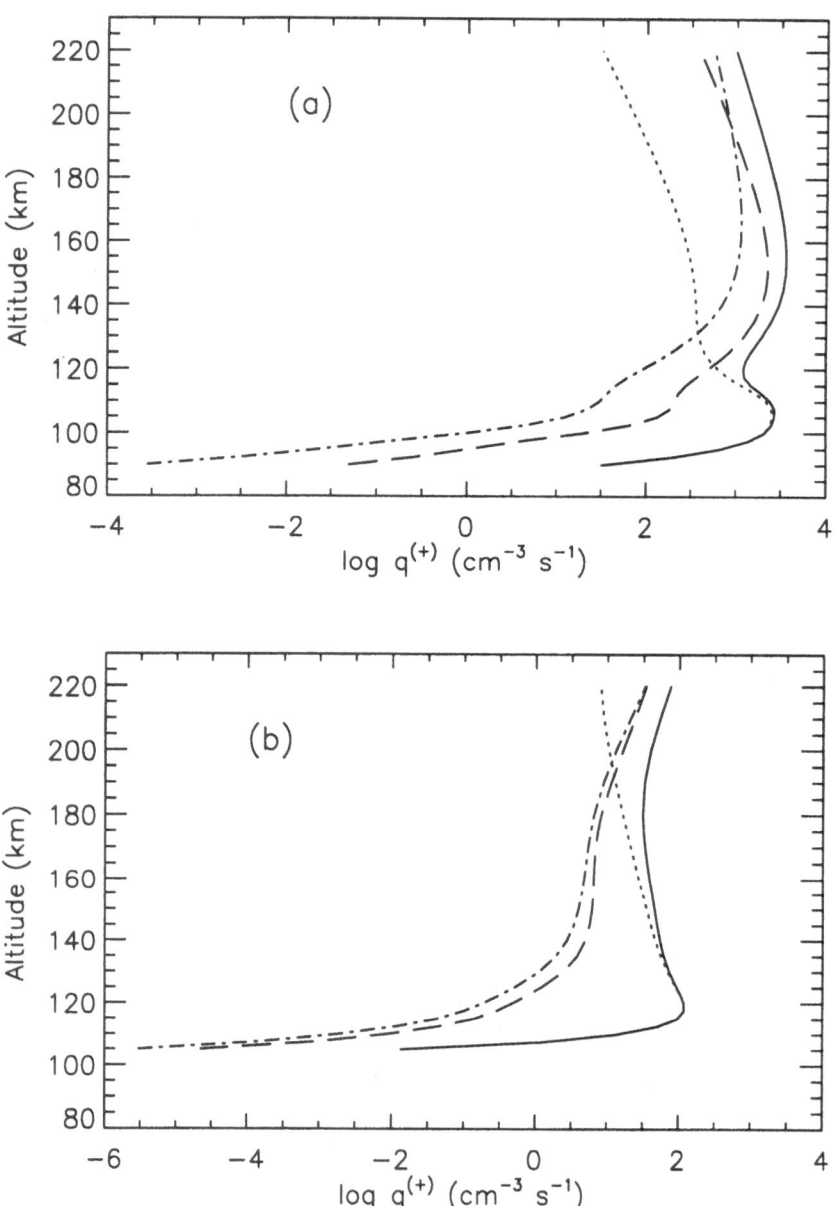

Figure 7.1. Total (solid line) and partial (O_2 – dotted line; N_2 – dashed line; O – dashed-dotted line) photoionization rates in the lower thermosphere for low ($F_{10.7} = 70$) solar activity and for different local solar times (a – $\theta_\odot = 5°$; b - $\theta_\odot = 90°$).

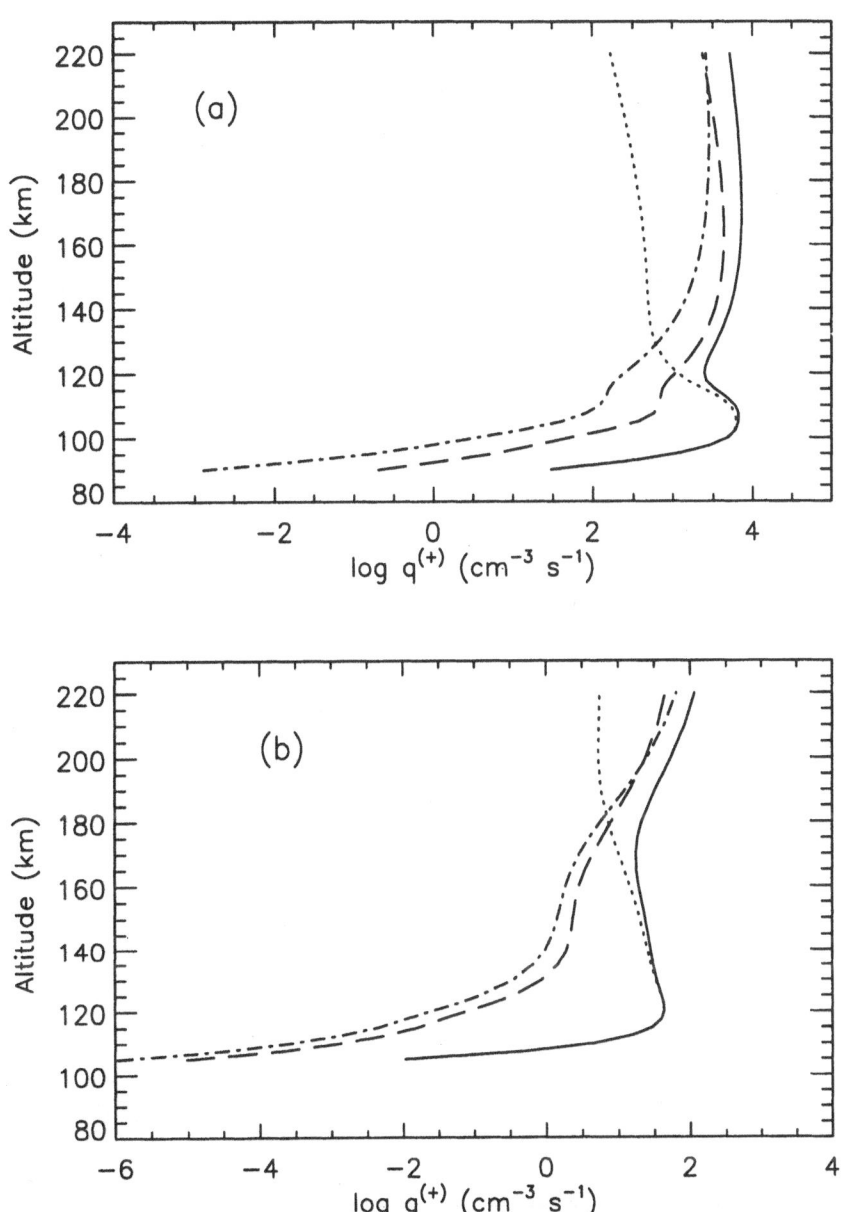

Figure 7.2.   Same as Fig. 7.1 for high ($F_{10.7} = 245$) solar activity.

processes of dissociation and ionization of atmospheric components;

- high-energy electrons with energies $\sim$ 200 to 800 eV formed by direct, dissociative and K-shell ionization of O_2 and N_2 in the lower thermosphere

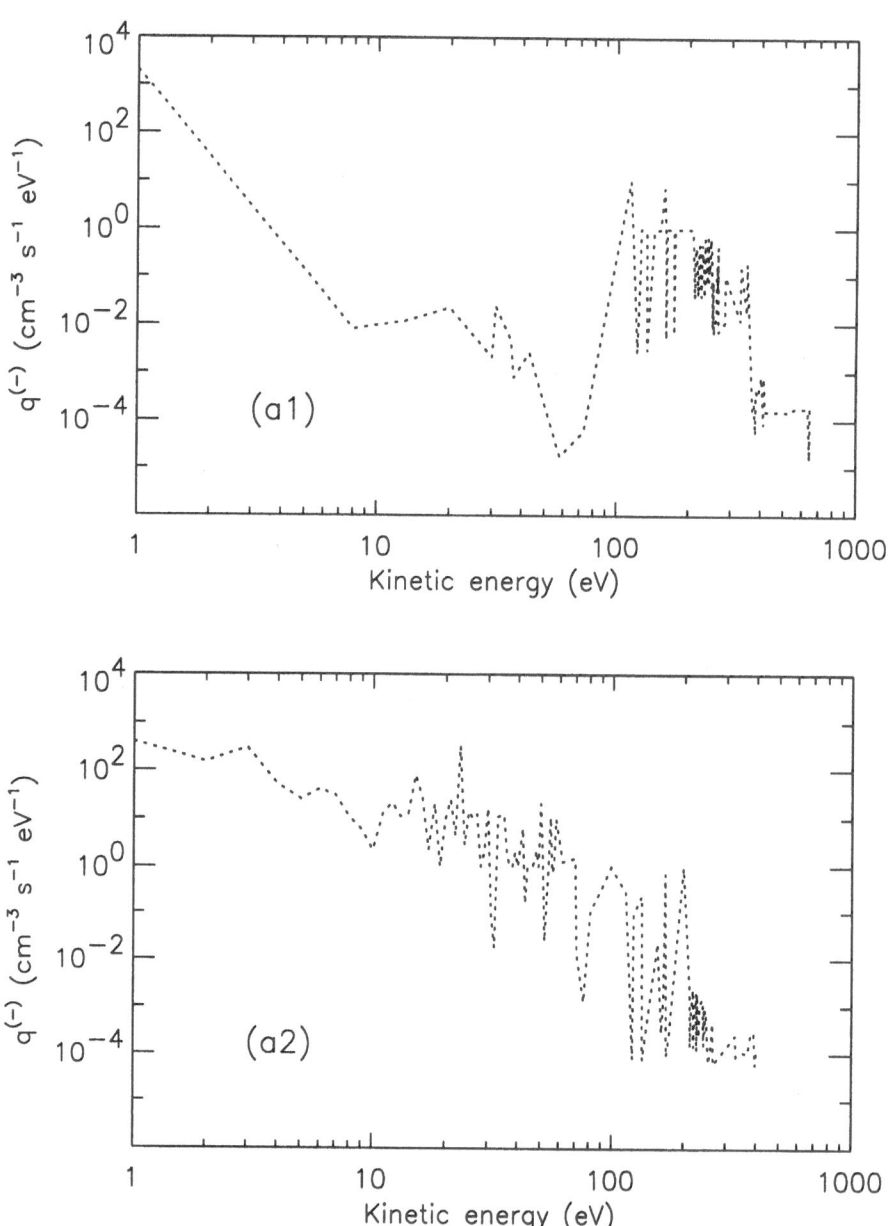

Figure 7.3. Kinetic energy spectra of primary photoelectrons at 100 (a1), and 200 (a2) km for low solar activity ($F_{10.7}$=70.)

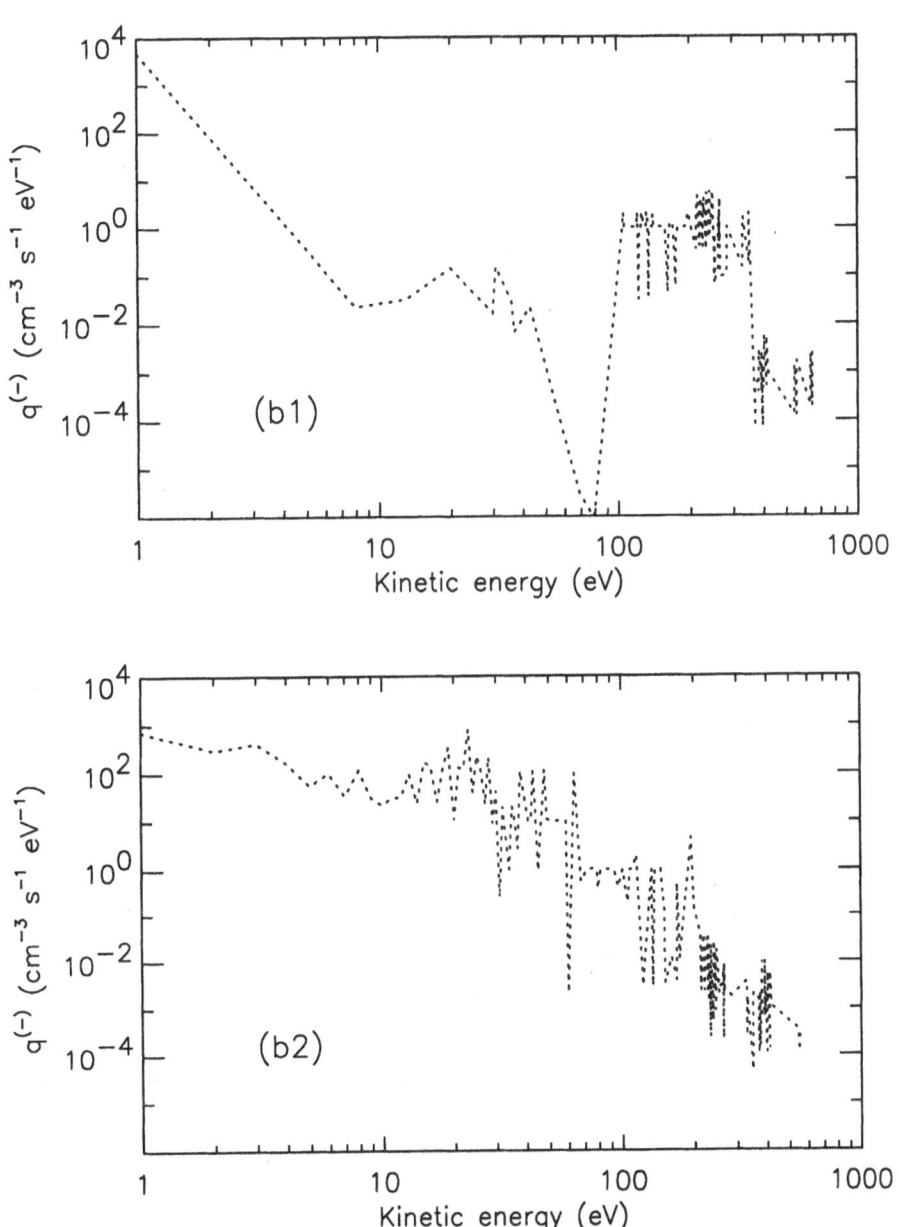

Figure 7.4. Kinetic energy spectra of primary photoelectrons at 100 (b1), and 200 (b2) km for high solar activity ($F_{10.7}$=245.)

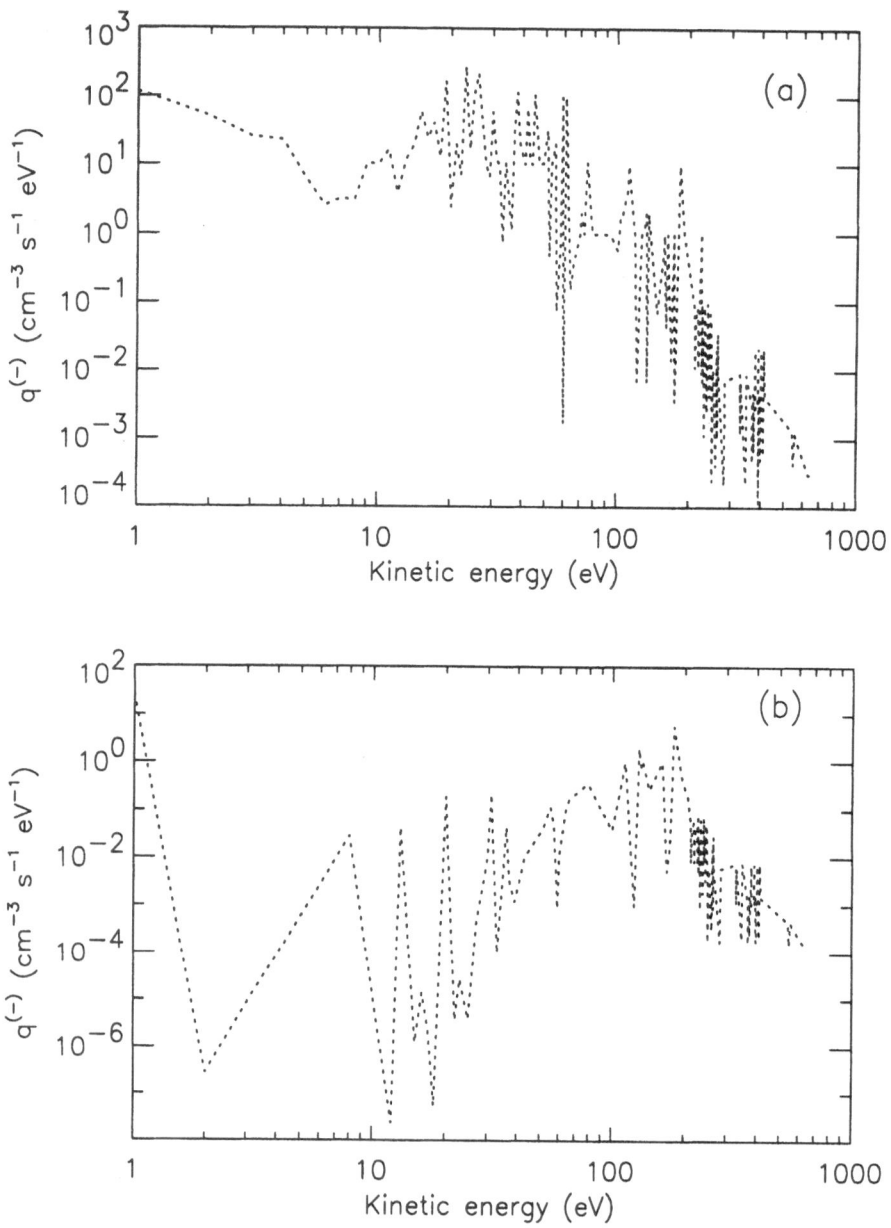

Figure 7.5. Kinetic energy spectrum of primary photoelectrons at 160 km for different local solar times [a - t=16^h ($\theta_\odot = 60°$); b - t=18^h ($\theta_\odot = 90°$)].

by hard EUV radiation and by soft X-ray fluxes and causing cascade dissociation and ionization of the thermospheric components.

With increasing solar zenith angle $\theta_\odot \rightarrow 90^o$, considerable attenuation of the EUV radiation in the higher layers of the atmosphere occurs and consequently, the role of soft X-ray fluxes in the formation of primary photoelectrons also increases (*Fig. 7.5a, b*).

These primary photoelectrons are an additional source of ions and radicals, formed in the thermosphere by impact effects on atmospheric gas (Krinberg, 1978; Stamnes and Rees, 1983; Richards and Torr, 1988; Buonsanto *et al.*, 1992; Titheridge, 1996). In order to analyze in detail the kinetics of these high-energy photoelectrons, we use the kinetic numerical model of cascade degradation of the photoelectron flux. This model considers the energy spectrum up to 10^3 eV and, consequently calculates the kinetic rates of impact electron effect at heights of the lower thermosphere ≤ 120 km, and at large solar zenith angles $\theta_\odot \rightarrow 90^o$ as well. In the numerical realization of the model the energetic spectra of primary photoelectrons were calculated (see *Figs. 7.3* and *7.7.4*), where the total and partial cross-sections of excitation, dissociation and ionization of atmospheric components due to the impact effects of energetic electrons (Rees, 1989) were adopted.

The results of calculations of the total (summarized by the main atmospheric species) kinetic rates of excitation, dissociation, direct and dissociative ionization (3.4) - (3.6) by photoelectron fluxes are shown in *Figs. 7.6a, b* and *7.7a, b* at low and high levels of solar activity, respectively. It follows from these results that at the heights ≥ 120 km the impact effect of photoelectrons represents an additional source (reaching ~ 10 to 30 % of the photolytic effect by UV radiation) of superthermal, excited and ionized particles in the thermosphere. The results of previous studies (Krinberg, 1978; Stamnes and Rees, 1983; Richards and Torr, 1988; Buonsanto *et al.*, 1992; Titheridge, 1996) are in agreement with these values. In the lower thermosphere (≤ 120 km) the impact effect of high-energy photoelectrons formed by the absorption of hard EUV and soft X-ray radiation, becomes comparable or even exceeds the rates of photodissociation (for N_2) and ionization. Hence, in the lower thermosphere where molecular nitrogen is the dominant component, this effect must be especially important. Indeed, the comparison of photolytic and electron impact ionization rates

$$r_\alpha^{(+)} = {}^{(e\nu)}q_\alpha^{(+)} / {}^{(\nu)}q_\alpha^{(+)}$$

of molecular nitrogen (*Fig. 7.8a, b*) shows that their ratio is strongly increased at these heights. The impact effect of energetic photoelectrons formed by soft X-ray absorption results in the cascade ionization of N_2 and O_2, and, consequently, the ionization rate by electrons becomes greater than that by photons. The relative role of different ionizing agents such as hard EUV and soft X-ray ($\lambda \leq 100$Å) radiation, EUV radiation ($100 \leq \lambda \leq 911$ Å), and photoelectrons is shown in *Fig. 7.8b*.

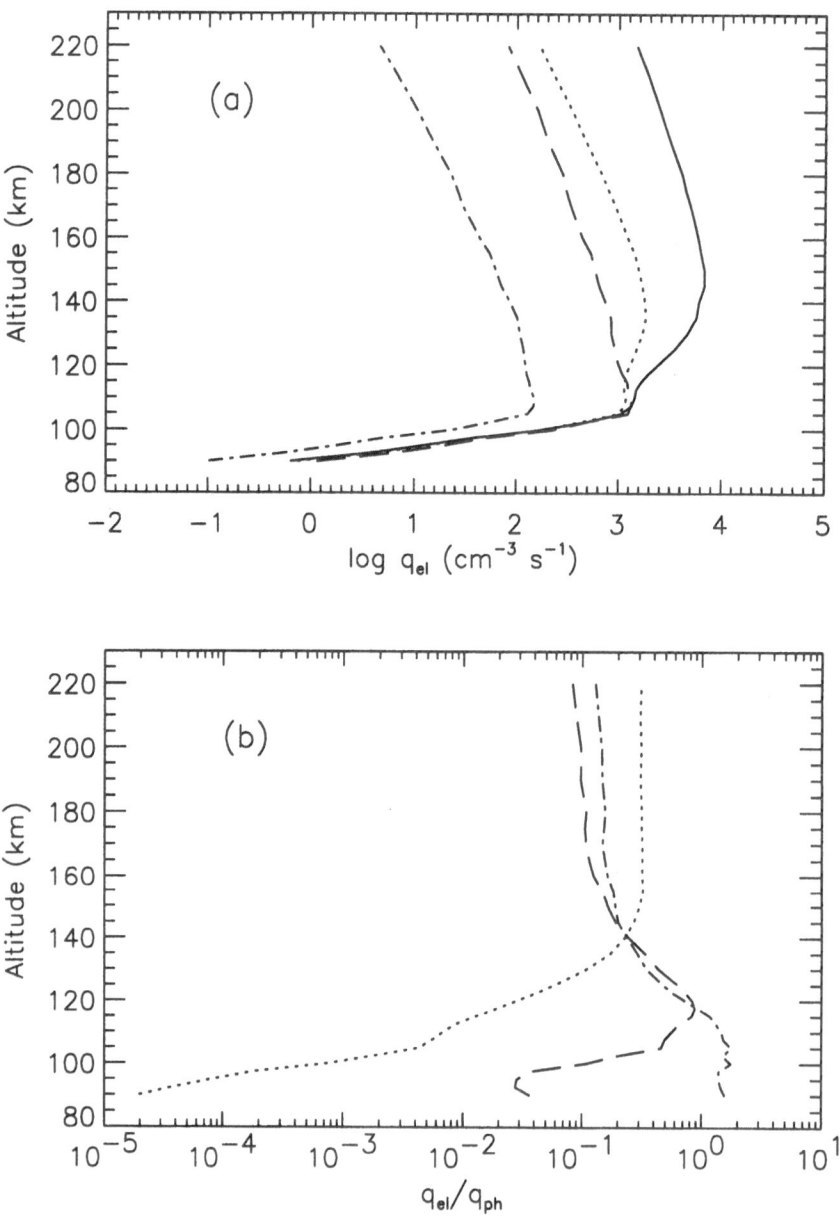

Figure 7.6. (a) Total rates of excitation (solid line), dissociation (dotted line), direct (dashed line) and dissociative (dashed-dotted) ionization of thermospheric components by photoelectron flux at low $F_{10.7}=70$ solar activity. (b) The rate ratio of electron impact and photolytic effects on thermospheric gas.

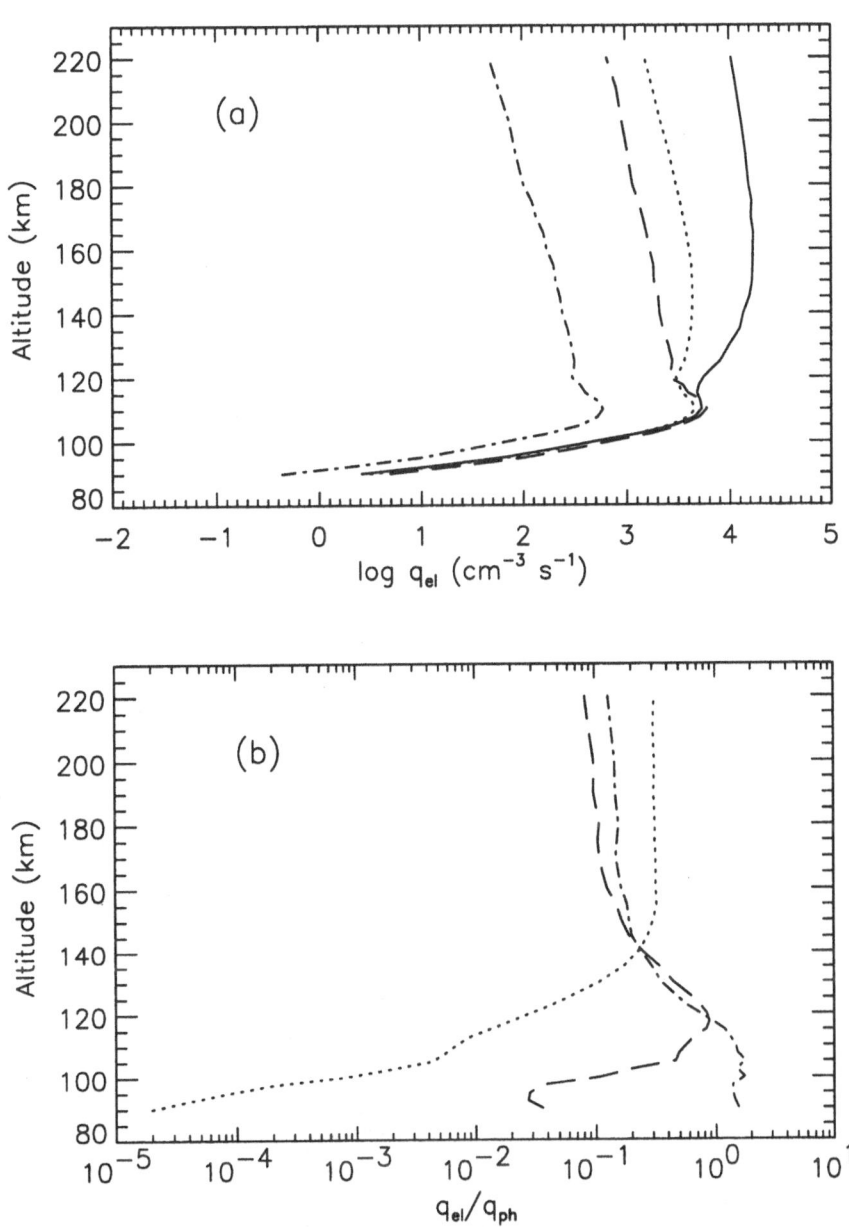

Figure 7.7. Same as *Fig. 7.6* for high solar activity.

These results bring support to the original idea on very important role of photoelectrons in the photochemistry and energetics of the lower thermosphere, and therefore, their kinetics deserves to be studied in more detail.

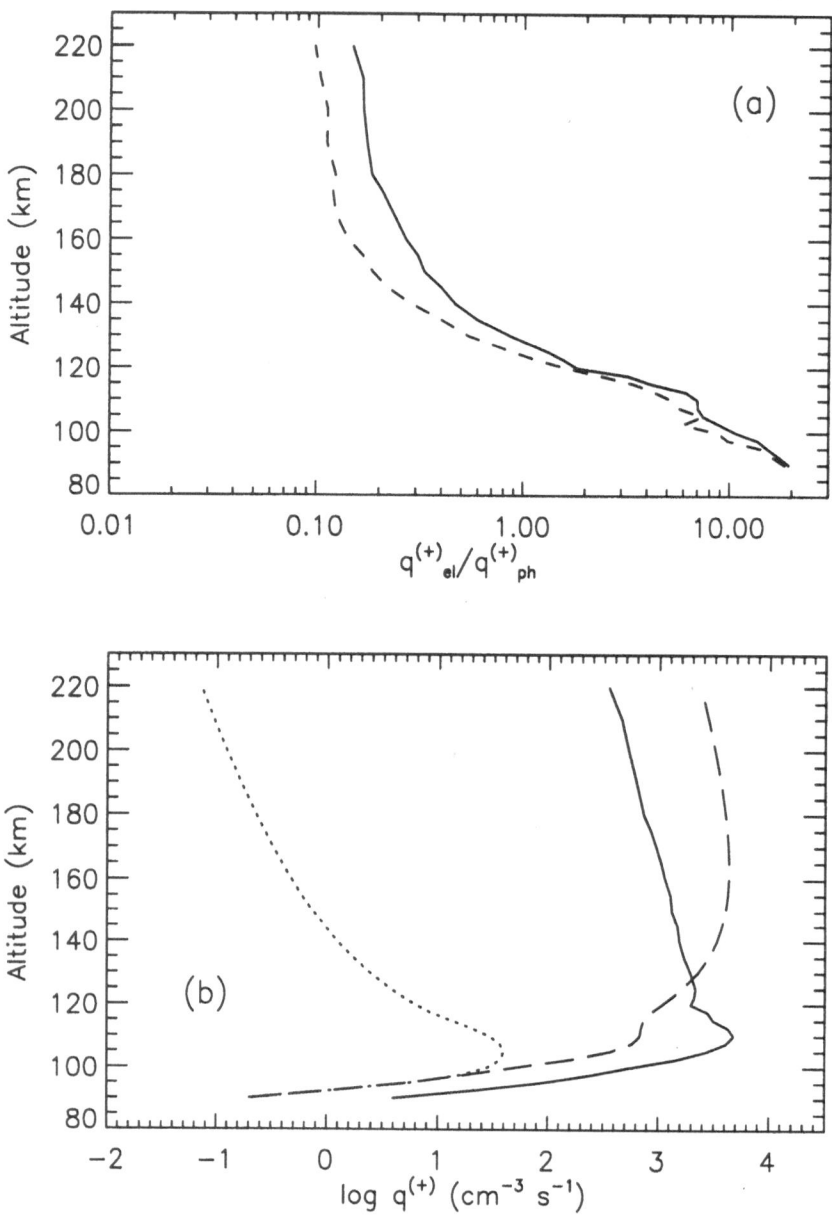

Figure 7.8. Ionization of N_2 in the lower thermosphere: (a) - altitude profiles of the ratio of impact- and photoionization rates for low $F_{10.7} = 70$ (solid line) and high $F_{10.7} = 245$ (dashed line) solar activity; (b) - ionization rates due to soft X-rays and hard EUV radiation (dotted line), EUV radiation (dashed line), and photoelectron impact (solid line).

From the numerical realization of the respective kinetic model the steady-state energy distribution functions of superthermal photoelectrons were obtained. These distributions at low and high solar activities are presented in *Fig. 7.9a, b* at 100, 160 and 200 km. *Fig. 7.10a, b* shows the steady-state distribution functions of photoelectrons at 160 and 200 km depending on the solar zenith angle $\theta_\odot = 5^o$ and 90°, i.e. for local solar time t $= 12^h$ (noon) and 18^h (sunset).

The analysis of the steady-state distributions of energetic photoelectrons clearly demonstrates that they are strongly in nonequilibrium as compared to the distributions of the thermal ionospheric electrons. In the energy interval [0, 100 eV] these distributions coincide well with the results of satellite measurements, as well as with some other model calculations for the Earth's thermosphere (Krinberg, 1978; Richards and Torr, 1983, 1984, 1985, 1988). Obviously, the results of the calculated nonequilibrium steady-state energy distributions may be used for the evaluation of the contribution of photoelectrons to the excitation of atmospheric emissions and to the heating rate of neutral, ionized and thermal electron components of the upper atmosphere (Roble *et al.*, 1987).

It is also important to note that processes of photo- and impact electron direct ionization are responsible for the formation of ions both in ground and in electronically excited states. These ions, participating in exothermic ion-molecular reactions (3.7) to (3.11) and in the processes of dissociative recombination with thermal electrons (3.12) - (3.14), result in the formation of chemical sources of superthermal and excited particles in the lower thermosphere. The products of dissociative ionizations are an additional source of superthermal particles.

7.4. Kinetic Rates of Photodissociation and Electron Impact Dissociation

Dissociation of the molecular atmospheric components O_2 and N_2 in the Earth's upper atmosphere is caused by the photodissociation by solar UV photons and by the impact effect of photoelectrons. The schemes of dissociation of these molecules differ in some details. Some electronic states appearing in the process of dissociation of molecules of O_2 and N_2, are illustrated in *Fig. 7.11a, b*.

All prominent excited states of the O_2 molecule : $A^3\Sigma_u^+$, $B^3\Sigma_u^-$, etc. are bound ones, although their potential minima occur at different internuclear distances. Accordingly, the excitation times of these states are small in comparison with the time scales of vibrations or relative motion of the nuclei. In accordance with the Frank-Condon principle, these transitions take place at the initial nuclear separation and cause the excitation of

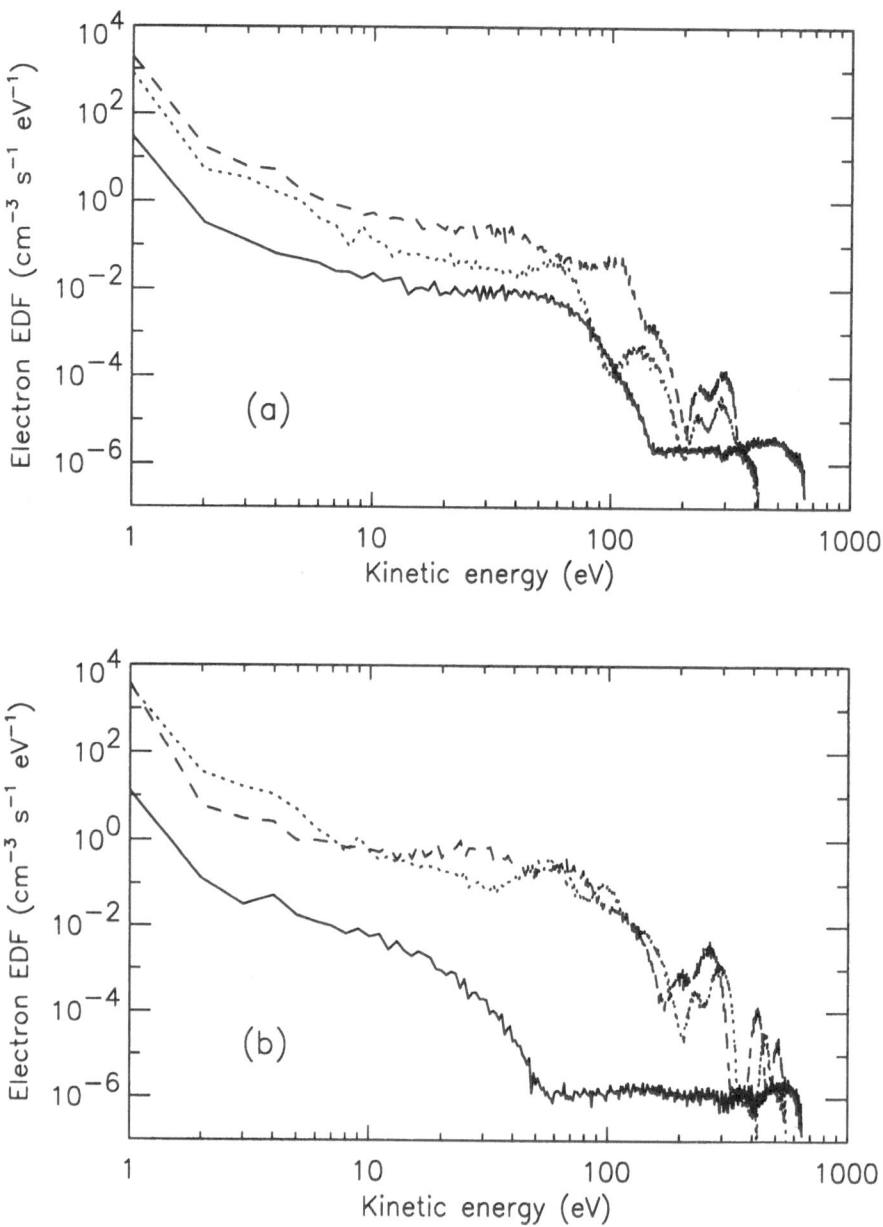

Figure 7.9. Steady-state energy distribution functions of superthermal photoelectrons at 100 (solid line), 160 (dashed line), and 200 (dotted line) km for low $F_{10.7}=70$ (a) and high $F_{10.7}=245$ (b) solar activity.

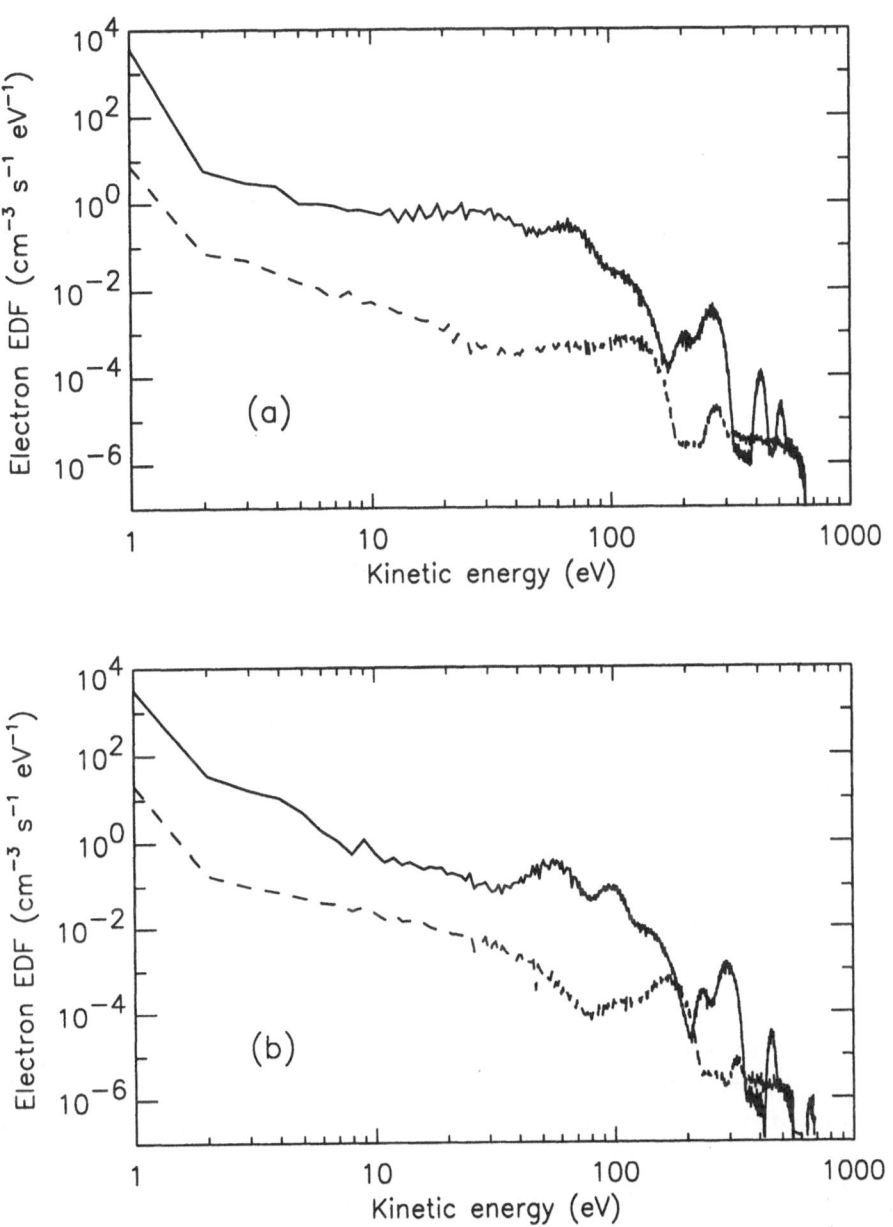

Figure 7.10. Steady-state energy distribution functions of superthermal photoelectrons at 160 km (a) and 200 km (b) for high $F_{10.7}=245$ solar activity and for two different solar zenith angles (solid line - $\theta_\odot = 5°$; dashed line - $\theta_\odot = 90°$).

vibrational levels of higher electronic states with energies greater than the dissociation threshold.

Dissociation via the $A^3\Sigma_u^+$ state yields two O atoms in the ground state, and via $B^3\Sigma_u^-$ it yields the excited atom $O(^1D)$ and ground state atom $O(^3P)$. The transition $X^3\Sigma_g^- - B^3\Sigma_u^-$ is optically permitted, while the transition $X^3\Sigma_g^- - A^3\Sigma_u^+$ violates one of the selection rules for dipole transitions (i.e. this transition is optically forbidden). Consequently, dissociation via the A state is considerably weaker than via the B state. This circumstance is one reason for the considerably stronger absorption of UV radiation in the Schumann-Runge continuum ($1250 \div 1750$ Å) than in the Herzberg continuum (≤ 2422 Å).

Unlike O_2, the photoabsorption spectrum of the N_2 molecule has no continuum connected with dissociation, and only absorption bands in the range $800 \div 1000$ Å exist. The absence of a dissociative continuum shows that there are no electronic states with transitions populating the vibrational levels with energies above the dissociation threshold. However, the nitrogen molecule has several high lying electronic states (in $^1\Pi_u$ and $^1\Sigma_u^+$ electronic configurations), which can be excited by photons or electrons through dipole allowed transitions. These states are refered to as predissociation states, because the probability of radiationless transition to states that lead to dissociation is much higher than that leading to radiation. This scheme is illustrated in *Fig. 7.11b* for the predissociation state $b^1\Pi_u$. The vertical transition shown in *Fig. 7.11b* results in the excitation of level $v' = 4$ of state $b^1\Pi_u$. This state dissociates to excited atoms $N(^2D)$ and $N(^4S)$ by means of intersection with the shallow state $C'^3\Pi_u$. However, it should be noted that the dissociative ionization of N_2, as well as of O_2, occurs directly without the formation of intermediate states.

The processes of molecular dissociation are one of the major sources of superthermal atoms O and N in the Earth's thermosphere. The dissociation process (7.4) is accompanied by the excitation of molecule $a^\alpha(E^*)$ levels with energies greater than the dissociation threshold $E^* > E_\alpha^{(0)}$. After disintegration of these excited molecules (in accord with the schemes described above) dissociation products in various states of inner excitation are formed and their thermal states are characterized by an excess of energy (7.6).

We present here some results of a detailed quantitative study of the dissociation kinetics involving the numerical modeling of the production rates and energetic spectra of O_2 and N_2 dissociation products (Shematovich, 1987; Shematovich *et al.*, 1991b). The total rates of O_2 and N_2 photodissociation at low ($F_{10.7} = 70$) and high ($F_{10.7} = 245$) levels of solar activity for the equatorial thermosphere are shown in *Fig. 7.12a, b*. The O and N atom production rates due to photodissociation at high solar activity de-

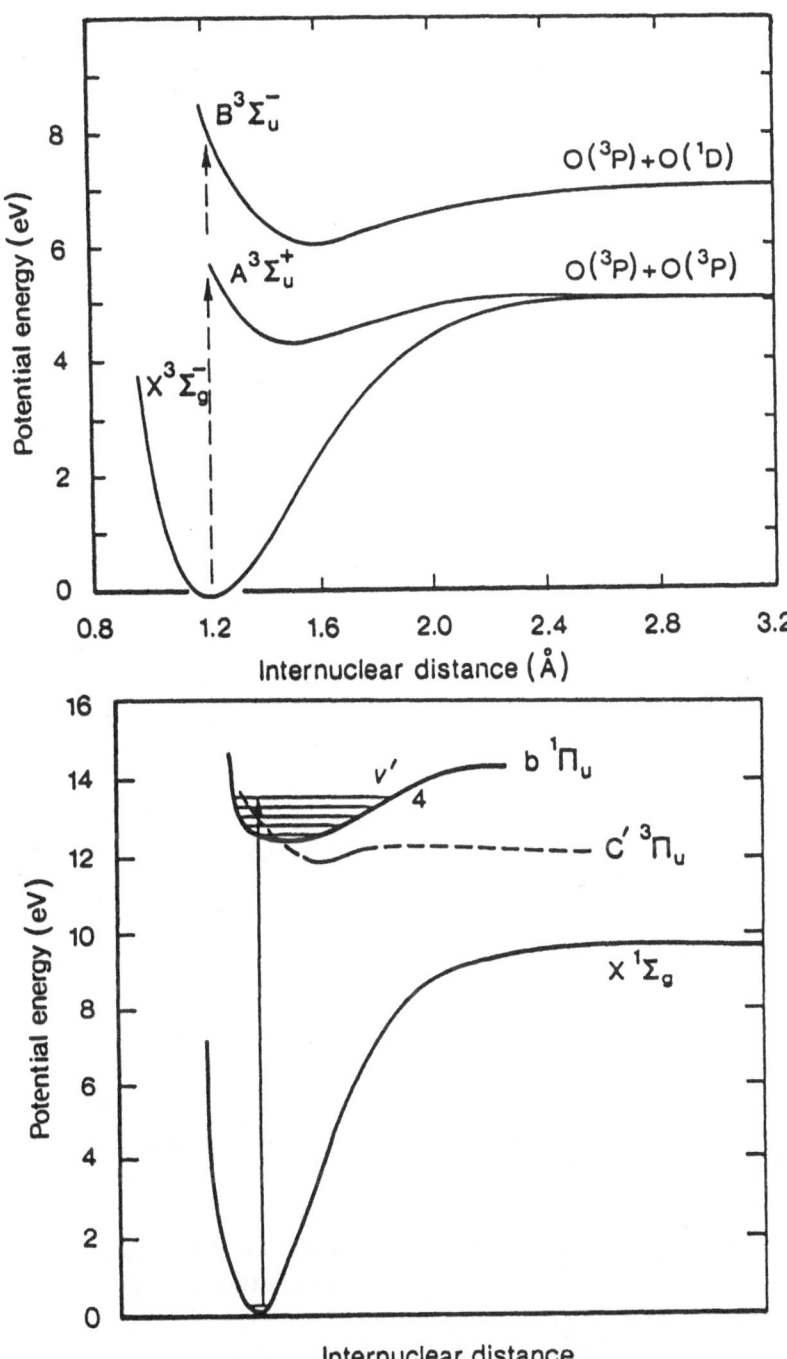

Figure 7.11. (a) Partial potential curve diagram for O_2 showing the principal states contributing to dissociation. (b) Partial potential diagram for N_2 showing states involved in the dissociation of N_2 by excitation of a predissociation level (Rees, 1989)

pendence on local time is shown in *Fig. 7.13a, b*. These results exhibit a general dependence of the photodissociation processes of the atmospheric molecular constituents on the solar- and geophysical conditions. The maximum production rates of superthermal O and N atoms are observed in the lower thermosphere. This location is defined by the geometry of the penetration of photolytically active solar radiation into the atmosphere.

As it was argued in the previous section, another important source of superthermal atoms $N(^4S, ^2D)$ is the impact dissociation of N_2 by energetic photoelectrons. Calculations of the production rate of hot N atoms due to this process is shown in *Fig. 7.14a, b* for both low and high levels of the solar activity. We note that at altitudes ≤ 130 km, photoelectron impact is the major source of hot N atoms. This is due to the fact mentioned earlier, that in this region of the lower thermosphere only soft X-ray and EUV radiation ≤ 100 Å penetrates. This radiation predominantly ionizes N_2 molecules and consequently, produces a flux of high-energy photoelectrons and Auger electrons with $E \sim 200 \div 600$ eV. As a result of the local relaxation of this photoelectron flux, high N_2 dissociation rates ($\sim 1 \div 3 \times 10^3$ cm^{-3} s^{-1}, depending on the level of solar activity) are observed.

In these calculations we have used the stochastic model of photodissociation, i.e. all dissociative collisions were simulated. This approach gives rise to the energetic spectra of superthermal products of dissociation. The calculated energy spectra (7.8) of $O(^3P, ^1D, ^1S)$ and $N(^4S, ^2D)$ at 100 km are presented in *Fig. 7.15* and *Fig. 7.16* and at the heights of 160 km in *Fig. 7.17* and *Fig. 7.18a, b*, respectively. The energetic spectra of O and N atoms depending on local solar time at high solar activity are shown in *Fig. 7.19* and *7.20*.

Analysis of the calculated energy spectra of $O(^3P, ^1D)$ atoms shows that these atoms are formed in the energy interval $0 \div 1$ eV by the absorption of radiation in the Schumann-Runge continuum, while the atoms with energies ≥ 1 eV are formed by absorption of radiation in the wavelength 700 to 1000 Å. The maximum production rates are observed for the formation of hot O atoms with energies ~ 0.1-0.2 eV, as well as in the energy range 1.3-1.4 eV. These atoms have strongly nonequilibrium initial kinetic energy distributions.

The structure of energy spectra of N atoms at the heights $90 \div 130$ km is generally defined by the impact effect of photoelectrons, since in this region photodissociation is due only to the absorption of radiation in a line CIII 977 Å. At the heights ≥ 130 km absorption in the wavelength 800 to 1000 Å contributes to the photodissociation process in addition to CIII line (see *Fig. 7.18a*). At these heights, the photodissociation rate becomes comparable or exceeds the rate of impact dissociation.

The impact effect of electrons results in the excitation of predissociation

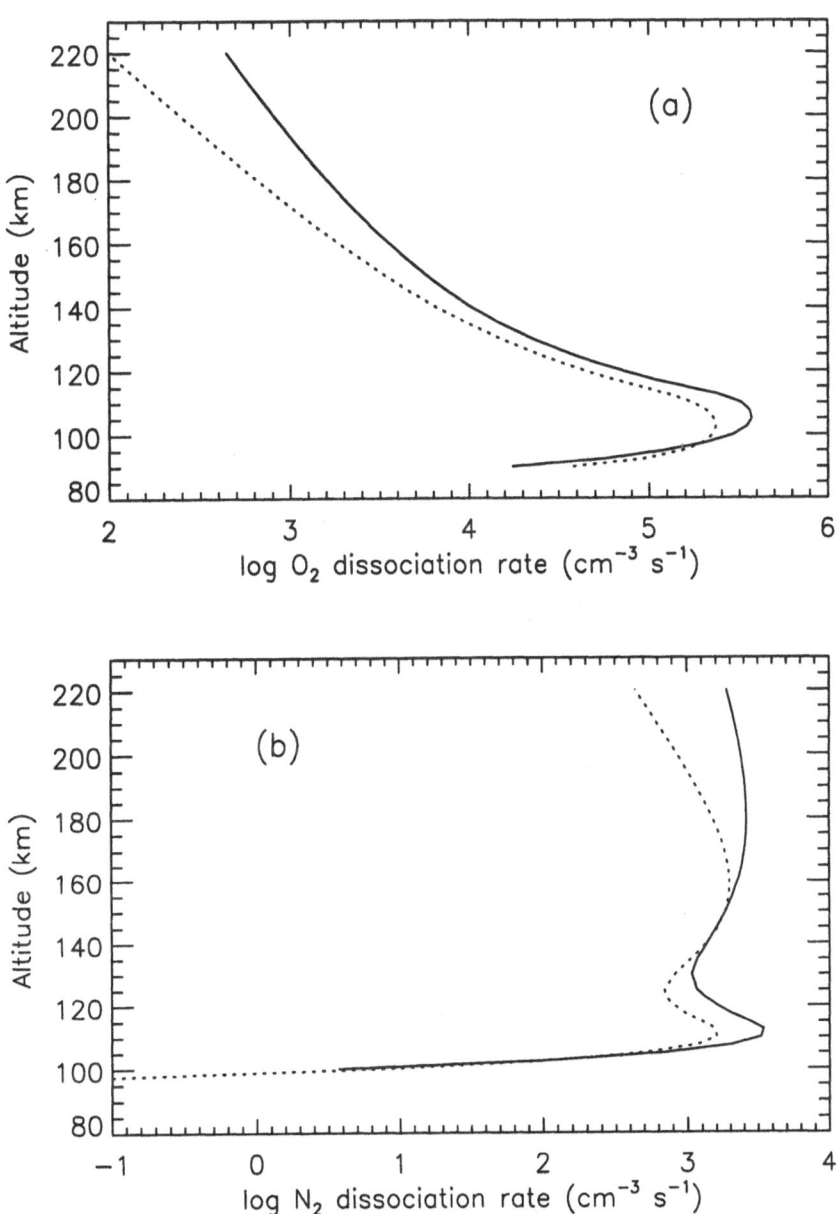

Figure 7.12. O_2 (a) and N_2 (b) photodissociation rates in the equatorial lower thermosphere for low $F_{10.7}=70$ (dotted line) and high $F_{10.7}=245$ (solid line) levels of solar activity. The local solar time is 12^h.

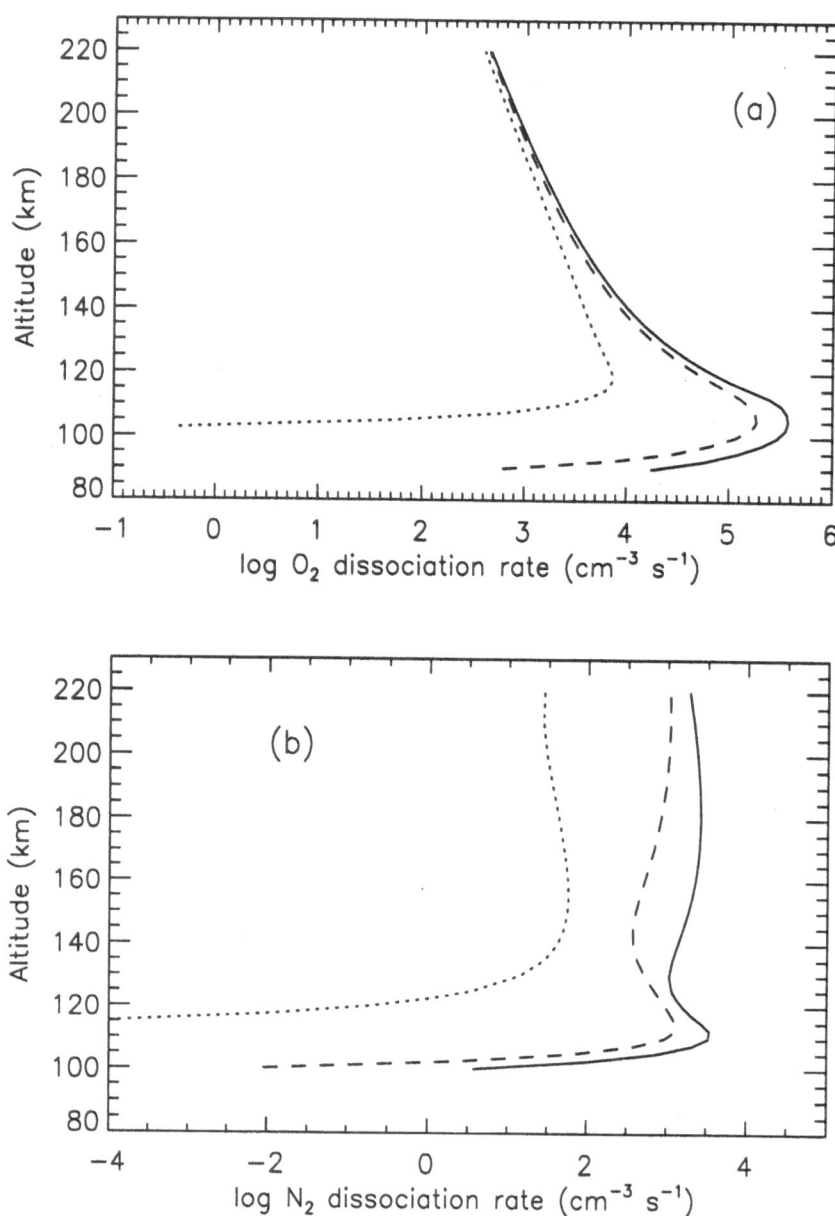

Figure 7.13. O_2 (a) and N_2 (b) photodissociation rates for different local solar times $\theta_\odot = 5°$ (solid line), $60°$ (dashed line), and $90°$ (dotted line) at high solar activity.

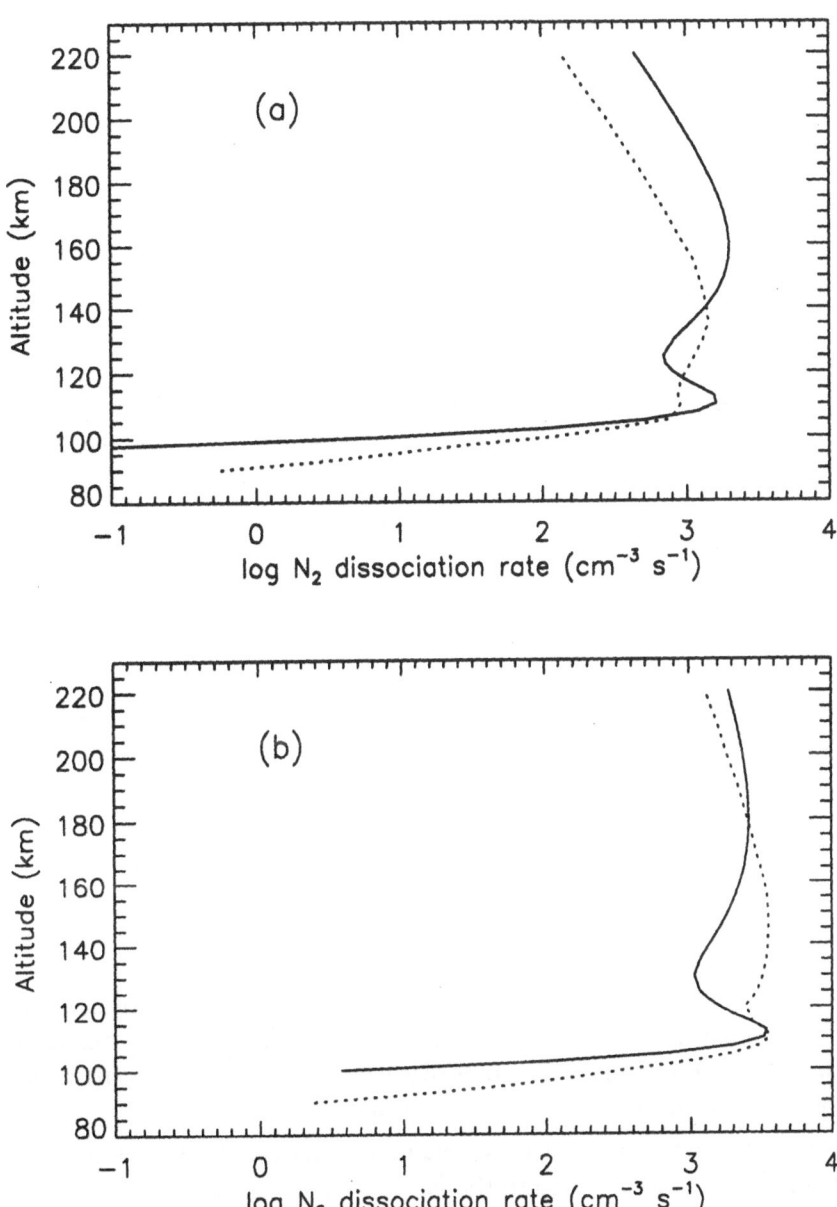

Figure 7.14. N_2 dissociation rates by photon (solid curve) and photoelectron impact (dotted curve) in the lower thermosphere for low $F_{10.7}=70$ (a) and high $F_{10.7}=245$ (b) solar activity.

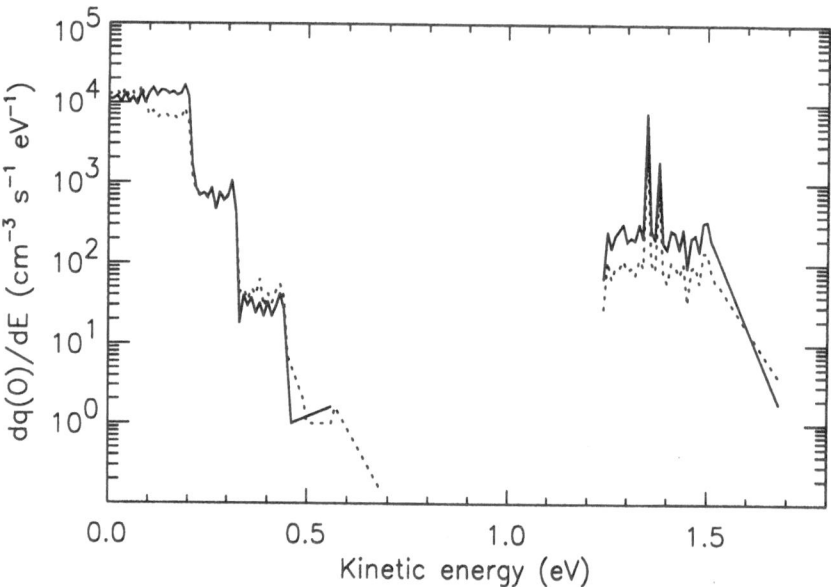

Figure 7.15. Kinetic energy spectra of superthermal O atoms produced by O_2 dissociation at 100 km for low $F_{10.7}=70$ (dotted line) and high $F_{10.7}=245$ (solid line) solar activity.

states $^1\Pi_u$ and $^1\Sigma_u^+$ in various vibrational levels (Zipf and McLaughlin, 1978) with energies exceeding the dissociation threshold. As a result, the hot N atoms with energies $\sim 0.6 \div 1.24$ eV are formed by decay of these states. The energy distribution of these particles is strongly nonequilibrium as well.

The dependence of the energy spectra of hot O and N atoms on solar and geophysical parameters results from the analysis of the data presented in *Figs. 7.15-7.20.* For example, at sunset $\theta_\odot = 90^o$, the optical depth of the atmosphere becomes larger, and the photodissociation rates decrease. Under these conditions the photodissociation rate of molecular nitrogen becomes small even at the heights ≥ 130 km (*Fig. 7.13*).

We may conclude that the typical kinetic rates of photodissociation and electron impact dissociation in the Earth's upper atmosphere are high. The above examples of the calculated kinetic characteristics of the processes of dissociation show that these processes are important sources of superthermal particles which must be included in the models of atmospheric chemistry.

In summary, the processes induced by electromagnetic and corpuscular solar radiation are responsible for the formation of significant sources of

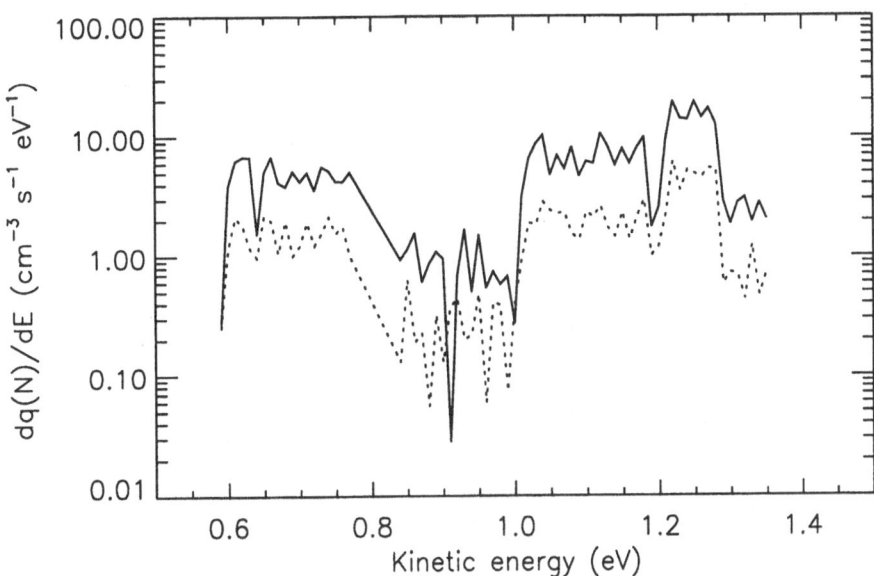

Figure 7.16. Kinetic energy spectra of superthermal N atoms produced by N_2 dissociation at 100 km for low $F_{10.7}=70$ (dotted line) and high $F_{10.7}=245$ (solid line) solar activity.

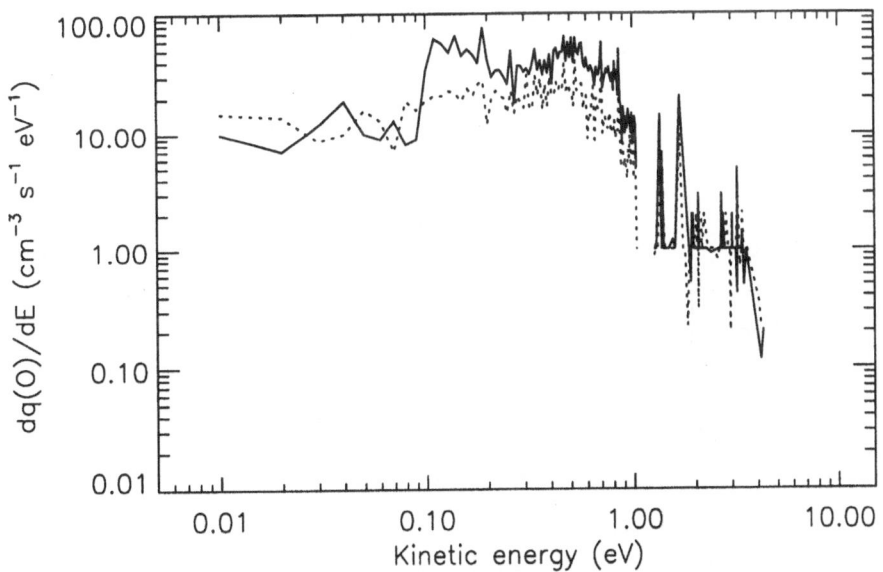

Figure 7.17. Same as *Fig. 7.15* at 160 km.

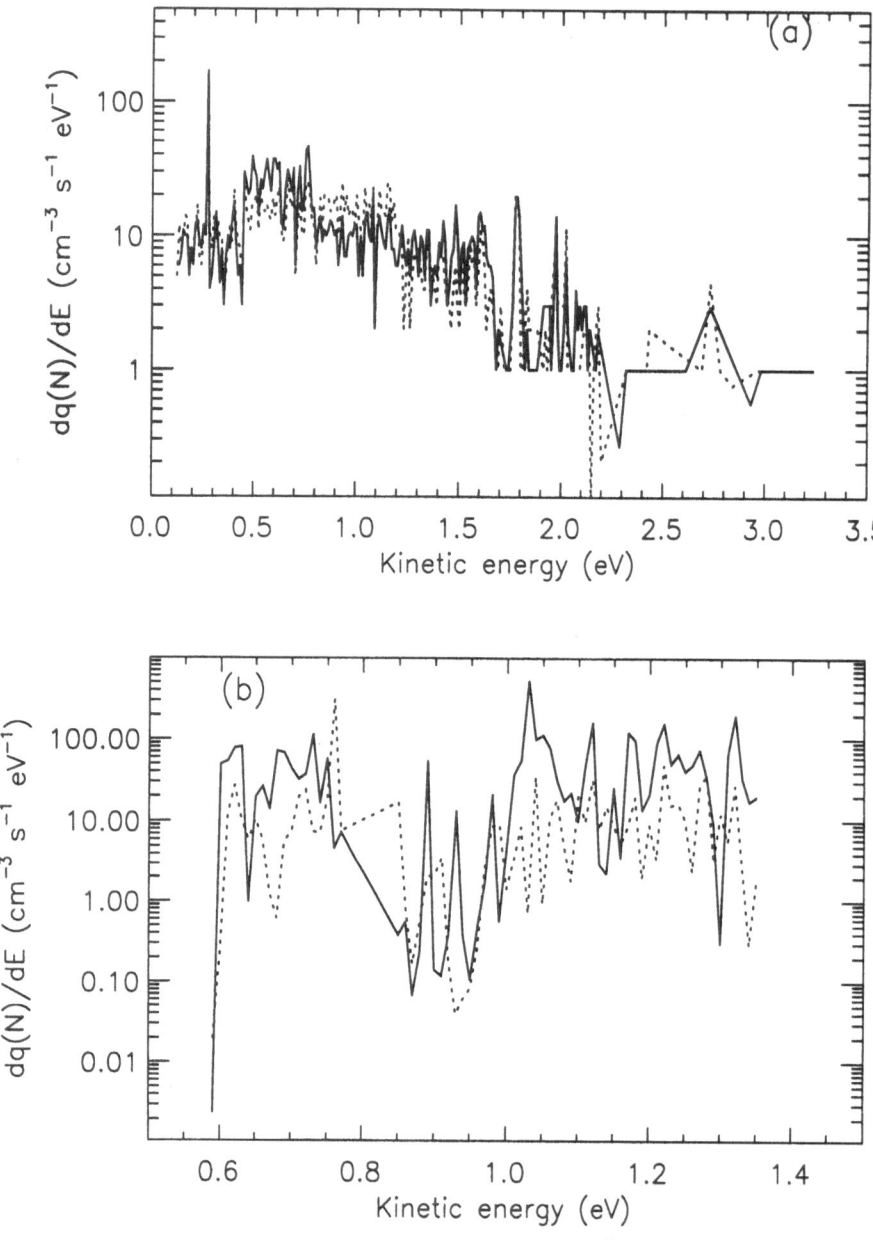

Figure 7.18. Same as *Fig. 7.16* at 160 km (a - due to photodissociation; b - due to electron impact dissociation).

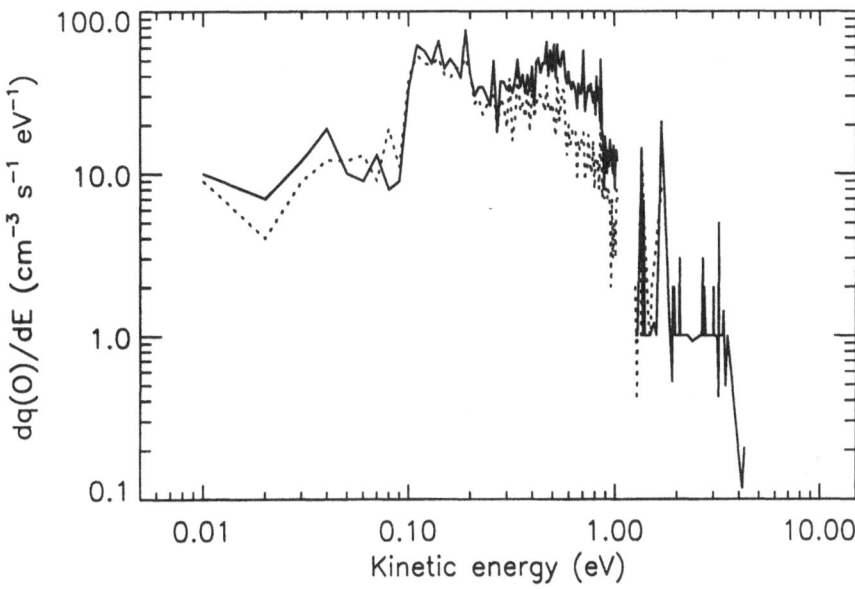

Figure 7.19. Kinetic energy spectra of superthermal O atoms produced by O_2 dissociation at height 160 km for different solar zenith angles $\theta_\odot = 5°$ (solid line) and 90° (dotted line) at high solar activity.

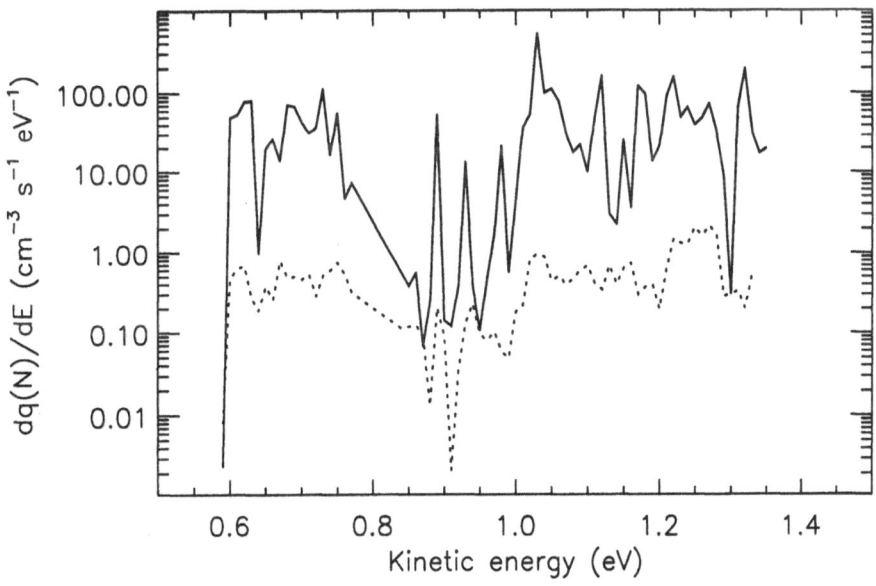

Figure 7.20. Same as *Fig. 7.19* for N atoms

nonthermal particles. These particles play an important role in the physical and chemical processes occurring in the upper atmosphere and their study in the framework of mathematical models contributes to progress in planetary aeronomy. In this chapter only direct effects of the solar incident radiation affecting the atmospheric gas were considered. At the same time it is well known that the interaction of solar wind with planetary atmospheres can lead to the formation of additional sources of nonthermal particles. In particular solar-planetary (and specifically solar-terrestrial) interactions are responsible for the auroral precipitation of high energy charged particles of magnetospheric origin (electrons, protons, and heavier ions). These additional sources are discussed in more detail below, as a part of the evaluation of specific aeronomy problems, for which the nonequilibrium processes induced by the fluxes of these particles dominate or at least can not be ignored.

CHAPTER 8

NONEQUILIBRIUM CHEMISTRY OF ODD NITROGEN IN THE EARTH'S THERMOSPHERE

The odd nitrogen thermospheric constituents $N(^2D)$, $N(^4S)$ and NO are produced by the dissociation and ionization of N_2 by photons, photoelectrons and energetic auroral particles. Nitric oxide plays an important role as a source of ionization in D region (Nicolet and Aikin, 1960), as a source of chemical heating in the thermosphere (Roble, 1995), and as a cooling term in the lower thermosphere (Kockarts, 1980; Gordiets et al., 1982; Gérard and Roble, 1986, 1988; Roble, 1995). It also controls the NO^+ / O_2^+ ratio in the E region and its downward transport in the polar night stratosphere or production by energetic solar protons generates the catalytic recombination of ozone in the middle atmosphere.

Photochemical-transport models of nitric oxide were developed for the comparison with measurements and prediction of the NO distribution. As observational and numerical techniques improved, it became apparent that efficient sources of 'active' nitrogen atoms are produced with a high efficiency by N_2 dissociation and ion-molecular reactions. Quite recently, however, satellite, rocket and ground-based obervations brought evidence that, in the present state of knowledge of the odd nitrogen chemistry, classical processes cannot explain the NO density observed in the thermosphere. One possible solution of this problem is the role played by hot nitrogen atoms as an additional source of nitric oxide. In this chapter, we first describe the 'classical' odd nitrogen photochemistry and the one- and multi-dimensional models developed in order to calculate the NO and N distributions. The role played by nonthermal $N(^4S)$ and their effect on the NO production rate and density is then discussed. Finally, observational evidence of nonthermal N atoms in the NO distribution and infrared emission spectrum is reviewed.

8.1. 'Classical' Photochemistry of Odd Nitrogen

The concept of odd nitrogen was introduced in the middle atmosphere aeronomy to distinguish a group of reactive nitrogen compounds having strong mutual chemical interaction. The thermospheric odd nitrogen (NO, $N(^4S)$ and $N(^2D)$) is addressed as a natural extension of this family. For example, atomic nitrogen and nitric oxide in the thermosphere are closely chemically coupled since nitrogen atoms are both source and a loss of nitric oxide.

The role of nitric oxide in aeronomy was recognized by Nicolet (1945) who first suggested ionization by Lyman-α as a source of ions in the atmosphere. The importance of this and other processes responsible for the formation of the ionospheric D-region was demonstrated by Nicolet and Aikin (1960). In 1964, Barth first observed the presence of the NO - γ bands in the ultraviolet spectrum of the Earth's day airglow (Barth, 1964). He then obtained nitric oxide densities of the order of 10^7 - 10^8 cm^{-3} in the lower thermosphere, a hundredfold larger than had been predicted. The altitude, latitude, seasonal and solar activity dependence of thermospheric NO has been extensively studied since then with increasingly complex and sensitive instruments on board rockets and satellites.

Initially, the main source of NO in the thermosphere was thought to be

$$N(^4S) + O_2 \rightarrow NO + O \ . \tag{8.1}$$

However, the strong temperature dependence of this reaction ($k = 4.4 \times 10^{-12} e^{-\frac{3220}{T}}$ cm^3 s^{-1}) considerably decreases its efficiency in the lower thermosphere and makes it insufficient to explain the NO density peak observed near 110 km if only thermal nitrogen atoms are taken into consideration. This lead Norton and Barth (1970) to suggest that metastable $N(^2D)$ atoms may be a dominant source of nitric oxide in the lower thermosphere where the reaction

$$N(^2D) + O_2 \rightarrow NO + O \ ,$$

with $k = 5 \times 10^{-12}$ cm^3 s^{-1} is about several orders of magnitudes faster than coefficient of the reaction (8.1) at 300 K, as it had been claimed by the laboratory measurements.

This idea has drawn considerable interest for the metastable $N(^2D)$ atom chemistry and the associated [NI] $^2D^o \rightarrow \ ^4S^o$ forbidden doublet transition at 5198.5–5207 Å usually observed as a single feature at 5200 Å. It also indicated that the NO-$N(^2D)$ - $N(^4S)$ system is closely coupled, since $N(^2D)$ atoms deactivate radiatively (lifetime $\sim$ 26 hrs) or collisionally to the ground-state ^4S level:

$$N(^2D) \ (+M) \rightarrow N(^4S) \ (+M) \ ,$$

where M designates atomic oxygen or thermal electrons. At all altitudes, the mutual destruction of N and NO:

$$NO + N(^4S, \ ^2D) \rightarrow N_2 + O$$

is an internal process limiting the total concentration of the odd nitrogen. Nitric oxide is also destroyed by charge transfer with O_2^+ and predissociation near 1908 Å:

$$NO + O_2^+ \rightarrow NO^+ + O_2$$
$$NO + h\nu \rightarrow N(^4S) + O$$

A first source of nitrogen atoms is the direct dissociation of N_2 by collisions of photons or photoelectrons with N_2:

$$N_2 + \{^{h\nu}_e\} \rightarrow N(^4S) + N(^4S, {}^2D, {}^2P) \quad (+e) \quad .$$

Additional indirect dissociation is generated by ionization of N_2:

$$N_2 + \{^{h\nu}_e\} \rightarrow N_2^+ + e \quad (+e')$$
$$N_2 + \{^{h\nu}_e\} \rightarrow N^+ + N + e \quad (+e')$$

followed by charge transfer of N_2^+ with O:

$$N_2^+ + O \rightarrow NO^+ + N(^2D)$$

and dissociative recombination of ionized nitric oxide and nitrogen:

$$NO^+ + e \rightarrow O + N(^2D, {}^4S)$$
$$N_2^+ + e \rightarrow N(^4S) + N(^2D, {}^4S)$$

Fig. 8.1 shows the main processes controlling the production and loss of odd nitrogen in the Earth's thermosphere. Table 8.1 lists the important ion-neutral and neutral-neutral chemical processes for terrestrial atomic nitrogen and NO. Current estimates of the relevant reaction coefficients are also given.

We now focus on the results of observations of the $N(^4S)$ and $N(^2D)$ density distributions and the current status of one- and multi-dimensional modeling of odd nitrogen in the Earth's upper atmosphere.

8.1.1. GROUND-STATE N(^{4}S) ATOMS

The most reliable direct measurements of the $N(^4S)$ thermospheric density were carried out with the neutral mass spectrometer on board the *Atmospheric Explorer (AE)* satellites. The signal peaked at 30 atomic mass units gives a direct evidence of the nitric oxide density produced by recombination of O and N atoms inside the instrument. Consequently, this technique may be applied in a straightforward manner above 200 km where $N(^4S)$ number density exceeds by far that of nitric oxide (Mauersberger *et al.*,

TABLE 8.1. Important odd nitrogen reactions used in the model.

Number	Reaction	Rate coefficient ($cm^3 s^{-1}$)
R1	$N_2 + e \rightarrow N + N + e$	$f(N(^2D))=0.5 - 0.6$
R2	$N_2 + h\nu \rightarrow N(^4S) + N(^2D)$	$f(N(^2D))=0.5$
R3	$NO^+ + e \rightarrow N + O$	$4.2 \times 10^{-7}(300/T_e)^{-0.85}$; $f(N(^2D)=0.78$
R4	$N_2^+ + O \rightarrow N(^2D) + NO^+$	$1.4 \times 10^{-10}(300/T_i)^{0.44}$; $f(N(^2D))=1$
R5	$N_2^+ + e \rightarrow N(^4S) + N(^2D)$	$1.8 \times 10^{-7}(300/T_e)^{-0.39}$; $f(N(^2D))=0.5$
R6	$N(^2D) + O_2 \rightarrow NO + O$	6×10^{-12}
R7	$N(^4S) + O_2 \rightarrow NO + O$	$4.4 \times 10^{-12}\exp(-3220/T)$
R8	$N_f(^4S) + O_2 \rightarrow NO + O$	see Section 7.2
R9	$N(^4S) + NO \rightarrow N_2 + O$	3.4×10^{-11}
R10	$O_2^+ + NO \rightarrow NO^+ + O_2$	4.4×10^{-10}
R11	$N(^2D) + NO \rightarrow N_2 + O$	6.7×10^{-11}
R12	$N(^2D) + O \rightarrow N(^4S) + O$	6.9×10^{-13}
R13	$N(^2D) + e \rightarrow N(^4S) + e$	$6.0 \times 10^{-10}(T_e/300)^{0.5}$
R14	$N(^2D) \rightarrow N(^4S) + h\nu$	$1.07 \times 10^{-5} s^{-1}$
R15	$O^+ + N_2 \rightarrow N(^4S) + NO^+$	1.57×10^{-12}
R16	$N^+ + O_2 \rightarrow N(^4S) + O_2^+$	4.0×10^{-10}
R17	$NO + h\nu \rightarrow N(^4S) + O$	
R18	$NO + h\nu \rightarrow NO^+ + e$	

1975). Below this altitude the NO concentration measured independently with the UV spectrometer on board the same spacecraft must be first reconciled with the mass spectrometer data (Engebretson *et al.*, 1977a). These data were used to develop an empirical $N(^4S)$ model (Engebretson *et al.*, 1977b) and later incorporated into the $MSIS-86$ model (Hedin, 1987) of the thermosphere and its subsequent revised version to describe thermospheric distribution of N. A simple altitude dependence was used, based on the two-dimensional odd nitrogen model by Gérard *et al.* (1984). The dependence of N density on geomagnetic activity was studied by Engebretson and Mauersberger (1983).

An other experimental approach, potentially applicable for remote sensing technique, is based on the chemiluminescent two-body recombination of N and O atoms :

$$N(^4S) + O(^3P) \rightarrow NO^* (C^2\Pi, v = 0)$$
$$NO^* \rightarrow NO + h\nu,$$

followed by subsequent rise to the v' = 0 progressions of the δ and γ NO systems (Groth *et al.*, 1971; Feldman and Takacs, 1974; Du and Dalgarno, 1990). This method was used to obtain nighttime $N(^4S)$ vertical distribution

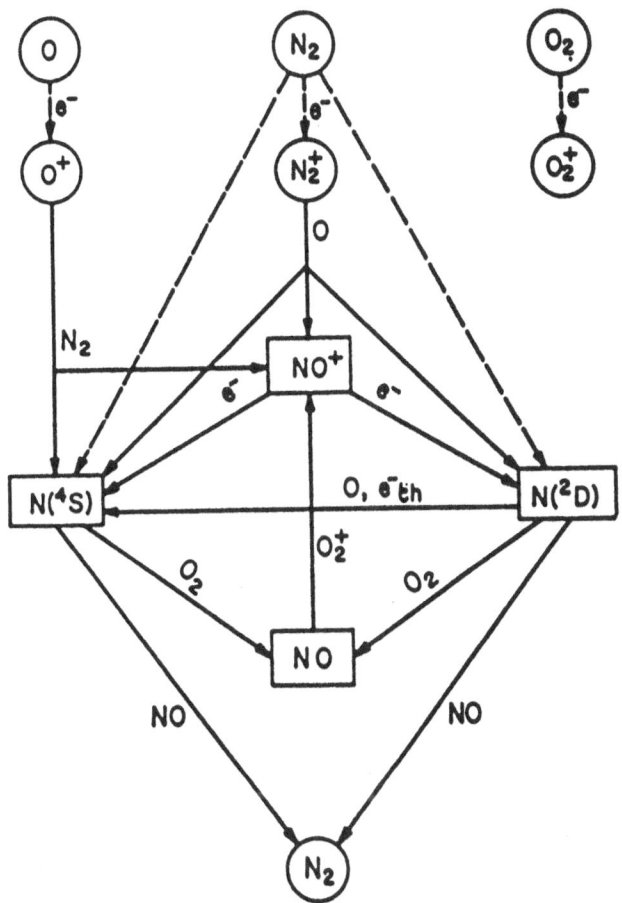

Figure 8.1. Simplified diagram of the odd nitrogen photochemistry in the Earth's thermosphere without involvement of fast nitrogen atoms.

from rocket observation of the NO recombination airglow (cf. review by Tennyson *et al.*, 1986). The $N(^4S)$ densities and their observed variations deduced from these measurements are in good agreement with the modelled local time and solar cycle variations.

The [NI] ^4S-^4P resonance transition at 1200 Å may be, in principle, helpful to obtain $N(^4S)$ abundances in the sunlit thermosphere. However, as discussed by Meier (1991), a not clearly defined optical depth of the transition, together with an uncertainty of the relative photoelectron impact on $N(^4S)$ and N_2 and photodissociative excitation of N_2, have hindered so far the utilization of the 1200 Å intensity profiles to derive the atomic nitrogen distribution.

8.1.2. N(^{2}D) METASTABLE ATOMS

Metastable N(^{2}D) atoms are produced by a variety of photochemical and ionospheric processes (see Table 8.1). They are destroyed by reaction (R5) and by deactivation in collisions with O and electrons in addition to spontaneous radiative transition to the N(^{4}S) state. Measurements of the [NI] transition at 5200 Å thus provide a direct determination of the N(^{2}D) column density. The small probability transition of about $1 \times 10^{-5} s^{-1}$, as it was recalculated by Godefroid and Froese-Fischer (1984) and Zeippen (1982), basically confirmed the original Garstang's (1956) value.

Hays *et al.* (1988) reviewed the results of interpretation of the 5200 Å airglow which was extensively measured with the *Visible Airglow Experiment (VAE)* on board the *AE* satellites. These *VAE* observations were also analyzed in detail by Frederick and Rusch (1977) for the dayglow and Rusch and Gérard (1980) for the aurora. They concluded that the various sources of N(^{2}D):

$$N_2 + \{^{h\nu}_e\} \rightarrow N(^2D,^4S) + N(^2D,^4S) + (e)$$
$$NO^+ + e \rightarrow N(^2D, \,^4S) + O$$
$$N_2^+ + O \rightarrow N(^2D) + NO^+$$
$$N_2^+ + e \rightarrow N(^2D) + N$$

need to proceed with maximum efficiency of the N(^{2}D) production rate compatible with the selection rules. They also estimated that the quenching coefficient by atomic oxygen fitting best the observations was on the order of 4.5×10^{-13} cm^3 s^{-1}. Until recently, experimentally determined deactivation coefficients exceeded this value, varying from 1×10^{-12} cm^3 s^{-1} (Piper, 1989) to 1.8×10^{-12} (Davenport *et al.*, 1976) and even 2×10^{-11} (Jusinski *et al.*, 1988). This high rate was shown, however, to be incompatible with the current understanding of the odd nitrogen chemistry (Fesen *et al.*, 1989; Siskind *et al.*, 1989) and with observations of the [OI] emission at 6300 Å (Bates, 1989a). Both the nitric oxide density and the 5200 Å airglow calculated with this quenching rate failed by far to match the observations. In spite of modeling efforts, it was not possible to reconcile the laboratory and the aeronomically determined rates until Fell *et al.* (1990) measured this coefficient more carefully and reported the value of $6.9(+0.7 - 1.1) \times 10^{-13}$ cm^3 s^{-1}, fairly close to that required by the odd nitrogen modeling. New information on this problem was provided by a compilation of the *AE-C* odd nitrogen data base by Rusch *et al.* (1991). It contains profiles of NO, N(^{2}D), the major positive ions, O and N$_2$ densities, as well as the neutral, ion and electron temperatures measured during

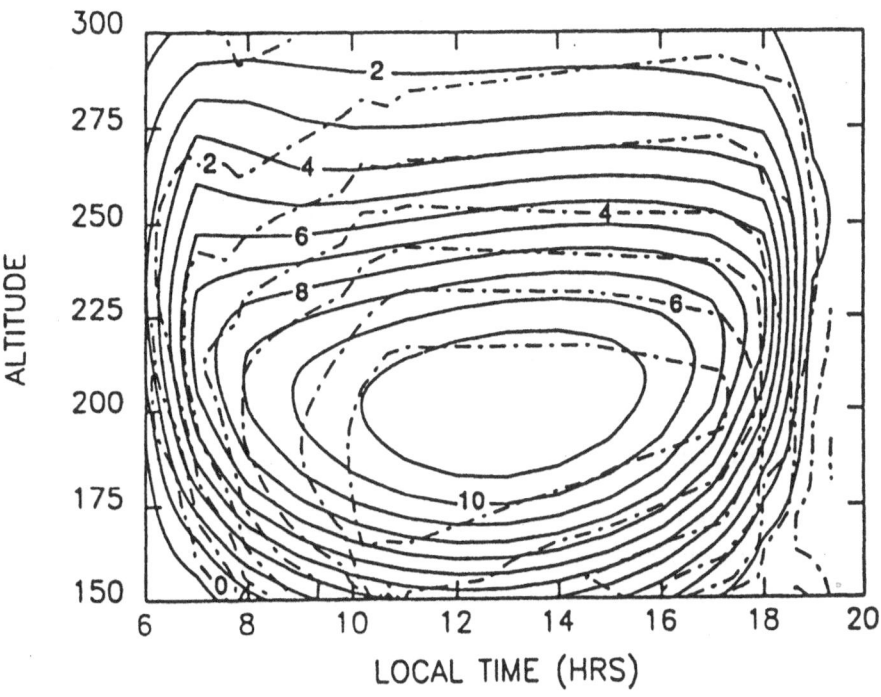

Figure 8.2. Comparison between metastable $N(^2D)$ atom distribution calculated by one-dimensional photochemical model (solid line) and observed with the *VAE* experiment on board the *AE-C* satellite (dashed line) (Rusch *et al.*, 1991).

11 weeks from August to October 1974. During this period, the satellite perigee ($\sim$ 135 km) drifted slowly in latitude but more rapidly in local time, covering 18 hours of local time during the selected periods. In order to avoid contamination of these data by the processes involving energetic particles, the latitude motion of the perigee was restricted to vary between 45°N and 50°S. *VAE* observations of the 5200 Å airglow were inverted to volume emission rates using the Abel solution of the integral equation. When compared with the one-dimensional calculations made for geophysical and solar activity conditions similar to the observations (*Fig. 8.2*), the measurements exhibit similar local time variations and absolute emission rates. They confirm the low value of the $N(^2D)$ deactivation coefficient needed to account for the observed 5200 Å emission rate and also show that, used in the model of odd nitrogen photochemistry, this quenching rate provides reasonably good simultaneous agreement with the measured NO and $N(^2D)$ vertical distributions.

8.1.3. N(^{2}P) METASTABLE ATOMS

N(^{2}P) atoms are produced at a smaller rate in the reactions R1 and R2, followed by destruction in the reactions with O_2 and collisional deactivation by O and electrons. However, because of much shorter radiative life time (~ 13 s) the N(^{2}P) number density is considerably less than N(^{2}D) density. An analysis of their sources and sinks in the aurora was made by Zipf *et al.* (1980) and Torr and Torr (1982). Zipf *et al.* (1980) suggested that N(^{2}P) quenching results primarily in production of N(^{2}D) atoms, so that the net N(^{2}D) yield is the sum of the primary N(^{2}D) and N(^{2}P) yields. Bates (1989b) estimated the rate coefficient of 1×10^{-10} cm^3 s^{-1} for the N(^{2}P)+O $\rightarrow$ N(^{2}D)+O reaction. The reaction with O_2 to form NO proceeds with the coefficient of 3.5×10^{-12} cm^3 s^{-1}.

8.1.4. NITRIC OXIDE

Nitric oxide has been extensively measured since the above mentioned pioneering work of Barth (1964). Most NO density determinations are based on the resonance scattering technique. In this method, the ultraviolet γ-bands of NO are observed by spectrometer or photometer on board a space platform. The NO column density $\eta = g \int [NO]ds$ is directly related to the column emission rate by the relationship:

$$\phi = g\eta \ ,$$

where g is the emission rate factor estimated as 6.4×10^{-6} photon/s$\times$molec. for optically thin emission. When the column density exceeds a critical value, an optical thickness correction must be applied and g is no longer constant (Eparvier and Barth, 1992). More recently infrared emission in the (1-0) fundamental band at 5.3 μm has been used to obtain NO densities from limb observations in emission (Wise *et al.*, 1995) or absorption features by solar occultation technique (Barth *et al.*, 1996). They yield values in good agreement with those derived from fluorescence scattering.

A large number of rocket measurements based on the technique of resonance scattering by solar radiation near 2000 Å have provided "snapshots" of the vertical distribution of NO in the equatorial, mid-latitude and high-latitude thermosphere. The global distribution of thermospheric NO has been measured with UV spectrometers using the same method on various satellites: *OGO-4* (Rusch and Barth, 1975; Gérard and Barth, 1977), *AE - C and D* (Cravens and Stewart, 1978) and *Solar Mesosphere Explorer (SME)* (Barth *et al.*, 1988; Fesen *et al.*, 1990), Gérard *et al.*, 1990). Spectral and spatial images were also obtained from the *Space Shuttle* by Torr *et al.* (1995).

A substantial database has been therefore accumulated during the last 20 years and, despite of its considerable variability, a global picture of the general morphology, local time and altitude dependence of NO has emerged. Common features of these observations include a dependence on solar and geomagnetic activity, a latitudinal gradient with larger values near the auroral oval, and peak densities between 110 and 115 km varying within $6 \times 10^6 \div 3 \times 10^7$ cm^{-3} at low latitudes.

Large variations of the nitric oxide peak density were observed with the *SME* satellite from 1982 to 1985 during the declining phase of the previous solar cycle (Barth *et al.*, 1988; Fesen *et al.*, 1990; Barth, 1990). These resonance scattering measurements showed a decrease by a factor of 7.5 as solar activity dropped during the 4-year period of the observations. Using *SME* data collected only during days of low geomagnetic activity ($A_p \leq 15$), Gérard *et al.* (1990) and Fesen *et al.* (1990) derived a peak density decrease by a factor of 2.5 to 4, depending on the season and latitude taken from the same database. In addition, variation by nearly the factor of 2 was observed with a 27-day period during the high solar activity phase in 1982. It was interpreted (Barth *et al.*, 1988; Gérard *et al*, 1993; Fuller-Rowell, 1993) as a consequence of a solar rotation-induced change of the XUV and soft X-ray radiation interacting with the E region and the lower thermosphere.

The *SME* data set was complemented by a set of rocket measurements made in Japan by the University of Tokyo team using the same observational technique. These observations were made at a fixed latitude of 31° N at sunset between 1981 and 1987 (Kuze and Ogawa, 1988) and covered $F_{10.7}$ solar indices ranging from 71 to 259. They also showed a solar activity dependence of the maximum NO density near 110 km. However, the increase from low to very high activity conditions was only by the factor of 3.5. In addition, their absolute NO densities turned out larger than the *SME* values, possibly as a consequence of the uncorrected effect of the optical thickness in the *SME* limb viewing geometry and upgraded value of the g-factor they used.

Ground-based microwave observations were also reported by Clancy *et al.* (1992). The measured vertical nitric oxide values were used to determine approximated NO peak densities. These values are systematically larger than those derived from other methods. This point will be further discussed in a subsequent section.

The large solar cycle and solar rotation dependence of the NO peak provoked Barth *et al.* (1988) and Siskind *et al.* (1989) to emphasize the role of the solar soft X-rays ($\lambda < 200$ Å) in the solar control of nitric oxide in the lower thermosphere. Indeed, this part of the solar spectrum varies strongly with solar activity (and rotation) and plays an important role in the production of the precursors of nitric oxide. Empirical models synthetizing the NO peak density observations were developed by Smith *et al.* (1993) and Titheridge (1996).

8.1.5. ONE-DIMENSIONAL ODD NITROGEN MODELING

Models of thermospheric odd nitrogen have been developed for the comparison with rocket and satellite observations of odd nitrogen species. One-dimensional chemical-diffusive models were built first. They could not ensure predictions of the global effects because lacked to take into account the horizontal transport, which may be important in the presence of strong density gradients. However, they still remain useful tools to investigate the detailed photochemical processes and their impact on the calculated odd nitrogen distribution. An advantage of one-dimensional models is also in the fact that they account for the local time variations, in contrast to two-dimensional zonally averaged models which run the solution to steady state for average values of the diurnal solar illumination.

As it was earlier mentioned, nitric oxide is produced in the thermosphere by the reaction of O_2 with atomic nitrogen (reactions R5 and R6 of Table 8.1), the first one being strongly temperature-dependent (DeMore *et al.*, 1990). The second reaction has no measurable activation energy and is faster at temperature prevailing in the lower thermosphere compared to the first one. Metastable $N(^2D)$ atoms are produced by a series of processes involving direct and indirect dissociations of N_2 molecules (see Table 8.1). Since the main loss process of NO in the thermosphere is the mutual destruction with nitrogen atoms:

$$NO + N(^4S) \rightarrow N_2 + O$$

the NO steady-state concentration in the lower thermosphere, where $N(^2D)$ is a dominant source, depends critically on the ratio of the production rates of $N(^2D)$ and $N(^4S)$ atoms. For most sources, the quantum yield of these states has been determined either experimentally, or deduced from theoretical considerations based on physical properties of the nitrogen molecule (Zipf and McLaughlin, 1978; Oran *et al.*, 1975).

A key question has been posed on the relative yield of ground and excited nitrogen atoms produced by the impact of energetic electrons on N_2:

$$N_2 + e \rightarrow N + N + e \ .$$

Assuming that excited atoms are created whenever energetically possible, Zipf and McLaughlin (1978) calculated the dissociation cross sections for the production of $N(^4S)$, $N(^2D)$, and $N(^2P)$ and N^+ fragments for electron energies up to ~ 500 eV. For example, they found that impact of 100 eV electrons yields the three nitrogen atom states in the 0.46 : 0.35 : 0.19 ratio.

It is reasonable to assume that all nitrogen atoms which are not found in the ground state cascade into the $N(^2D)$ state. Therefore, the effective $N(^2D)$ branching ratio for reaction R1 (Table 8.1) at 100 eV is close to 0.54.

These models are based on the coupled set of continuity equations

$$\frac{\partial n_i}{\partial t} = P_i - \nu_i n_i - \frac{\partial \phi_i}{\partial z}$$

and the vertical flux equations :

$$\phi_i = -D_i\left(\frac{n_i}{H_i} + \frac{\partial n_i}{\partial z} + n_i\frac{\partial \ln T}{\partial z}\right) - K\left(\frac{n_i}{\bar{H}} + \frac{\partial n_i}{\partial z} + n_i\frac{\partial \ln T}{\partial z}\right)$$

for NO and $N(^4S)$. In these equations, n_i is the NO or $N(^4S)$ concentration; P_i and ν_i are the photochemical production and loss rates, respectively; D_i is the molecular diffusion coefficient; H_i is the diffusive scale height; and $\bar{H}$ is the background atmospheric scale height. Photochemical equilibrium is usually assumed for $N(^2D)$, positive ions and electrons. The background neutral densities and the temperature profile for the major thermospheric constituents are usually taken from the $MSIS$ model.

This system is solved numerically with finite differences methods on a discrete altitude grid with appropriate boundary conditions. The local time dependence is explicitly calculated or the model is running to steady state, until an equilibrium is reached. One-dimensional time dependent models of thermospheric NO were developed and described by Strobel et al. (1970), Rusch (1973), Kuze and Ogawa (1988) , Cleary (1986), Siskind et al.(1989, 1995), Gérard and Taieb (1986). The globally averaged models were used by Roble and Emery (1983) and Roble et al. (1987) to investigate the mean energetics of the processes involved. Besides, Fuller-Rowell (1993) studied the effect of the solar cycle variations on the NO density.

8.1.6. MULTIDIMENSIONAL ODD NITROGEN MODELING

Quite recently, as satellite global measurements of nitric oxide became available, multi-dimensional models of odd nitrogen were developed what allowed to compare the observational data sets with chemical-dynamical models. Since the thermospheric NO density is strongly controlled by the atmospheric temperature and background gas chemical composition, the N_2, O_2, O densities, temperature and odd nitrogen fields are simultaneously calculated in framework of these models.

Two-dimensional models of odd nitrogen for solstice conditions were developed by Gérard et al. (1984) and Gérard and Roble (1986, 198o).

These models incorporated the dynamical-chemical calculations of Roble and Kasting (1984) which provided the background neutral densities, temperature and vertical and meridional winds with a complete odd nitrogen photochemical scheme. Consequently, the $N(^2D)$, $N(^4S)$ and NO zonally averaged distributions were calculated consistently for various levels of solar and geomagnetic activity. These studies demonstrated that the summer-to-winter pole meridional circulation plays an important role in the redistribution of NO and $N(^4S)$ accross the solar terminator. In particular, winds carry NO into the polar night region where, in addition to the local production by auroral ionization and dissociation of N_2, nitric oxide is carried downward by vertical transport into the mesosphere. Comparison with global satellite observations of NO showed that the model is able to reproduce the main features of the NO latitudinal distribution measured with the *AE-C* (Gérard *et al.*, 1984), and *SME* (Fesen *et al.*, 1990) satellites.

The role of thermospheric cooling by the NO 5.3 μm radiation was studied by Kockarts (1980) and Gordiets *et al.* (1982) and in more detail on the basis of a globally averaged model by Roble *et al.* (1988). Roble *et al.* (1988) found that at solar minimum activity conditions, the NO cooling makes a 88 K contribution to the total heat budget, though it may reach 725 K at solar maximum. Two-dimensional calculations (Gérard and Roble, 1986, 1988) investigated the influence of this infrared cooling term in the energy equation. It was found that the NO cooling reduced the summer-to-winter thermal contrast by 45 K at low solar activity ($F_{10.7} = 80$) and by 110 K at high solar activity ($F_{10.7} = 200$). It is important to note that quantitative conclusions on the role of the NO cooling strongly depends a correct prediction of the nitric oxide density distribution. An adequate description of the NO photochemistry and transport is therefore needed.

Some observed features of the global NO distribution obviously can not be explained by zonally averaged model calculations. For example, the longitudinal dependence observed by Cravens and Stewart (1978) in the NO peak concentration at a given geographic or geomagnetic latitude clearly involves the offset between the geographic and the magnetic poles in addition to horizontal transport. Cravens and Killeen (1988) suggested, on the basis of the consideration of neutral parcel trajectories calculated with the use of the *NCAR Thermospheric Global Circulation Model (TGCM)*, that longitudes associated with the geomagnetic poles are favoured for equatorward transport out of the auroral zones. These characteristics could explain the maximum NO abundance found near 70° W longitude in the North and 160^0 E longitude in the South.

Three-dimensional odd nitrogen fields were calculated with the *NCAR Thermospheric Ionosphere General Circulation Model (TIGCM)* and described by Roble *et al.* (1988), as well as its more recent time-GCM version

(Roble and Ridley, 1994) proceeded from dynamic meteorology and adapted to the physics of thermosphere and ionosphere. The total neutral and ion temperature, velocity vector, N_2, O_2 and O mixing ratios, positive ion and electron densities and $N(^2D)$, $N(^4S)$ and NO concentrations were revealed consistently between 97 km and the exosphere in a time-dependent way. The photochemistry was identical to that in the one-dimensional global mean thermospheric model by Roble *et al.* (1987) and Roble and Ridley (1994). $N(^2D)$ was assumed to be in photochemical equilibrium but $N(^2D)$ and NO were transported by winds and diffusion. Parameters defining the external forcing on the thermosphere include the solar UV and EUV radiation, the total auroral particle power release and the cross-polar cap potential drop driving the ionospheric convection pattern. Results showing the NO global distribution for minimum solar activity conditions were presented by Roble *et al.* (1988).

Fig. 8.3 shows a south polar view of the distribution of nitric oxide calculated with the *TIGCM* near 120 km for medium solar activity. The NO densities increase from low to high latitudes as a result of auroral particle precipitation. The complexity of the latitudinal-longitudinal-local time factors is apparent. At low latitudes, the local time and chemical composition factors are dominant and the maximum density is calculated in the early evening region. The minimum occurs near sunrise as a result of the NO night time loss by chemical destruction and transport. At high latitudes, the NO structure reflects the combined effect of local time, solar control and particle precipitation morphology. This is clearly seen in *Fig. 8.3*. The offset between the geographic and geomagnetic poles, the complex morphology of the auroral precipitation and convection patterns combine to generate a complex structure. The largest NO densities occur in the auroral oval with a local minimum in the polar cap. The pattern of NO concentration has a distorted shape with a maximum of about 3.2×10^7 cm^{-3}. The longitudinal dependence of the electron energy characteristics also produces a marked longitudinal structure in the nitric oxide distribution. A clear longitudinal asymmetry is also apparent. It is related to asymmetries in the neutral composition, temperature and wind vortices driven by the magnetospheric convection. At this altitude, the temperature and circulation structures are also strongly influenced by tides propagating upward.

The $N(^4S)$ atoms (*Fig. 8.4*) are in near photochemical equilibrium at this altitude. Maximum concentrations of $\sim 7 \times 10^6$ cm^{-3} are obtained in the region of maximum N_2 dissociation. The morphology reflects the structure of the $N(^4S)$ sources similar to that of $N(^2D)$, and its main photochemical sink, due to reaction with NO. This strong coupling between the two species explains in part the complex morphology. The high latitude pattern exhibits a horseshoe shape with an opening in the afternoon sector. $N(^2D)$ is in

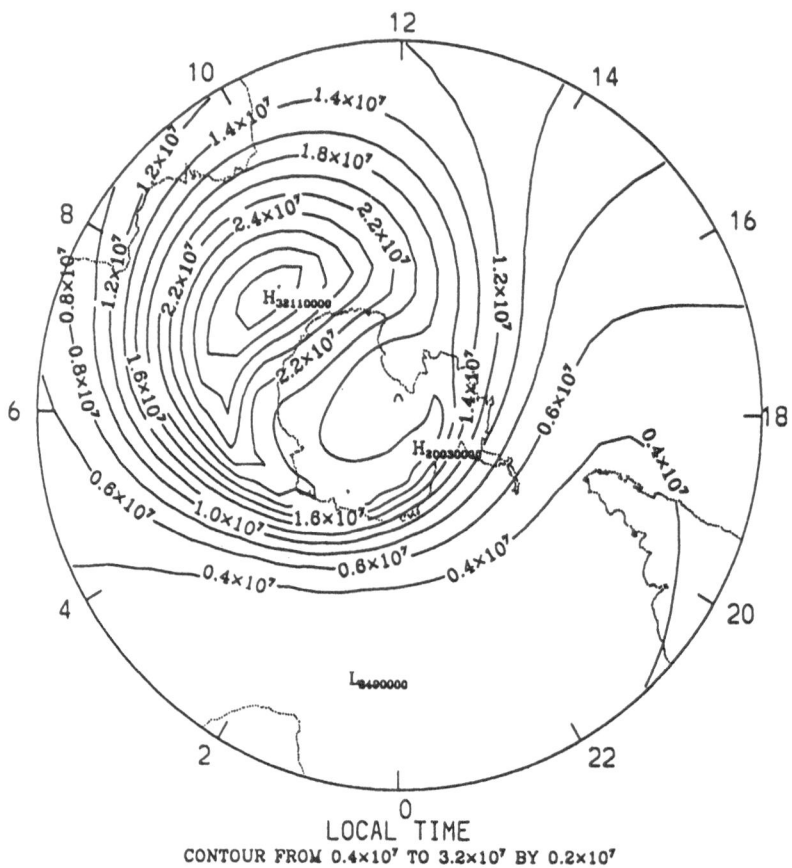

LOCAL TIME
CONTOUR FROM 0.4×10⁷ TO 3.2×10⁷ BY 0.2×10⁷

Figure 8.3. South polar view of the nitric oxide distribution at 120 km calculated with the *NCAR TIGCM* for equinox medium solar activity conditions ($F_{10.7} = 150$), 0:00 UT, cross-tail potential of 70 keV and global particle precipitation power of 16 GW (R.G. Roble, 1991; private communication).

photochemical equilibrium in the thermosphere and its distribution exhibits a maximum solar produced component at local noon. Its density quickly drops off along the solar terminator and is significantly enhanced in the auroral oval and magnetospheric cusp.

Comparison of model calculations based on the descibed odd nitrogen photochemistry lead to the abundances of N(^{2}D) that are in good agreement with the observed airglow values. However, the calculated NO density near 110 km tends to be on the low side of the measured quantities. This is especially the case when compared with the recent microwave measurements giving systematically higher peak values. This discrepancy was removed somewhat arbitrarily in models by using branching ratios of the

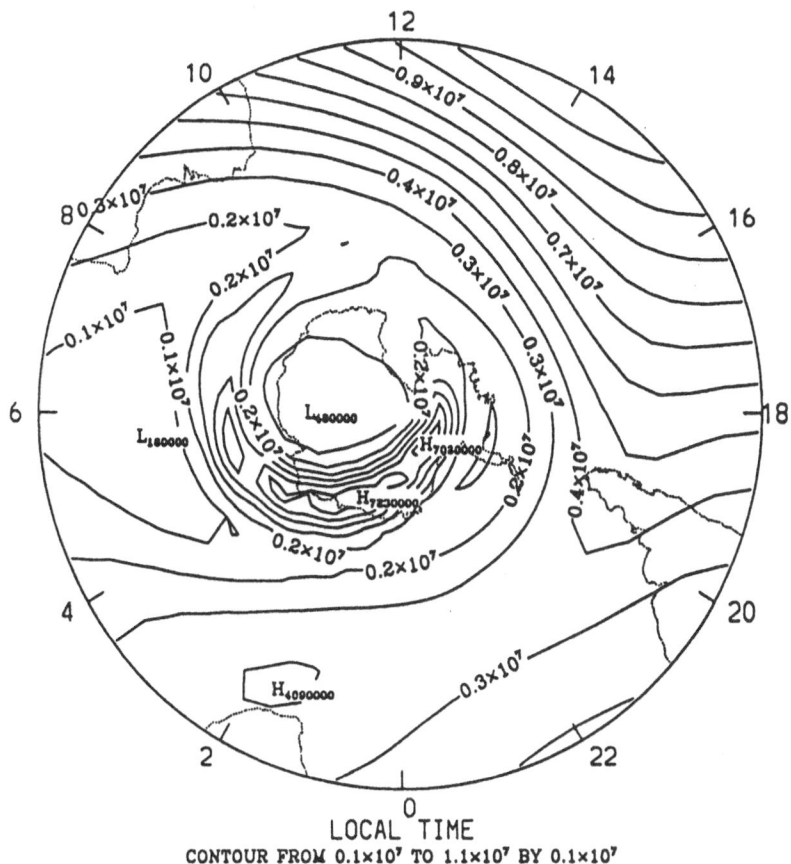

CONTOUR FROM 0.1×10⁷ TO 1.1×10⁷ BY 0.1×10⁷

Figure 8.4. High southern latitude distribution of N(^{4}S) near 120 km plotted in polar coordinates centered on the geographic pole and calculated for the same conditions as *Fig. 8.3* (R.G. Roble, 1991, private communication).

N(^{2}D) production rates higher than those predicted by laboratory measurements or *a priori* molecular calculations, or by adopting quenching rates of N(^{2}D) lower than it was measured. For example, the N(^{2}D) branching ratio employed for reaction R1 has frequently been used as a fitting parameter, with values ranging from 0.5 (Oran *et al.*, 1975) to 1 (Frederick and Rusch, 1977). The most widely used value in recent works, however, is 0.6 (McCoy, 1983; Cleary, 1986; Gérard *et al.*, 1990; Siskind *et al.*, 1989, 1995) which gives the best fit to the NO profiles when adopting the N(^{2}D) quenching coefficient by O on the order of $4 \div 7 \times 10^{-13}$ cm^3 s^{-1}, in agreement with the aeronomical and more recent laboratory determinations. Another quantity used as a fitting parameter is the solar EUV flux and its dependence on the solar activity level. Barth et al. (1988) closely related to the short wavelengths (far UV-, soft X-rays) of the solar emission spectrum. Siskind et al. (1989, 1990) used an arbitrary scaling factor to impose the large variability of this flux, what is necessary in their model to account for the rocket observations of NO vertical distribution at solar maximum.

This situation prompted a search for additional sources of nitric oxide. One such possibility is the role played by nonthermal hot N atoms which may substantially increase the efficiency of the reaction of $N(^4S)$ with O_2. The sources of translationally hot N and its effects on the NO production and density will be discussed in the next section.

8.2. Nonthermal Photochemistry of Odd Nitrogen in the Lower Thermosphere

We have seen that in the thermosphere nitrogen molecules can be splitted into two nitrogen atoms by photodissociation, electron impact and ion chemical processes, producing photochemically active species: N and NO (odd nitrogen). Nitrogen atoms are produced in various electronically excited states (principally ^2D and ^2P) as well as in the ^4S ground state and are characterized by a thermal energy excess. These nonthermal (non-maxwellian) N atoms may play a significant role in the odd nitrogen photochemistry. Obviously, their energy distribution function serves as the milestone for calculations of the $N(^4S)$ atoms abundance in the Earth's thermosphere and their dependence on altitude and solar activity. The results of this modeling [Shematovich et al. (1991b, 1992) and Gérard et al. (1991, 1993, 1995, 1996)] are based on the kinetic approach described in detail in Chapters 6 and 7.

8.2.1. SOURCES OF SUPERTHERMAL $N(^4S)$ ATOMS

The photodissociation of N_2 by solar photons

$$N_2 + h\nu \rightarrow N(^4S) + N(^4S,^2D)$$

in the wavelength between 800 and 1000 Å yields translationally excited (hot) $N(^4S)$ atoms with kinetic energies in the range $0.1 \div 3$ eV. The branching ratio between $N(^4S)$ and $N(^2D)$ in the reaction R2 is usually assumed to be 0.5 : 0.5 (Zipf and McLaughlin, 1978).

Another source of hot $N(^4S)$ is electron impact dissociation:

$$N_2 + e^* \rightarrow N(^4S) + N(^4S,^2D,^2P) + e$$

Photoelectron impact on molecular nitrogen is accompanied by the excitation of predissociation levels $^1\Pi_u$ and $^1\Sigma_g^+$ of N_2 with energies above the dissociation threshold (Zipf and McLaughlin, 1978). These excited levels preferably dissociate and form $N(^4S)$ atoms with energies $\sim 0.1 \div 1.4$ eV.

The hot $N(^4S)$ photochemistry in the lower thermosphere is driven by the solar EUV flux in the different wavelength intervals (soft X rays below the 50 Å, XUV below 250 Å, and EUV below 1100 Å). Fast photoelectrons (with energies less than 1000 eV) formed by photoionization of N_2 also play a significant role in the ionization and dissociation of N_2. A soft X ray flux extended down to 11 Å was also used, based on the experimental data by Donnally and Pope (1973). This extension leads to additional sources of hot $N(^4S)$ atoms and high energy electrons produced by N_2 dissociative ionization and Auger ionization processes:

$$N_2 + h\nu \;\; (\text{soft X ray}) \rightarrow \begin{cases} N_2^+ + e_\nu \\ (N_2^+)^* + e_\nu \rightarrow N_2^{++} + e_{\text{Auger}} + e_\nu \\ N^+ + N(^4S) + e_\nu \end{cases} \quad (8.2)$$

$$N_2 + e_\nu \;\; (E < 1000 \text{ eV}) \rightarrow \begin{cases} N_2^+ + e_\nu' + e \\ (N_2^+)^* + e_\nu' + e \rightarrow N_2^{++} + e_\nu' + \\ \qquad\qquad\qquad\qquad + e + e_{\text{Auger}} \\ N^+ + N(^4S,\,^2D,\,^2P) + e_\nu' + e \end{cases} \quad (8.3)$$

Our model also includes the Auger ionization of atomic and molecular oxygen. The consideration of K-shell ionization (Moddeman et al., 1971) results in the formation of high energy Auger electrons in addition to the secondary photoelectron. Consequently, the electron impact dissociation and ionization rates of N_2 increase in the lower thermosphere, leading to additional production of odd nitrogen.

Cross-sections for N_2, O_2 and O for solar flux down to 18 Å are partially taken from Conway (1988) and they are very close to those published by Richards et al. (1994). For the 11–17 Å interval, values from Siskind et al. (1995) were adopted. The rate coefficient for $N(^2D)$ quenching by atomic oxygen is 6.7×10^{-13} cm^{-3} s^{-1}, the most recent laboratory determination of this important quantity (Fell et al., 1990). Molecular and eddy vertical transport are considered for NO and $N(^4S)$ as described by Gérard and Taieb (1986) and photochemical equilibrium is assumed for $N(^2D)$, ions and electrons.

In the model four main exothermic chemical sources of $N_f(^4S)$ were considered :

$$\begin{array}{llll} NO^+ + e & \rightarrow & N(^4S) + O & +2.75\,eV \\ O^+ + N_2 & \rightarrow & NO^+ + N(^4S) & +1.09\,eV \\ N^+ + O_2 & \rightarrow & O_2^+ + N(^4S) & +1.72\,eV \\ N(^2D) + O & \rightarrow & O + N(^4S) & +2.38\,eV \end{array} \qquad (8.4)$$

All these sources have strongly energy dependent cross-sections (see e.g. Rees, 1989). Therefore, a numerical Monte Carlo model was developed to calculate the initial energy distribution of the hot $N(^4S)$ atoms. This code provides the initial hot $N(^4S)$ distribution for given cross-sections and exothermic reaction energy releases. Theoretical studies (Whipple, 1974; Sharma *et al.*, 1996) of the kinetics of bimolecular chemical reactions indicate that the products of exothermic reactions have energy distributions different from the thermal ones. This fact leads to additional broadening of the initial energy distribution functions (EDF) of the reaction products and, consequently, can influence their kinetics. This broadening was taken into consideration in the calculation of the $N(^4S)$ EDF. Let us note that the analytical solution for the initial EDF of the reaction products is known only for a few simple cases when the reaction cross-sections are independent on energy.

Additional production of hot $N(^4S)$ is due to dissociative ionization of N_2 by XUV photons and by high energy electrons. The energy distributions of $N(^4S)$ produced by photons were derived from the experimental studies by Gardner and Samson (1975). The EDF of $N(^4S)$ produced by electron impact was adopted from Locht *et al.* (1992) and Van Zyl and Stephen (1994). The main processes leading to the production of fast $N(^4S)$ atoms and odd nitrogen are sketched in *Fig. 8.5*.

The influence of hot $N(^4S)$ on the odd nitrogen photochemistry is strongly dependent on the rates of two processes: i) the reaction of hot nitrogen with O_2 to produce NO, and ii) the cooling of the hot particles in relaxation (elastic and inelastic) collisions with the ambient gases (N_2, O_2, O). In this model we used the energy dependent cross-section for reaction of hot $N(^4S)$ with O_2 after Polak *et al.* (1984), which is consistent with the recent results of Duff *et al.* (1994). The situation concerning elastic cross-sections is more complex. As a first approximation, gasdynamical estimates of cross-sections deduced from the consideration of diffusive processes (Morgan and Schiff, 1964) could be used. However, the possibility that elastic cross-sections are smaller for high energy relaxation collisions was taken into account by decreasing its value by an arbitrary factor. Following this approach, we have included recent results about the relaxation cross-section of $N(^4S)$

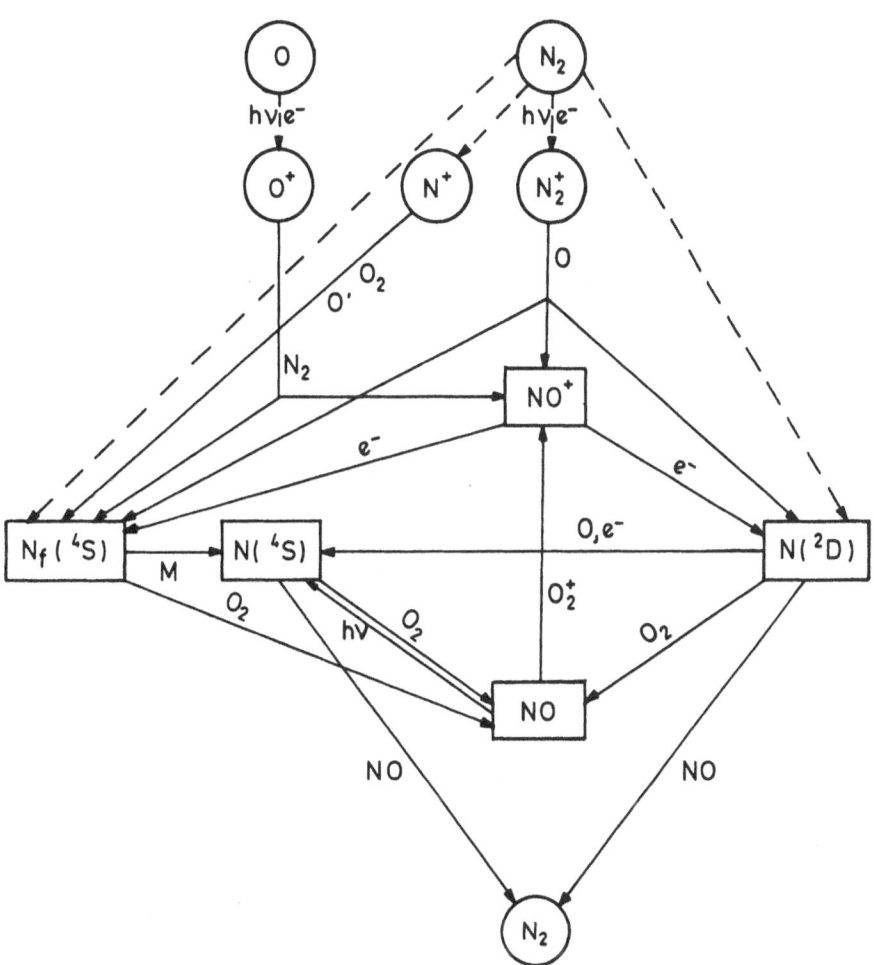

Figure 8.5. Schematic diagram of photochemical processes leading to the production of nonthermal $N(^4S)$ atoms and controling nitric oxide in the thermosphere. Dotted lines indicate ionization and dissociation of N_2 by photons and photoelectrons. M denotes elastic collisions.

in N_2 and applied these new energy dependent values to the calculation of $N(^4S)$ energy distribution.

The NO peak density critically depends on the relative production rate of ground state $N(^4S)$ atoms (which are mostly responsible for a sink of NO) and $N(^2D)$ metastable atoms (which are a source of NO). On the basis of physical properties of the N_2 molecule, Zipf and McLaughlin (1978) concluded that the photodissociation of N_2 yields $N(^4S)$ and $N(^2D)$ atoms with the same efficiency. The electron impact dissociation of N_2 slightly favors the production of $N(^2D)$, as do the NO^+ dissociative recombination and the $N_2^+ + O$ charge transfer (reactions R3 and R4 in Table 8.1).

To investigate the role of hot $N_f(^4S)$ in odd nitrogen photochemistry, the microscopic energy distribution function F_f of these particles must be first of all determined. Since we are mostly focused on the lower thermosphere, where free path lengths of active particles are much less than atmospheric density scale height, physical and chemical processes involving active particles can be considered in a local approach, i.e. without transport effects.

An important aspect of this calculation concerns the thermalization processes of $N_f(^4S)$. The main sinks of $N_f(^4S)$ are elastic and inelastic collisions with the ambient atmospheric constituents. The kinetics of $N_f(^4S)$ atoms is defined by the balance between the main sources and sinks and may be correctly described by the Boltzmann kinetic equation (Shematovich *et al.*, 1991b).

The calculations of XEUV absorption by the main atmospheric constituents were made using a compilation of recent laboratory measurements of photoabsorption and photoionisation cross-sections of O, O_2 and N_2 (Conway, 1988). The calculation of electron impact on the main constituents was made using the analytical approximations of excitation, dissociation and ionization cross-sections given by Jackman *et al.* (1977) and Peterson *et al.* (1973). For the electron impact dissociation of N_2, the partial and total cross-sections of Zipf and McLaughlin (1978) were adopted.

The collisional relaxation rate of nonthermal atoms was determined by the elastic cross-sections with major gases. Estimates of these cross-sections for dissociated gases with a large kinetic energy are complicated (Mason and Marrero, 1970). Basically, the cross-sections for relaxation of the translationally hot atoms by collisions with the ambient atmosphere are the most uncertain quantity in the evaluation of contribution of fast $N(^4S)$ atoms to the NO formation. It is reasonable, in a first approximation, to assume that the value $\sigma_{el} = 3.5 \times 10^{-15}$ cm^2 (Morgan and Shiff, 1964), derived from molecular diffusion coefficients, represents an upper limit to the actual cross-section for hot N atoms. Theoretical treatments of hot particle collisions (Polak *et al.*, 1984) indicate that elastic collision cross-sections are smaller than the respective values at room temperature. For example, Schmitt *et al.* (1981) obtained a value of 1.5×10^{-15} cm^2 for the thermalization of fast $O(^1D)$ atoms in the thermosphere, in agreement with the calculations by Yee and Dalgarno (1987). In our model, we adopted $\sigma_{el} = 1 \times 10^{-15}$ cm^2 as an approximation to an energy-independent value.

Although there are several theoretical examples of the relaxation of fast atoms in atomic collisions (Yee and Dalgarno, 1985; Tharamel *et al.*,

TABLE 8.2. Relaxation cross-sections by N_2 for fast $N(^4S)$ atoms.

E_T	σ_{in}	σ_{el}	$< \Delta f_T >$	N_{col}	σ_{eff}
(eV)	(10^{-16} cm^2)	(10^{-16} cm^2)			(10^{-16} cm^2)
0.075	14.6	203.9	0.059	2	109.3
0.125	19.2	178.1	0.058	11	17.9
0.25	23.5	148.3	0.056	23	7.4
0.50	25.2	124.3	0.054	35	4.3
0.75	25.8	112.1	0.052	43	3.2
1.0	25.7	104.5	0.051	48	2.7
1.5	25.3	94.7	0.048	56	2.1
2.0	24.7	88.6	0.046	62	1.8
2.5	24.5	83.9	0.044	67	1.6
3.0	24.1	80.4	0.042	70	1.5

1995), little is known about the relaxation of fast atoms in molecular collisions (i.e., where internal degrees of freedom of the colliding partner may be important). Previous estimates of relaxation cross-sections for $N+N_2$ based on diffusion coefficients have ranged from 10 Å^2 to 35 Å^2. In order to assess the effect of internal degrees of freedom and provide a better description of the relaxation of fast $N(^4S)$ atoms by molecular N_2, the quasiclassical trajectory (QCT) method (Truhlar and Muckerman, 1979) was used to compute the translational energy dependence of the energy transfer cross-sections and the final N_2 vibrational/rotational state distributions. The interaction energy for $N+N_2$ collisions was represented by LEPS potential energy surface with parameters chosen in such a way that a barrier to reaction is approximately correct (Laganà et al., 1987). Although an accuracy of the potential energy surface for the energy transfer calculations is currently unknown, the results obtained by J. Duff provide a semiquantitative description of the relaxation process (Gérard et al., 1996). Standard Monte Carlo techniques were used to compute the energy transfer cross-sections as a function of the initial translational energy E_T and the initial N_2 vibrational-rotational states which are selected from a 500 K Boltzmann distribution.

The results of relaxation calculations are summarized in Table 8.2 for the range of initial translational energies (in the laboratory coordinate system) from 0.075 to 3 eV. Elastic scattering is ineffective in the process of $N(^4S)$ relaxation, making an important contribution only for $E_{T'} \neq E_T$ (large impact parameter processes). The efficiency translation energy transfer is illustrated by considering the average fraction of translational energy transferred per collision Δf_T, which is approximately 5 % as shown in Table 8.2. The QCT result is significantly smaller than the value 44 % predicted by a hard sphere model assuming isotropic scattering (Libby, 1947), which was used by Logan and McElroy (1977) in the study of energetic O atoms and Solomon (1983) in an estimate of significance of fast N atoms in the odd nitrogen chemistry. Although the final translational energy distribution function could be used directly in the kinetic model, a simpler approach involves an estimate of the effective relaxation cross-section based on the total cross-section and average translational energy transferred per collision.

The effective relaxation cross-section can be evaluated provided an average number of collisions necessary to relax atoms possessing initial energy E_T to the average final energy corresponding to translational temperature of 500 K, is known. This estimate can be done using the average fraction of translational energy transferred during $N+N_2$ collision listed in Table 8.2, as a function of energy. Since the transfer of translational energy is rather inefficient (approximately 5 % of the total value as shown in Table 8.2), as many as 70 collisions are necessary to relax $N(^4S)$ atom having initial energy 3 eV to final energy of ~ 0.06 eV, the collisional efficiency being similar to vibrational relaxation. The thermalization cross-section is then estimated as the total cross-section per collision. The energy dependence of these estimated cross-sections, which are by the factor of 5-25 smaller than the gas kinetic value the 3.5×10^{-15} cm^2 and are comparable to the $N(^4S)+O_2$ reaction cross-section (Duff et al., 1994), are given in Table 8.2. The influence of significantly smaller relaxation cross-sections on the NO formation is discussed below.

For reaction R8 the energy-dependent cross-section $\sigma(E)$ for energies $E > 0.85$ eV obtained by Polak et al. (1984) was used, whereas for the energy interval from the reaction threshold to 0.85 eV the best estimate for the reactive cross-section was derived from the relation between the thermal reaction rate coefficient $k(T)$ and $\sigma(E)$ (averaged over a Maxwellian distribution):

$$k(T) = (\pi\mu)^{-0.5}(2kT)^{1.5} \int_0^\infty \sigma(E)E \exp(-E/kT)\, dE \quad ,$$

where μ is the reduced mass of the N - O$_2$ system. This means that the value of $\sigma(E)$ used for low values of E is in good agreement with the thermal reaction rate coefficient.

8.2.2. EFFECT ON NITRIC OXIDE PRODUCTION AND CONCENTRATION

These calculations were made using the following procedure:

a) The processes of XUV and soft X ray absorption and the subsequent kinetics of fast photoelectrons were calculated in detail using the kinetic model described in Chapters 6 and 7. This model includes excitation, dissociation, and direct, dissociative, and Auger ionization processes induced by the photon and electron fluxes. For the photoelectron flux calculations, an energy range stretching from thermal energies up to 1000 eV was used and the energy deposition of the cascade electrons was taken into account;

b) The results obtained at the previous step, including the rates of XUV and soft X-ray photon and fast photoelectron impact processes when affecting thermospheric gases, were used in the one-dimensional chemical - diffusive odd nitrogen model (Gérard and Taieb, 1986; Gérard et $al.$, 1991). Vertical steady state profiles of the odd nitrogen species were calculated in this approach. They are referred to hereafter as standard case (no hot $N(^4S)$ contribution);

c) The production rates and initial kinetic energy distribution functions of hot $N(^4S)$ were calculated in steps a) and b) by the following nonstandard codes: i) at the step (a) for N_2 dissociation and dissociative ionization by photons and photoelectrons; ii) at the step (b) for exothermic chemical reactions (8.4). These values were applied to the hot $N(^4S)$ kinetics code (also nonstandard) which calculates the relaxation of hot atoms by elastic collisions with the ambient atmospheric gases and by reactive collisions with O_2. This kinetic model provides the steady state energy distribution functions of $N(^4S)$ at different altitudes and the height dependent fraction of hot $N(^4S)$ reacting with O_2 (reaction efficiency);

d) Using the reaction efficiency and the total production rate of hot $N(^4S)$, the odd nitrogen species profiles were calculated with the use of another nonstandard photochemical code. These profiles show the influence of hot $N(^4S)$ on the NO concentration in the lower thermosphere. These profiles are referred to hereafter as hot N cases.

The hot nitrogen production rates calculated using the described procedure are presented in $Fig.$ 8.6. It is seen, that the quenching of $N(^2D)$ by atomic oxygen is the largest source of $N(^4S)$ for altitudes above 130 km. At lower altitudes, the main sources of hot nitrogen are the dissociation reactions R1 and R2 (Table 8.1). Low-latitude profiles of nitric oxide observed with rockets and satellites consistently show a peak occurring below 120 km (Barth et $al.$, 1988). This means that the odd nitrogen chemistry makes an important contribution to the lower thermosphere energetics at these altitudes, and consequently, an in-depth study of all possible sources of odd nitrogen in this region is very important.

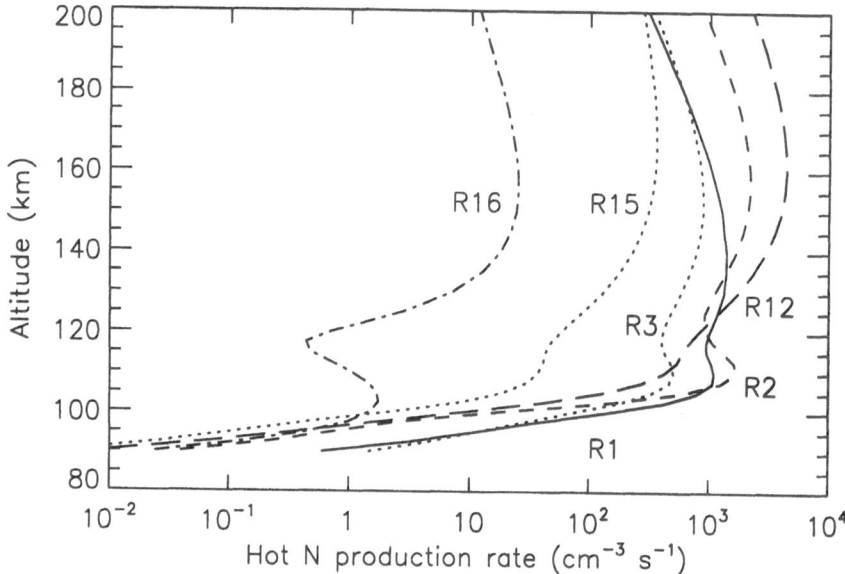

Figure 8.6. Hot nitrogen atom production rates by direct N_2 photon- and electron impact dissociation, and by exothermic chemical reactions for low solar activity conditions.

Basically, the kinetic calculations (Shematovich *et al*, 1991b, 1992, Gérard *et al.*, 1995, 1996) confirm that soft X-rays are the main source of hot nitrogen atoms below 110 km. Indeed, since the cross-sections of N_2, O_2 and O are smaller at the shortest wavelengths (Conway, 1988), soft X-rays penetrate to the lower thermosphere where they produce a large increase in the N_2 ionization rate, i.e. an increase in the high energy photoelectron flux, and consequently, the main source of $N_f(^4S)$ at these altitudes is the photoelectron impact dissociation of N_2. As a result of these calculations, the $N_f(^4S)$ production spectrum was obtained for all altitudes.

The steady-state translational energy distribution for $N_f(^4S)$ atoms after relaxation by elastic and inelastic collisions was calculated for a given solar spectrum and neutral atmosphere as a function of altitude. *Fig. 8.7* shows examples of this energy spectrum at altitudes 110 km and 140 km. Calculations show that this distribution function for hot nitrogen atoms remains nonequilibrium. The production rates of hot atoms and elastic collision frequencies have a complex altitude dependence, and the steady-state population of nonthermal atoms increases at high altitudes.

For the aeronomical purposes it is very useful to calculate the fraction f of hot $N_f(^4S)$ atoms reacting chemically with O_2 (reaction R8) rather than atoms thermalized by elastic collisions. This value is taken from the

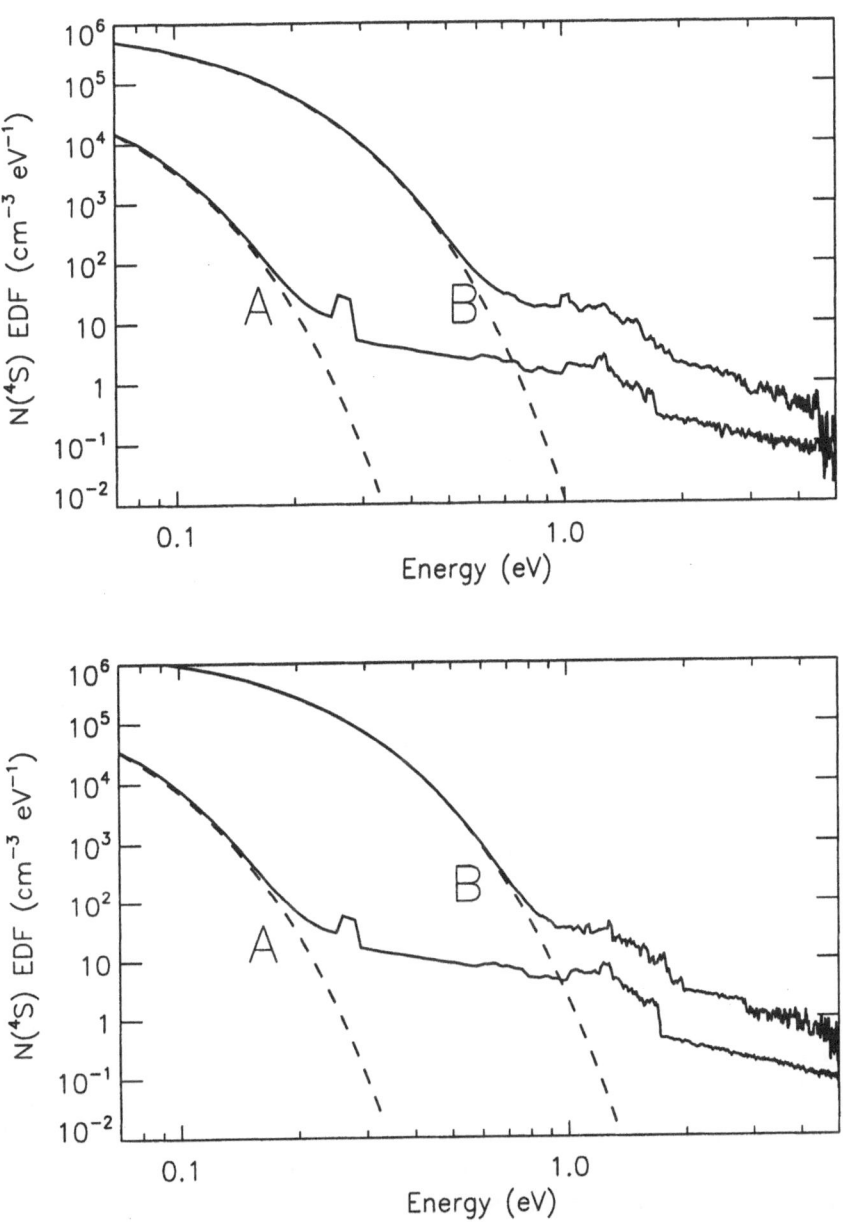

Figure 8.7. Steady state energy distribution function of $N(^4S)$ at 110 km (A) and 140 km (B). Solid lines show the calculated EDF and dashed curves are the local Maxwellian EDFs for gas temperatures of 209 K (curve A) and 575 K (curve B). (a) : low solar activity case (($F_{10.7} = 71$), (b) : high solar activity case ($F_{10.7} = 245$)

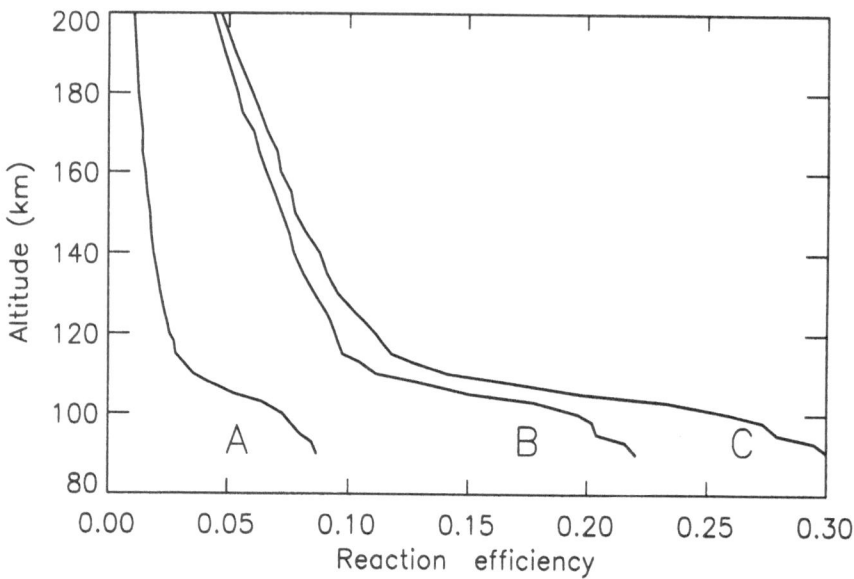

Figure 8.8. Altitude dependence of the $N_f(^4S) + O_2$ reaction efficiency f for high solar activity and different models of elastic cross-sections. A: hard sphere approximation ($\sigma_{el} \simeq 10^{-15}$ cm^2); B: hard sphere approximation for $N_f(^4S) + O_2$, O collisions and energy dependent cross-section for $N_f(^4S) + N_2$. C: energy dependent cross-section for collision of $N_f(^4S)$ with O_2, O and N_2.

lated steady-state velocity distributions of N atoms. For the quiet solar and geomagnetic conditions quoted above and for equatorial, the altitude dependence of this fraction is presented in *Fig. 8.8*. It is evident that f is very sensitive to the value of elastic cross-section for high energy collisions, that is unfortunately only poorly known. As it was discussed earlier, we adopted the value $\sigma_{el} = 1 \times 10^{-15}$ cm^2 as a first approximation (case A of *Fig. 8.8*). The results of calculations of f with the previously described energy dependent relaxation cross-section are also presented. This latter approach is applied to the $N(^4S)$-N_2 collisions only in the case B. The same energy dependence is assumed for collisions with O and O_2 in the case C. The rate coefficient of reaction R8 was calculated in a self-consistent manner, and all relaxation collisions of secondary and higher order were included in this kinetic approach. Clearly enough, a correct evaluation of the reactive fraction f requires a detailed calculation of the steady state $N_f(^4S)$ energy distribution. Numerical simulations of the hot $N_f(^4S)$ kinetics show that these atoms are present in significant amounts in the Earth's thermosphere and play an important role in the odd nitrogen chemistry and energetics of this atmospheric region.

All calculations discussed in this section were made for the equatorial lower thermosphere under quiet geomagnetic conditions and for both low ($F_{10.7} = 71$) and high ($F_{10.7} = 243$) solar activity. The results can be briefly summarized as follows. The vertical distributions of the major constituents and temperature were taken from the $MSIS - 86$ model (Hedin, 1987). A one-dimensional model of thermospheric odd nitrogen similar to those described in the previous section (Gérard and Taieb, 1986; Gérard et al., 1990, 1991) was used to investigate the role of fast N atoms on the steady-state odd nitrogen distribution at low latitudes. Unlike the standard photochemistry, reaction R8 is added as an additional source of nitric oxide. The fast N atom production rate and reactive fraction f were calculated. The role of the effective branching ratio for the production of $N(^2D)$ atoms by photoelectron dissociation (R1) was investigated in detail (Shematovich et al., 1992) and the experimentally determined ratio 0.54:0.46 was adopted. The $N(^4S)$ density is not directly affected by consideration of the hot N chemistry but, since the $N(^4S)$ atoms are quickly destroyed by the NO molecules in the lower thermosphere (reaction R9), their density decreases when the NO concentration increases. The $N(^2D)$ concentration increases as a result of the larger $N(^2D)$ quantum yield, but it is unaffected by the hot $N(^4S)$ atom photochemistry. In contrast, the NO peak density increases when reaction (R8) is taken into account, but the shape and altitude of its maximum are also modified by the effect of fast atoms.

This last point can be best understood by considering Fig. 8.9 where the altitude dependence of the three sources of NO molecules is shown. As expected for solar minimum conditions, if the exospheric temperature remains low (730 K in this example), the $N(^2D) + O_2$ reaction dominates at all altitudes. This is a well known consequence of the large temperature dependence of the coefficient of reaction R7, which is very small at temperatures less than 1000 K in comparison with R6. Consequently, below 120 km, R8 is the second most important source of NO. In particular, soft X-rays produce a significant population of energetic photoelectrons which generate hot N atoms near and below 110 km. It is therefore quite obvious that the additional NO production due to the reaction of $N_f(^4S)$ with O_2 plays the most important role in the lower thermosphere.

The NO loss rate distribution is shown in Fig. 8.10. Obviously, the mutual destruction of $N(^4S)$ and NO is the major sink of NO at all altitudes. Below 160 km, the charge transfer reaction with O_2^+ is also important, although it is not responsible for a net loss of odd nitrogen but for a redistribution among the family members. At higher altitudes, the $N(^2D) +$ NO reaction is also an important mechanism of NO sink. In turn, predissociation of NO in the δ-bands and photoionization are minor loss processes.

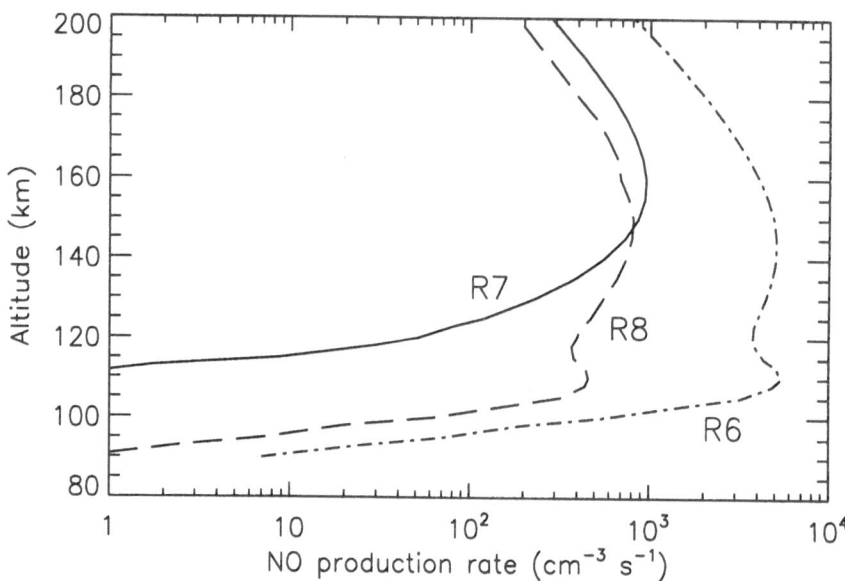

Figure 8.9. NO production rates in the lower thermosphere for solar minimum activity.

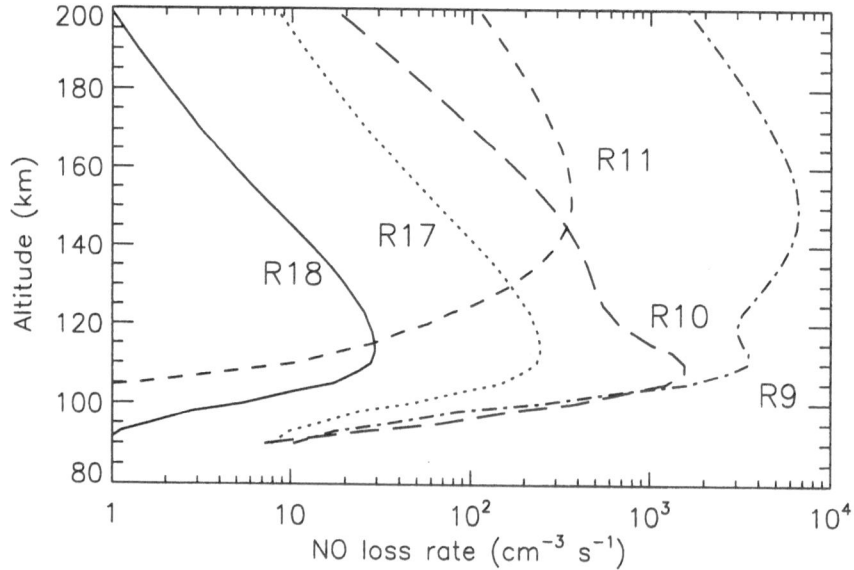

Figure 8.10. NO loss rates in the lower thermosphere for solar minimum activity.

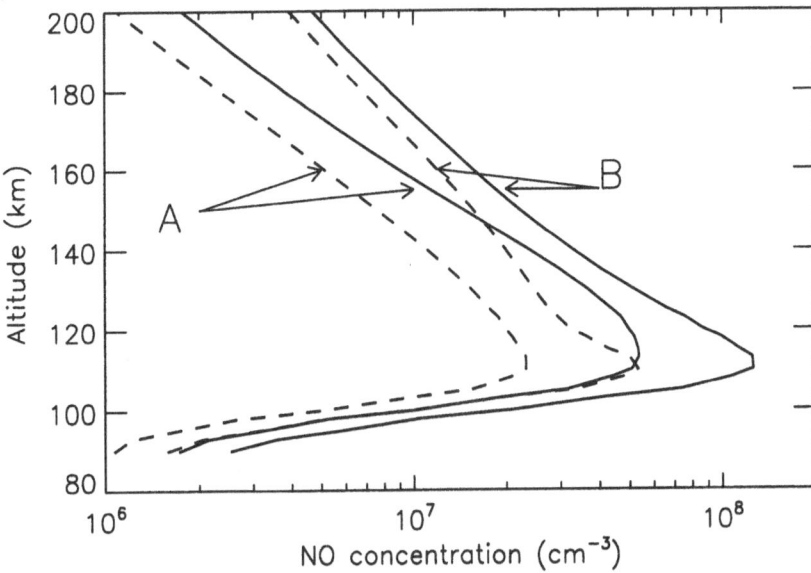

Figure 8.11. Nitric oxide density profiles for low (curves A) and high (curves B) solar activity levels. Dashed lines show the values for the standard NO photochemistry (no fast N), solid lines are for the hot N cases

The steady state nitric oxide density distribution exhibits a peak near 110 km, in agreement with rocket and satellite observations (*Fig. 8.11*). The importance of the hot $N(^4S)$ contribution is shown to compare the two cases. In the first one, the full model (including hot N) is used, and the NO peak reaches 1.3×10^8 cm^{-3} (Table 8.3) for the high solar activity conditions ($F_{10.7} = 243$) adopted in this model run. In the second case, the model was developed under identical conditions except for the effect of reaction R8 that was neglected. In this latter case, the NO maximum drops to 5.3×10^7 cm^{-3} (Table 8.3) as a consequence of the removal of the hot N source of NO.

The presence of hot $N(^4S)$ also results in a significant increase of the NO number densities. The hot N cases shown in *Fig. 8.11* are calculated for the case when the elastic cross-sections of the hot $N(^4S)$ collisions with N_2, O_2 and O include the energy dependence from Table 8.2. This run corresponds to larger values of the reaction efficiency, i.e. to curve C in *Fig. 8.8*. Add-on of hot $N(^4S)$ in the model leads to significant additional increase of the NO densities. For both low and high solar activities, this effect gives a more than twofold increase in comparison with the standard case (Table 8.3).

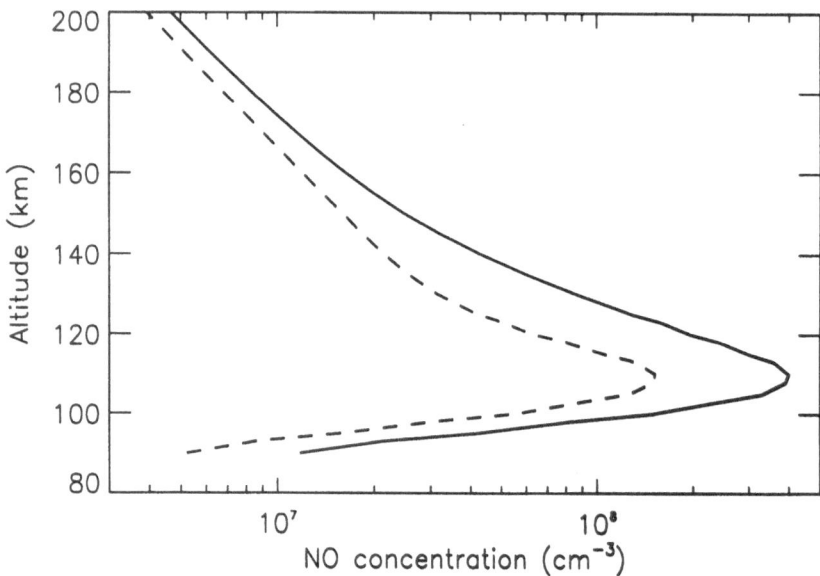

Figure 8.12. Nitric oxide density profiles for high solar activity conditions using the scaled version of the soft X-ray solar flux. The dashed line is for the standard NO photochemistry, the solid line is for the hot N case

A few additional runs have been conducted to test the sensitivity of the model to different uncertain parameters. To assess the influence of the elastic cross-sections, the NO height profiles for high solar activity were calculated for three different cross-sections approaches (see *Fig. 8.8*). We found that the NO peak density increases by 29 % in the standard simplest case of the hard sphere approximation. For the intermediate case the increase is 92 %, and it is 136 % for the last case of the energy dependent relaxation cross-section. This last case is clearly most justified as the idea of an elastic cross-section energy dependence for N-N_2 collisions is based on the well supported grounds. The consideration of the energy dependence of the N-O_2 relaxation cross-sections, which is presently not available, would additionaly improve the model, though it is clear that it is unlikely to significantly modify the calculated values of NO increase, remaining between 92 % and 136 %.

Also of interest is the sensitivity of the model to the XUV and soft X ray solar fluxes (*Fig. 8.12*). The previous runs used the values of these fluxes deduced from the observations. There are indications, however, that the soft X ray flux may exhibit larger variations with solar activity than Tobiska's values (Siskind *et al.*, 1995). Hence, an additional run was undertaken using

TABLE 8.3. Calculated nitric oxide peak density (cm^{-3}).

Solar activity	Standard	Hot N
SOLMIN	2.3×10^7	5.4×10^7
SOLMAX	5.3×10^7	1.3×10^8

the same scaling factor of the soft X ray flux as Siskind *et al.* (1995) to fit the calculated NO densities to the high values of nitric oxide obtained in microwave measurements by Clancy *et al.* (1992). For the interval 18–50 Å, the baseline values of *SC21REFW* were increased by the scaling factor of 100, and for the interval 11–17 Å the measured value was increased by the factor 75. The NO density profiles for the standard case and for case with hot N(^{4}S) contribution are shown in *Fig. 8.12*. As it can be seen, for the standard case, NO peak density is about 1.5×10^8 cm^{-3}, and for the hot N case it is about 4.0×10^8 cm^{-3}.

In conclusion of this section, the results of our numerical simulations show that the dissociation of N_2 by EUV solar photons and photoelectrons, as well as other exothermic chemical processes, yield atomic nitrogen species with excess kinetic energy. At steady-state, a non-Maxwellian population of ground state N atoms occurs in the thermosphere. Some of these hot atoms have sufficient kinetic energy to exceed the activation energy of the N(^{4}S) + O_2 → NO + O reaction and produce additional nitric oxide in comparison with the standard "thermal" photochemistry. The shape of NO vertical distribution in the ionospheric E region is significantly modified by the contribution of the fast N atoms. The introduction of a superthermal particle component into the chemical model allow to reconcile the laboratory and theoretical studies of the N(^{2}D) yield with the observed NO distribution in the lower thermosphere. Therefore, this model is greatly advantageous compared to the classical models based on the chemical reaction kinetics involving only thermal nitrogen atoms, which yield NO densities lower than the major part of the observational data. Unlike arbitrarily modifying the values of reactions coefficients or *ab-initio* calculated efficiencies or branching ratios, the involvement of nonthermal nitrogen atoms and hot atom in photochemical models removes the constrains placed on the efficiency of the thermal NO photochemistry. In addition, there are indications that the rovibrational distribution of intensity in the infrared emission spectrum of nitric oxide bears signature of the presence of nonthermal N atoms in the thermosphere as well. This problem of relevance of the anomalies in the intensity distribution to the presence of hot N atoms has been thoroughly debated for years and will be studied in detail in the next section.

8.3. Observational Signature of Nonthermal Nitrogen in the Earth's Atmosphere

We learnt that because the thermospheric gas density is low enough, the $N(^4S)$ kinetic energy distribution function becomes non Maxwellian above ~ 0.3 eV. However, no direct measurements of EDF of the hot N atoms are possible and thus the ideas of EDF's potential effects on the N-NO photochemistry remain rather speculative. Nevertheless, observations of high resolution spectra in the NO infrared bands offer an opportunity to identify a signature of hot N atoms.

8.3.1. NO INFRARED BAND FEATURE

Observations of thermospheric infrared airglow from 130 to 190 km were obtained for the NO fundamental band $\Delta v = 1$ at 5.3 μm by the *CIRRIS-1A* interferometer on board the *Space Shuttle*. High rotational level transitions exceeding $J = 50$ were identified such as (1-0) and (2-1) bandheads (Smith and Ahmadjian, 1993; Armstrong *et al.*, 1994). These data cannot be explained by classical sources such as

$$\begin{array}{ll} NO(v = 0) + O \rightarrow NO(v = 1) + O & (a) \\ N(^2D) + O_2 \rightarrow NO(v, J) + O & (b) \end{array} \quad . \tag{8.5}$$

Indeed, process (8.5a) does not provide rotationally excited $NO(v, J)$ highly enough while $N(^2D)$ number density is too low for reaction (8.5b) to contribute significantly to the process. A new candidate for NO^* production was suggested to explain the presence of bandheads (Sharma *et al.*, 1993):

$$N_f(^4S) + O_2 \rightarrow NO(v, J) + O \quad . \tag{8.6}$$

Since this mechanism has an activation energy ~ 0.3 eV and populates high rotational levels very efficiently at higher energies (Duff *et al.*, 1994), the presence of fast ground state nitrogen atoms $N_f(^4S)$ can increase the production rate of the rotationally excited NO to explain the bandheads. It is also possible that sufficiently hot thermal N atoms carry energies above the 0.3 eV threshold to quickly react with O_2 and produce $NO(v, J)$. These N atoms would belong to high energy tail of the Maxwellian distribution whose extent critically depends on the local temperature and thus on the altitude in the thermosphere.

Both possibilities were investigated by Hubert *et al.* (1996). The detailed production of $NO(v, J)$ by reaction (8.6) involving both hot Maxwellian and superthermal $N(^4S)$ atoms has been evaluated and it was attempted to proceed to the resulting IR NO spectrum. In particular, synthetic NO spectra have been generated to determine an altitude range where superthermal atoms can produce the observed bandheads.

A model was developed to calculate the $NO(v, J, W)$ production rates for $v = 0$ to 13; $J = 0.5$ to 149.5; and $\Omega = 1/2, 3/2$ by the process (8.6). All different levels were considered separately and their energy $E(v, J, \Omega)$ was taken from Amiot (1982). We used reaction constants $k(E, v, J)$ for the process (8.6) calculated by Duff $et\ al.$ (1994) where E is kinetic energy of the reacting species in the center of mass frame of reference (ECM), v and J being respectively the vibrational and rotational quantum numbers of the $NO(v, J, \Omega)$ product. Duff's calculations provide, for every energy and v level, a reaction constant and a rotational temperature from which $k(v, J)$ was calculated. Dependence of the production rate versus Ω was set to a fraction $f_\Omega = 2/3$ for $\Omega = 1/2$ and $f_\Omega = 1/3$ for $\Omega = 3/2$, according to Lipson $et\ al.$ (1994). The total production rate $P(v,J,\Omega)$ of $NO(v, J, \Omega)$ by reaction (8.6) was obtained by collecting the contributions of collisions at all energies, i.e.

$$P(v, J, \Omega) = \int_0^\infty dE f_{CM}(E)k(E, v, J)f_\Omega[N(^4S)][O_2] \ , \qquad (8.7)$$

where $f_{CM}(E)$ is the distribution function of the E_{CM}, the latter being the probability of a collision between $N(^4S)$ and O_2 which have an ECM between E and $E + dE$, and $[M]$ is concentration of M. Reactions (8.5a, and b) were also included in the calculation of the total chemical production rate $P_c(v, J, \Omega)$ of excited NO. The rate coefficient of reaction (8.5b) and its dependence on v were taken from Kennealy $et\ al.$ (1978) and a Boltzmann distribution was assumed for the NO rotational levels. The synthetic spectrum for $\Delta v = 1$ and 2 was calculated at steady state and convolved with an instrumental Gaussian function of appropriate width. The kinetic energy distribution f of $N(^4S)$ described in the Section 7.2 was used to calculate the production rate given in (8.7). The kinetic energy distribution of O_2 was assumed to be Maxwellian. The O_2 and O concentrations and the local temperature were taken from the $MSIS$-90 model (Hedin, 1991), and [NO] from the rocket measurement by Siskind $et\ al.$ (1995) made at high solar activity ($F_{10.7} = 233$). Concentrations of $[N(^4S)]$ and $[N(^2D)]$ were accepted as the model values matching these above observations. Finally, the column intensity was also calculated for high solar activity conditions similar to those occurring during the $CIRRIS$-$1A$ measurements ($F_{10.7} =165$) by integrating the local emission rates along the line of sight of 128 km tangent height. This spectrum, showed in $Fig.\ 8.13$, agrees well with the spectrum presented by Armstrong $et\ al.$ (1994) in their $Fig.\ 1a$. Here QR/(1-0) de-

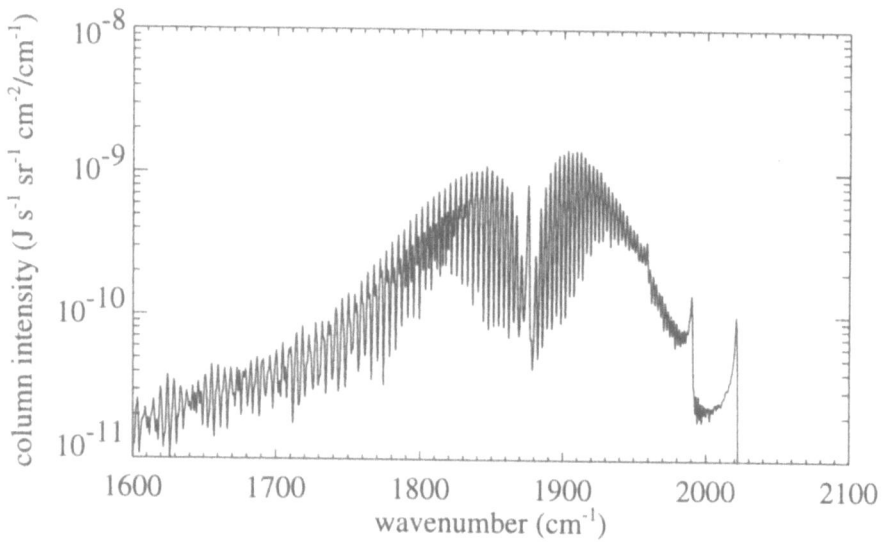

Figure 8.13. Column intensity for a tangent height of 128 km at high solar activity. The N(^{4}S) energy distributions used in this synthetic spectrum are non-Maxwellian. This spectrum must be compared with the spectrum of Armstrong *et al.*'s *Fig. 1a* (1994). The first and second bandheads are clearly identified. Absolute intensity agrees well with the observation being accounted for uncertainties of the calculation.

notes the ratio of maximum intensity of the R branch to that of the (1-0) bandhead. For the observed spectrum, QR/(1-0) $\simeq$ 44, while for the calculated spectrum QR/(1-0) $\simeq$ 38, which is an acceptable value with the account for uncertainties of the concentrations.

This result tends to show that the two features, i.e. the R branch and the (1-0) bandhead, which are produced by two different mechanisms are self consistently evaluated. This ratio is also an indicator of the relative importance of the reaction (8.6) and its role as a source of highly rotationally excited NO. Armstrong *et al.* (1994) obtained a NO(v, J) number density of 1×10^3 cm^{-3} at 140 km. We calculated an effective rate coefficient for the reaction (8.6) of 5×10^{-14} cm^3 s^{-1} at this altitude. Therefore, our estimate of the volume production rate of the reaction (8.6) is 1.5×10^4 cm^{-3} s^{-1} which corresponds to a steady state concentration of 1.2×10^3 cm^{-3}, quite close to the observational value, though an order of magnitude higher than that contributed by the reaction (8.5).

8.3.2. HOT N SPECTRAL SIGNATURE

The contribution of reaction (8.6) to the production of excited NO depends on whether the $N(^4S)$ kinetic energy distribution is Maxwellian or non-Maxwellian, because of the dependence of the reaction coefficient on the reactants ECM. Both calculations were performed at three altitudes (see *Fig. 8.14*), and the results are as follows.

At 110 km, for a gas temperature $T = 288$ K, the spectrum obtained for an optically thin atmosphere is drastically different in the cases of Maxwellian (Mx) or the non-Maxwellian (NMx) energy distributions, but no data were published which refer to this altitude. At 130 km, ($T = 616$ K), the (1-0) bandhead is much brighter in the NMx than in the Mx case. The (1-0) bandhead was observed by *CIRRIS-1A* at as low as 128 km tangent height (Armstrong *et al.*, 1994). The Mx maximum intensity of the (1-0) bandhead would match the observed ratio (within a factor 2 or 3) only if the local temperature were 800 K, a value unacceptably high at this altitude. Fast nonthermal atoms are needed to accommodate the reaction (8.6) and provide (1-0) bandhead in the local emission rate at 130 km. At 170 km, where temperature is high (1035 K), both spectra became very similar and the contribution of the fast Maxwellian atoms is sufficient to give rise to (1-0) bandhead formation. Interestingly, when the production rates of reactions (8.5a,b) are removed, the spectrum at 170 km exhibits bandheads up to (3-2) in the NMx as well as in the Mx calculations. This means that at the rather high temperature at this altitude, collisions involving thermal fast nitrogen atoms $N(^4S)$ become the main contributors for the emission excitation.

The column intensity at line of sight for the tangent height of 128 km was obtained by integrating the emission rates along this line. The contribution of the layers above 130 km is still important, because the $[N(^4S)][O_2]$ and $[NO][O]$ products slowly decrease with altitude. Due to the steep temperature gradient above 130 km, the integrated spectrum is adjusted due to contribution from the hot thermal N atoms at higher altitudes where they yield the main input to the reaction (8.6). The QR/(1-0) ratio of the integrated NMx spectrum is 38. Accounting for uncertainties of the calculations, this value is still consistent with the observed ratio of 44 (Armstrong *et al.*, 1994). The Mx calculation gives QR/(1-0)=79 for the integrated spectrum, only a factor 2 higher than the ratio in the observed spectrum. Consequently, the input from the fast N atoms generates substantial differences in the spectra emitted locally at 130 km, though the *CIRRIS-1A* data available make it difficult to find out a definite spectral signature of superthermal $N(^4S)$. Local or geometrically deconvolved observations in the lower thermosphere are needed to clarify this problem.

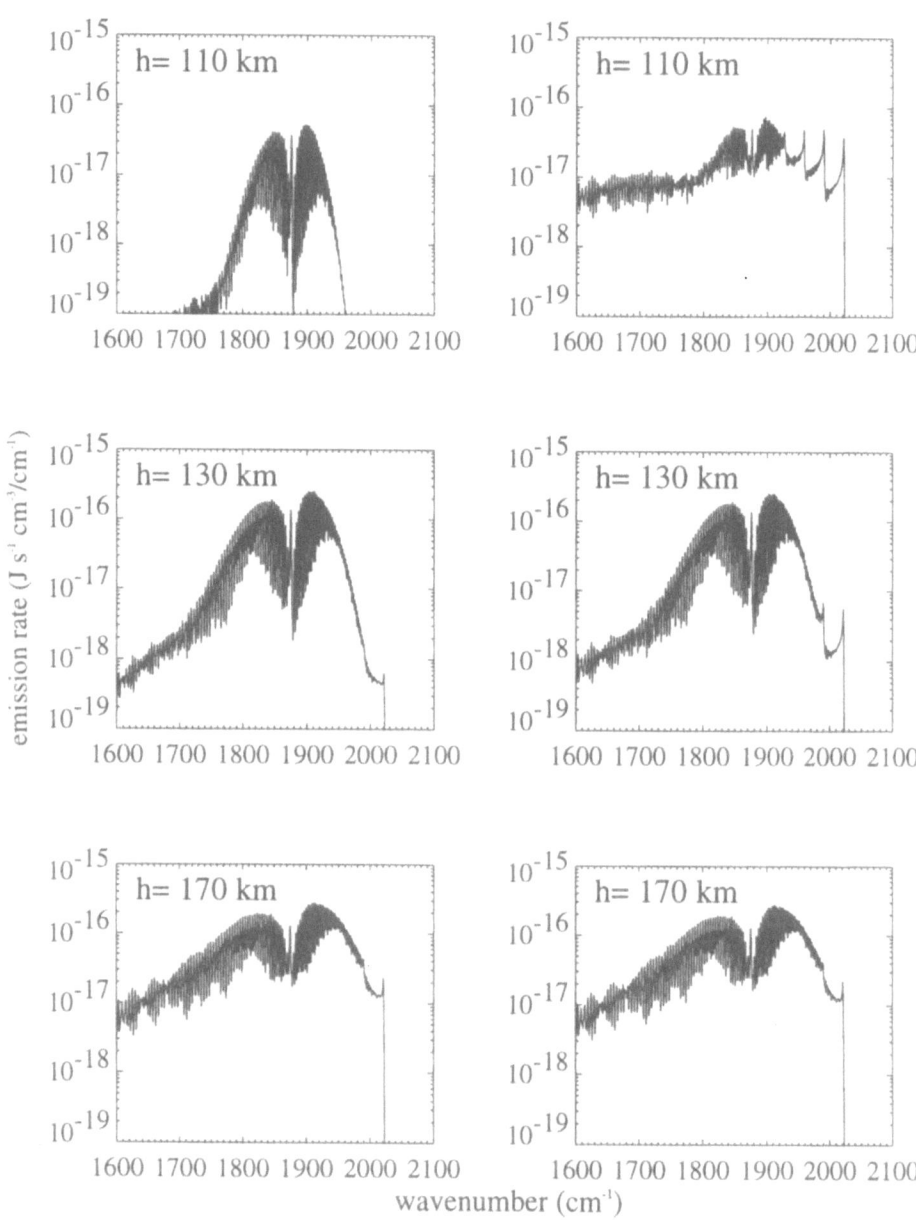

Figure 8.14. Volume emission rates calculated at various altitudes for a Maxwellian (left) and a non-Maxwellian N(^{4}S) EDF (right).

The results of our modeling reproduce quite well the main features of the observations, including the ratio between the maximum intensity of the R branch, resulting from the reaction (8.5a), and (1-0) bandhead, resulting from the reaction (8.6). Contribution of the reaction (8.5b) is negligible. As it was assumed in the observational data analysis at 130 km, the presence of hot superthermal $N_f(^4S)$ is required to produce the (1-0) bandhead by chemical excitation through the reaction (8.6). In the Maxwellian case, however, the temperature in the lower thermosphere is too low to efficiently produce rovibrationally excited NO by the reaction (8.6) and to form the bandhead. As altitude (and temperature) grows, hot thermal nitrogen atoms possess a sufficient energy to produce the (1-0) bandhead. Our calculated $NO(v, J)$ number density is also in good quantitative agreement with that obtained from the brightness of infrared spectrum because, though the concentrations we used were partially taken from empirical models combined with the observational data, their applicability is somewhat uncertain. Also a possibility remains that another mechanism could still be involved in the production of excited NO. An additional hot N atom source could modify the shape of the $N(^4S)$ EDF as well and thus to impact on the NO* production and the associated IR emission rates. Nonetheless, in any case, our model clearly indicates that fast nonthermal nitrogen atoms are a significant source of highly excited $NO(v, J)$ in the Earth's lower thermosphere.

The results of the study of hot nitrogen atoms in the Earth's upper atmosphere can be summarized as follows:

- fast nitrogen ground-state atoms are created fairly efficiently by photons and fast electron impact. Exothermic chemical reactions also contribute to substantial production of hot N in the Earth's thermosphere;
- these translationally hot atoms are thermalized by elastic and inelastic collisions with the ambient atmospheric constituents. However, at a steady state, an important nonthermal population exists, which is able to activate the production of nitric oxide through the $N(^4S) + O_2$ reaction. It results in an almost twofold increase of the steady state NO peak density near 110 km according to the model calculations, which are in better agreement with recently measured high NO densities;
- recent results also indicate that the energy dependence of the elastic collision cross-section of fast N atoms increases the potential role of hot nitrogen atoms in the odd nitrogen photochemistry;
- fast N atoms are responsible for specific spectral features in the 5.3 μm NO infrared band. The calculated NO infrared spectrum fits reasonably well the data available on infrared airglow observations from space.

CHAPTER 9

ROLE OF NONLINEAR PROCESSES IN THE FORMATION OF NEUTRAL PLANETARY CORONAS

Following the nomenclature given earlier, the top layers of the planetary atmosphere are characterized by vanishingly small values of gas particle number density and are referred to as exospheres. When the emphasis is put on its scattered properties, an exosphere is usually called a planetary corona (Chamberlain, 1978; Chamberlain and Hunten, 1987). Because the gas in planetary exospheres is very rarefied and strongly influenced by the solar electromagnetic and corpuscular radiation, only the methods of the kinetic theory are pertinent for an adequate mathematical description of this atmospheric region.

The problem of the development of an accurate mathematical model of the planetary exosphere is closely related to the fundamental problems of aeronomy and rarefied gas dynamics, in particular the transition regime from a continuous (gas dynamic) to a free-molecular (collisionless) gas flow which we discussed in detail in Chapter 6 (see also Ferziger and Kaper, 1972; Bird, 1976, 1994). From the viewpoint of the atmospheric evolution, consideration of the physical and chemical processes in the exosphere is extremely important, because they are connected with the loss of matter to space due to gas escape throughout the geological time scales (McElroy, 1972; Chamberlain, 1978; Fox and Dalgarno, 1980, Jokosky et al., 1994). Exospheric physics is also closely related to the photochemical and dynamical processes in the lower atmospheric regions - thermosphere and middle atmosphere and therefore, the knowledge of the state of the exospheric gas is very important for the development of adequate approaches in atmospheric modeling (Hunten, 1973; Chamberlain and Hunten, 1987; Marov and Kolesnichenko, 1987).

In this chapter we discuss the main mechanisms of formation of planetary coronas. The basic problem of exospheric physics is the determination of the particle velocity distributions, because these distributions include both fractions of resident atmospheric particles in the corona and particles escaping to outer space. We shall focus on different methods developed for the solution of this problem and the constraints involved. The main disadvantage of the traditional models of planetary geocoronas is that they are based on a linearized approach for the evaluation of the intrinsically a nonlinear kinetic processes. Here the results of nonlinear kinetic model developments and its application to the formation of a hot oxygen geocorona are presented.

9.1. Physical and Chemical Mechanisms of the Corona Formation

Obviously, in a planetary exosphere the thermalization rate is very small and superthermal particles infrequently interact with the ambient atmospheric gas. The superthermal particles at these altitudes are formed mostly by external forcing, though their production rates in the diluted exospheric gas are small (Chamberlain and Hunten, 1987). The main source of superthermal particles in the exosphere is connected with their transport from the transition (from collision-dominated to collisionless) atmospheric region (Jeans, 1923; Spitzer, 1952; Rohrbaugh and Nisbeth, 1973; Torr et al., 1974; Yee et al., 1980; Kozyra et al., 1982; Ishimoto et al., 1986; Hedin, 1989). As it was discussed in Chapter 4, in this region the gas flow regime is characterized by distribution functions varying on both microscopic and macroscopic scales. Superthermal particles formed in the transition region significantly disturb the thermal state of the ambient atmospheric gas. This is due to the fact that at these altitudes the production rates of superthermal particles are still a significant and the elastic thermalization rates strongly decrease. It results in significant amount of superthermal particles in this region, which define the parameters of the planetary corona. This was in particularly revealed by the analysis of experimental data for the Earth's (Bertaux, 1976; Yee et al., 1980; Hedin, 1989; Cotton et al., 1993), for Venus' (Takacs et al., 1980; Bertaux et al., 1981; Mahajan et al., 1992), and for Mars' (Nier and McElroy, 1977; Ip, 1990; Fox, 1993) atmospheres.

The particles populating the planetary coronas must have an energy excess allowing them to move upward to high altitudes in the planetary gravitational field. For given initial dynamic parameters, their trajectories in the collisionless approximation have conic sections of elliptical, parabolic or hyperbolic type. The energy and angular momentum conservation laws determine the trajectory type. The energy excess is available for particles in the tail of the Maxwellian energy distribution. Their amount is small and does not agree with experimental studies of different planetary corona populations (H, He, O, etc.), as reviewed by Chamberlain and Hunten (1987). This confirms the above statement that the main sources of corona particles are controled by physical and chemical processes, in the transition region which are responsible for the formation of superthermal particles.

The superthermal particles are produced in various nonthermal processes by chemically reactive, inelastic, or charge exchange collisions, their products having translationally excited states. If these hot particles are injected in the tail of the Maxwellian thermal distribution and are not thermalized quickly enough, a large fraction of them may disturb an initially Maxwellian distribution in the planetary corona. Chamberlain and Hunten (1987) summarized the main sources of nonthermal particles in planetary atmospheres, which are as follows: 1) charge exchange; 2) dissociative recombination; 3) photon and charged particle impact dissociation; 4) exothermic ion-neutral chemical reactions; 5) sputtering or knock-on.

The charge exchange process

$$A_{th} + B_{hot}^+(E) \rightarrow A_{th}^+ + B_{hot}(E' \leq E)$$

converts an ion with excess energy into an atom with approximately the same energy. Its collision partner becomes a thermal ion. The superthermal particles produced in this process are very energetically active (up to one hundred eV). This process is a resonant one and therefore, has a very large cross-section.

Dissociative recombination, photo- and impact dissociation, and exothermic chemical reactions

$$\begin{cases} AB^+ + e \rightarrow A_{hot}^* + B_{hot}^* \\ AB + h\nu(e_\nu) \rightarrow A_{hot}^* + B_{hot}^* + (e_\nu) \\ C + D \rightarrow A_{hot}^* + B_{hot}^* \end{cases}$$

typically yield energies of a few eV, a part of which may be stored in the excited inner degrees of freedom. For different celestial bodies the relative input of these sources are different and specific sets of these processes must be considered (Hunten, 1982; Chamberlain and Hunten, 1987).

The superthermal particles loose translational energy in elastic and inelastic collisions.

$$A_{th} + B_{hot}(E) \rightarrow A_{hot}^*(E'' < E) + B_{hot1}(E' < E).$$

In the case when the superthermal particles chemically differ from the ambient gas ($A \neq B$), these collisional processes lead to the thermalization of the primary energetically active particles. These processes are usually considered in the linear approximation assuming that the ambient gas is in thermal equilibrium and disturbances A_{hot}^* of its state are non taken into account (Hodges, 1994). However, if disturbances occur, the nonlinear approach should be used. In the case when $A = B$, the collisional processes become completely nonlinear, because the particles of secondary origin (with $E \gg kT$) are nonthermal, and their subsequent collisions with the ambient gas lead to cascade formation of new nonthermal particles.

Obviously, in this case only the nonlinear Boltzmann type kinetic equation
may adequately describe these processes.

As we discussed in the previous chapters, the nonequilibrium properties
of the particles formed with an excess of kinetic and internal energy make
it necessary to study these processes at the microscopic kinetic level. For a
rigorous self-consistent consideration of nonthermal particles and the role
they play in the corona formation, the system of nonlinear Boltzmann type
kinetic equations (4.4) must be solved. Some problems relating to the planetary coronas can be studied, however, in a simplified (linearized) approach.
In the next section, we discuss the essence of this linearized approach, including its main constraints.

9.2. Linearized Models of Planetary Coronas

Historically, the simplest exospheric models are based on the assumption
that no collisions occur. In this case, the distribution of exospheric particles
by velocities is provided by the model of collisionless free motion of particles
in a planetary gravity field above the exobase. The set of kinetic equations
pertinent for the evaluation of a steady-state, collisionless medium may
then be applied (Aamodt and Case, 1962)

$$\mathbf{c}\frac{\partial}{\partial \mathbf{r}}F_\alpha + \mathbf{S}\frac{\partial}{\partial \mathbf{c}}F_\alpha = 0 , \quad r > r_c . \tag{9.1}$$

The model uses the basic assumption of local equilibrium for the thermal
state of the gas at $r < r_c$, this state being determined from the equation

$$\sum_{\alpha'}\sum_{\beta}Q_\alpha(F_{\alpha'}, W_\beta) + \sum_{\alpha_1}J_{\alpha\alpha_1}(F_\alpha, F_{\alpha_1}) = 0 .$$

This approach was applied to the important problem of exospheric
physics: the study of the dissipation rate of atmospheric species into the
outer space. Jeans (1923) was the first who suggested that atmospheric
escape could be approximated as the rate of 'thermal evaporation' from
the exobase. In the exobase region light constituents (such as hydrogen
and partially helium atoms in the Earth's atmosphere) can have thermal
velocities exceeding the local escape velocity

$$c_{esc} = (2GM/r)^{\frac{1}{2}} , \tag{9.2}$$

and, consequently, the condition of dissipation is verified. Here M is mass of
the planet, G the gravitational constant and r the radial distance (distance
from the planet's center).

In collisionless theories of the exosphere, it is assumed that at the exobase level ($Kn \sim 1$) a local equilibrium distribution of atmospheric particles by thermal velocities prevails

$$F_\alpha^{(M)}(r_c) = n_{\alpha,c}[\frac{m_\alpha}{2\pi k T_c}]^{\frac{3}{2}} \exp[-\frac{m_\alpha c^2}{2kT_c}] \ , \tag{9.3}$$

except for particles which are moving downward with velocities exceeding c_{esc}. In accordance with this assumption, Jeans (1923) calculated the flux of escaping particles by thermal dissipation of a rarefied atmospheric gas

$$\bar{\Phi}_\alpha = \frac{n_{\alpha,c}}{2\sqrt{\pi}}\bar{c}_{\alpha,c}(1 + u_{\alpha,c}^2)\exp(-u_{\alpha,c}^2) \ , \tag{9.4}$$

where $\bar{c}_{\alpha,c} = (\frac{2kT_c}{m_\alpha})^{\frac{1}{2}}$, and $u_{\alpha,c}^2 = c_{esc}/\bar{c}_{\alpha,c}$. Two important features of Jeans' formula (9.4) should be emphasized. Firstly, at sufficiently high exospheric temperatures, when $u_{\alpha,c}^2 \leq 2$, thermal loss of matter from the exosphere occurs in the regime of gas dynamic outflow, rather than in the regime of thermal escape (Opik, 1963). This process of fast radial gas outflow can be considered as a planetary wind, by analogy with solar and polar winds (Fahr and Shizgal, 1983). Such a phenomenon can be most easily observed for planets or their satellites with relatively low accelerations due to gravity. Secondly, formula (9.4) defines the dissipation flux provided diffusion from the lower-lying layers of atmosphere is efficient, because the density $n_{\alpha,c}$ is determined from the condition of diffusive equilibrium. In other words, the flux of escaping particles is determined and controlled by diffusion processes at lower altitudes (Hunten, 1973; Hunten and Donahue, 1976).

Utilization of the collisionless models of exosphere is rather limited since they are based on the assumption that the gas flow near the exobase is not uniform and has a 'breakpoint' nature: above r_c the gas state is determined by the collisionless movement of particles, while below r_c collisions play the dominant role. In reality, a gradual transition from a collision-dominated regime of gas flows at low altitudes to collisionless flows at higher altitudes occurs in the planetary atmospheres. Nonetheless, this approach has been quite successfully used for the interpretation of some satellite observations of planetary corona (Bertaux, 1976; Meriwether et al., 1980).

Disadvantages inherent to the collisionless approach can be partially removed in the framework of collisional models of exosphere as reviewed by Fahr and Shizgal (1983) and Hodges (1994). Collisional models have been used to study a variety of effects such as (i) non-Maxwellian effects near the exobase and how they influence the reduction of the thermal escape flux (Brinkmann, 1971; Chamberlain and Smith, 1971; Fahr and Nass, 1978;

Shizgal and Lindenfeld, 1980); (ii) transition from collisional to collision-less regions of the upper atmosphere (Fahr, 1976; Lindenfeld and Shizgal, 1979); (iii) enchanced escape due to nonthermal processes (Hodges *et al.*, 1981; Hunten, 1982; Chamberlain and Hunten, 1987; Hodges and Breig, 1991, Hodges, 1994). All these models based on the linearized Boltzmann equation were mainly dedicated to the study of light atmospheric species because only in this particular case can the linear approximation yield reliable results.

As far as the first two problems are concerned, it is well known that the local equilibrium thermal distribution (9.3) is established at altitudes located below the exobase $r < r_c$, where $Kn \sim 0.1$. The low collision frequency in the transition region, where the Knudsen number varies within the limits 0.1 to 1, results in the violation of collision balance and causes energy transfer from high-velocity to slower moving particles, which eventually reduces the thermal dissipation.

As an example of the contribution of collisional processes let us consider the results of a kinetic simulation (Bisikalo and Shematovich, 1989) of the rarefied gas flow in the transition region of the Earth's upper atmosphere, where the Knudsen number $Kn \sim 0.1 \div 1$ corresponds to the altitude range $300 \div 600$ km for an average level of solar activity. Parameters at the lower boundary (where $Kn \sim 0.1$) of this region were defined from the *MSIS-86* model.

The radial flow of the rarefied gas consisting of oxygen and hydrogen (as light trace component) was studied under the assumption that the gas particles undergo only elastic collisions. The flux Φ of hydrogen atoms escaping from the Earth's transition region was determined from the calculated distribution functions. The maximum rate of thermal dissipation of light particles was found to occur just in the transition region and was explained by the fact that, on one hand, the gas density in this region is sufficient to maintain a rather high collision frequency, but on the other hand, it is low enough to ensure a large mean free path lengths in the vertical direction and hence to allow the particle escape. The escape of fast particles is responsible for the violation of the condition of local thermal balance for light atmospheric components in the transition region.

The results of evaluations of the ratio $\Phi/\bar{\Phi}$ depending on the level of solar activity are shown in *Fig. 9.1*, where the results of other calculations (Chamberlain and Smith, 1971; Brinkmann, 1971; Shizgal and Lindenfeld, 1980) are also plotted for comparison. The results of both numerical and analytical modeling are in good agreement and allow to estimate the deviation of the real velocity distribution of hydrogen atoms in the transition region from the local equilibrium Maxwellian distribution (9.3). It demonstrates a lower rate of thermal dissipation of atomic hydrogen in the Earth's

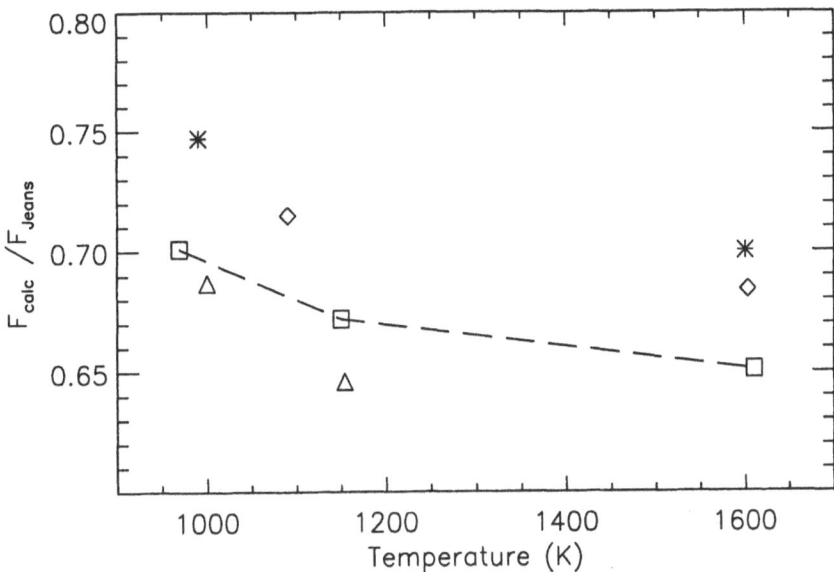

Figure 9.1. Thermal dissipation rate of atomic hydrogen in the Earth's upper atmosphere in comparison with Jeans' formula : boxes – Bisikalo and Shematovich, 1989; triangles – Shizgal and Lindenfeld, 1980; diamonds – Chamberlain and Smith, 1971; asterisks – Brinkmann, 1971.

upper atmosphere compared to Jeans' formula (9.4).

The next important step undertaken in the development of linear exospheric models was connected with the attempt to retain the equations additional terms accounting for the total dissipation rate of the planetary atmosphere, including nonthermal escape, radiative cooling, light pressure, etc. (see, e.g., Hodges *et al.*, 1981; Hunten, 1982; Fahr and Shizgal, 1983; Tinsley *et al.*, 1986; Hodges, 1994). These studies gave rise to significant improvement of the linearized models of planetary coronas by involving the thermosphere-exosphere-plasmasphere coupling, which in turn allowed to evaluate more correctly the total escape fluxes of light species. Application of these improved linearized models to the Earth's (see, e.g., Hodges, 1994), Venus'(see, e.g., Hodges and Tinsley, 1986; Nagy and Cravens, 1988; Hartle *et al.*, 1996), Mars' (see, e.g., Ip, 1988; Nagy *et al.*, 1990; Jokosky *et al.*, 1994), Titan's (see, e.g., Strobel, 1982), and Io's (see, e.g., Wilson and Scheider, 1994) exospheres showed that nonthermal processes play an important role and can even dominate in the escape flux evaluation. However, the role of nonlinear and nonequilibrium processes in the planetary corona formation was poorly defined in the framework of these models. It is worth mentioning that different numerical techniques involving the test particle

Monte Carlo method (see, e.g., Chamberlain and Smith, 1971; Brinkmann, 1971; Hodges, 1993, 1994) and the two or multistream approximations (see, e.g., Nagy and Banks, 1970; Nagy et al., 1990) were used for the kinetic equations solution underlying these models. More details with applications to different planets can be found elsewhere (see, e.g., Fahr and Shizgal, 1983; Chamberlain and Hunten, 1987; Nagy et al., 1990; Hodges, 1993, 1994).

9.3. A Nonlinear Kinetic Model of the Hot Oxygen Geocorona

The linearized approach is limited by a set of physical processes including only linear collisions. This approach is, in particular, valid when the coronal gas under consideration is a small admixture to the dominant atmospheric component and thus the collisions between particles of these species can be neglected. In addition, it is assumed that the ambient gas is in the state of thermal equilibrium. However, when the interaction of trace species with the disturbed component of the ambient gas, as well as the interaction between different nonthermal coronal species are taken into account, nonlinear effects become important (Hodges, 1994; Shizgal and Arkos, 1996).

In the case of the dominant coronal component, the strongly nonlinear system must be invoked in order to study interactions between particles of the same species. The reason is that, in an elastic interaction of superthermal particles with the thermal ambient gas consisting of the same atoms, the latter are disturbed and secondary nonthermal atoms are formed. In this case, both primary and secondary nonthermal atoms control the physical-chemical properties of the exospheric gas. The quantitative analysis of such systems is complicated because it involves the solution of nonlinear kinetic Boltzmann equations (4.4). As an example, we present here the development and the results of the nonlinear kinetic model for the hot oxygen geocorona.

A geocorona is considered as a typical object where nonlinear processes play an important role. Indeed, atomic oxygen is the dominant component in the transition region of the Earth's atmosphere. A detailed description of the hot oxygen geocorona formation requires to investigate how the ambient oxygen gas responds to the disturbances arising during collisions with superthermal oxygen atoms formed by external forcing and chemical reactions. These disturbances result in increase of translational energy of the ambient oxygen atoms followed by cascade production of nonthermal atoms of secondary origin. In order to study the processes involving kinetics and dynamics of primary and secondary superthermal particles, the stochastic simulation method described in Chapters 5 and 6 was utilized.

The existence of a geocorona of hot oxygen atoms was initially proposed theoretically by Rohrbaugh and Nisbet (1973). They suggested that it is formed by dissociative recombination of O_2^+ and NO^+ ions in the F region of the ionosphere. The excess kinetic energy carried by the fragments in exothermic recombination processes ranges from 0.79 to 6.95 eV for O_2^+ and 0.38 to 2.75 eV for NO^+, depending on the electronic excitation states of the O and N atoms. Following this study, Torr et al. (1974) suggested that energetic oxygen atoms can be also produced by precipitating energetic O^+ ions during magnetically disturbed periods. Such energetic O^+ fluxes have been reported by Shelley et al. (1972). Theoretical calculations of this effect based on linearized models were perfomed by Torr et al. (1974), Yee and Hays (1980), and Ishimoto et al. (1986, 1992). Recently, Richards et al. (1994b) demonstrated that there is a large number (a total of 22) of previously neglected sources of geocoronal hot oxygen. Richards et al. (1994b), Hickey et al. (1995) showed that exothermic reactions involving minor and metastable thermospheric species provide significant sources of translational energy to atomic oxygen that were previously overlooked. For the period of high solar activity in winter they predicted that quenching by atomic oxygen of O^+ (^2D), $O(^1D)$, $N(^2D)$, $O^+(^2P)$ and vibrationally excited N_2 provides a source of O kinetic energy about ten times higher than the 'classical' O_2^+ and NO^+ recombination sources.

Recently experimental evidence has accumulated to substantiate the presence of a hot O geocorona. These results are based on various techniques: mass spectrometric in situ measurements (Hedin, 1989), twilight airglow observations of the $O^+(^2P)$ 7319 Å emission (Yee et al., 1980) and oxygen ultraviolet airglow vertical distribution (Cotton et al., 1993). The hot O peak density derived from these observations range from $\sim 1 \times 10^4$ cm^{-3} for low to $\sim 10^6$ cm^{-3} for high solar activity conditions. The effective temperature obtained from these measurements proved to be also considerably higher than the kinetic temperature associated with the local oxygen Maxwellian distribution. Nearly simultaneously, the idea of the existence of hot oxygen coronas on other planets drew attention. Theoretical studies of a possible hot O corona on Mars, as a product of dissociative recombination of O_2^+, were performed by McElroy (1972), Knudsen (1973), Ip (1990), and Fox (1993). Theoretical and experimental work on the hot O corona of Venus was undertaken by Nagy et al. (1981), Nagy and Cravens (1988), Nagy et al. (1990), Mahajan et al. (1992).

The presence of a hot oxygen geocorona is potentially important, since hot O is capable to convert plasmaspheric H^+ ions into O^+, which could help to explain the high nighttime ionospheric concentrations. It can be also responsible for the increase of the high energy tail for light ions and therefore, the enhancement of the escape flux of light atoms such as helium, which is known to exceed the Jeans escape flux. It possibly also plays a role in the variation of the composition and energetics of the plasmasphere.

We shall first consider the importance of various sources of hot O in the formation and structure of the oxygen geocorona. It involves the evaluation of the inputs of 'classical' chemical sources, such as O_2 photodissociation, and O_2^+ and NO^+ dissociative recombination (Shematovich et al., 1994); the extended set of exothermic chemical reactions proposed by Richards et al. (1994) (Gérard et al., 1995); and O^+ ion precipitation (Bisikalo et al., 1995). We also evaluate the relative importance of these sources for high and low solar activity levels. We note that the chemical sources and those connected with O^+ ion precipitation have a different nature and their relative contribution is strongly dependent on the prevailing geomagnetic conditions. At quiet conditions, the chemical sources dominate, while in disturbed conditions the role of ion precipitation becomes more significant.

We now consider the three above mentioned chemical sources of superthermal O atoms in more detail. The following aeronomical reactions are mainly responsible for their production:

1. O_2 dissociation by solar UV radiation

$$O_2 + h\nu \rightarrow O(^3P) + O(^3P, {}^1D, {}^1S) \ . \tag{9.5}$$

Dissociatively produced O atoms have a continuous kinetic energy spectrum, in contrast to the dissociative recombination of molecular ions O_2^+ and NO^+ with thermal electrons:

$$\left\{ \begin{array}{l} O_2^+ + e \rightarrow \left\{ \begin{array}{l} O(^3P) + O(^3P) \ + 6.95 \text{ eV} \\ O(^3P) + O(^1D) \ + 4.98 \text{ eV} \\ O(^1D) + O(^1S) \ + 0.79 \text{ eV} \end{array} \right. \\ \\ NO^+ + e \rightarrow \left\{ \begin{array}{l} N(^2D) + O(^3P) \ + 0.38 \text{ eV} \\ N(^4S) + O(^3P) \ + 2.75 \text{ eV} \ . \end{array} \right. \end{array} \right. \tag{9.6}$$

In this case, superthermal O atoms are produced with an excess kinetic energy proportional to the energy output of exothermic reactions (9.6);

2. Chemical reactions proposed by Richards et al. (1994) who identified a total of 27 sources of hot oxygen. The most important reactions and their energy release are summarized in Table 9.1. Five of them are the 'classical' reactions (9.6). Amongst the new sources of hot O, the processes involving $O^+(^2D)$, $O^+(^2P)$, $N(^2D)$ and $O(^1D)$ quenching were found to be dominant in the upper thermosphere. The main contributors near 200 km are relaxation of vibrationally excited N_2 by collisions with O and O_2^+;

TABLE 9.1. Additional chemical sources of hot O and their exothermicities.

Number	Reaction	Rate coefficient ($cm^3 s^{-1}$)	ΔE_O (eV)
R1	$O + N(^2D) \rightarrow N(^4S) + O_f$	7.0×10^{-13}	1.11
R2	$O + O^+(^2P) \rightarrow O^+ + O_f$	4.0×10^{-10}	2.50
R3	$O + O^+(^2D) \rightarrow O^+ + O_f$	5.0×10^{-12}	1.65
R4	$N(^2D) + O^+ \rightarrow N^+ + O_f$	5.0×10^{-11}	0.68
R5	$O_2 + O^+ \rightarrow O_2^+ + O_f$	$2.1 \times 10^{-11} \; f(T)^{a)}$	1.03
R6	$N_2 + O^+(^2D) \rightarrow N_2^+ + O_f$	8.0×10^{-10}	0.85
R7	$N_2 + O(^1D) \rightarrow N_2 + O_f$	$2.0 \times 10^{-11} \; \exp(107.8/T)$	0.84
R8	$N(^2D) + O_2 \rightarrow NO + O_f$	6.0×10^{-12}	2.45
R9	$N(^2P) + O \rightarrow N + O_f$	1.7×10^{-11}	1.67
R10	$NO + N \rightarrow N_2 + O_f$	3.4×10^{-11}	2.07
R11	$N + O_2 \rightarrow NO + O_f$	$4.4 \times 10^{-12} \; \exp(-3220/T)$	0.90
R12	$N^+ + O_2 \rightarrow NO^+ + O_f$	2.0×10^{-10}	4.35
R13	$O^+(^2D) + O_2 \rightarrow O_2^+ + O_f$	7.0×10^{-10}	3.24
R14	$O^+(^2P) + N_2 \rightarrow N_2^+ + O_f$	4.8×10^{-10}	1.92
R15	$O(^1D) + O_2 \rightarrow O_2 + O_f$	$2.9 \times 10^{-11} \; \exp(67.5/T)$	1.31
R16	$O_2^+ + N \rightarrow NO^+ + O_f$	1.2×10^{-10}	2.74
R17	$N(^2D) + NO \rightarrow N_2 + O_f$	7.0×10^{-11}	3.58
R18	$N_2^*(v) + O \rightarrow N_2(v' = 0) + O_f$	McNeal $et\ al.$, 1974	0.19v

a) $f(T) = (T_n + 2T_i/3 * 300)^{-0.763}$ (Hickey $et\ al.$ (1995)).

3. The quenching reactions are an additional source of superthermal O_f atoms in the transition region. Superthermal O_f atoms in metastable states ^1D and ^1S produced in reactions (9.5), (9.6) can lose their internal excitation energy in the following spontaneous transitions and collisional quenching processes:

$$O_f(^1D) + \left\{ \begin{array}{l} \\ O \end{array} \right. \rightarrow \left\{ \begin{array}{l} O_f(^3P) + h\nu \\ O_{f'}(^3P) + O_{f'} + 1.97 \text{ eV} \end{array} \right. \tag{9.7}$$

$$O_f(^1S) + \left\{ \begin{array}{l} \\ O \end{array} \right. \rightarrow \left\{ \begin{array}{l} O_f(^1D) + h\nu \\ O_{f'}(^3P) + O_{f'} + 4.19 \text{ eV} \end{array} \right. . \tag{9.8}$$

Superthermal O_f atoms produced in the reactions (9.5)–(9.8) as well as those listed in Table 9.1 are characterized by initial nonequilibrium energetic spectra compared with the thermal state of the ambient oxygen atoms. These superthermal O_f atoms are thermalized in elastic collisions with atomic oxygen as the main atmospheric constituent. In elastic collisions between superthermal O_f atom having kinetic energy E and thermal O atom

$$O_f + O_{\text{th}} \rightarrow O_{f'} + O_{f'} \tag{9.9}$$

secondary nonthermal $O_{f'}$ atoms with kinetic energies $E' < E$ are produced. The subsequent fate of these secondary nonthermal $O_{f'}$ atoms depends on the ambient atmospheric density. They are fully thermalized at altitudes near the lower boundary of the transition region, while at higher altitudes they act as primary superthermal atoms and play an important role in the formation of the steady-state nonequilibrium distribution function of the exospheric gas particles.

Following the previous discussions, the physical system taking into account the production, elastic and inelastic relaxation, and transport processes for O atoms in the transition region may be evaluated from a system of nonlinear kinetic equations similar to (4.4)

$$
\begin{cases}
\frac{\partial}{\partial t} F_f^{3P} + \mathbf{c}\frac{\partial}{\partial \mathbf{r}} F_f^{3P} + \mathbf{S}\frac{\partial}{\partial \mathbf{c}} F_f^{3P} = Q_f^{3P} + J_{el}(F_f^{3P}, F_{th}) & (a) \\[2mm]
\frac{\partial}{\partial t} F_f^{1D} + \mathbf{c}\frac{\partial}{\partial \mathbf{r}} F_f^{1D} + \mathbf{S}\frac{\partial}{\partial \mathbf{c}} F_f^{1D} = Q_f^{1D} + W_f^{1D} + \\[1mm]
\qquad\qquad + J_{el}(F_f^{1D}, F_{th}) + J_q(F_f^{1D}, F_{th}) & (b) \\[2mm]
\frac{\partial}{\partial t} F_f^{1S} + \mathbf{c}\frac{\partial}{\partial \mathbf{r}} F_f^{1S} + \mathbf{S}\frac{\partial}{\partial \mathbf{c}} F_f^{1S} = Q_f^{1S} + W_f^{1S} + \\[1mm]
\qquad\qquad + J_{el}(F_f^{1S}, F_{th}) + J_q(F_f^{1S}, F_{th}) & (c) \\[2mm]
\frac{\partial}{\partial t} F_{th} + \mathbf{c}\frac{\partial}{\partial \mathbf{r}} F_{th} + \mathbf{S}\frac{\partial}{\partial \mathbf{c}} F_{th} = \sum_L J_{el}(F_{th}, F_L) + \\[1mm]
\qquad\qquad\qquad + \sum_N J_q(F_f^N, F_{th}) & (d)
\end{cases} \qquad (9.10)
$$

where F_{th} is the distribution function for the thermal atoms of the ambient atmospheric gas, and F^{3P}, F^{1D}, and F^{1S} are the distribution function for the nonthermal O_f atoms in the ^3P, ^1D and ^1S states. The right-hand side of the system (9.10) contains the source terms for superthermal O_f atoms due to chemical reactions, as well as the sink terms for spontaneous transitions W, and elastic J_{el} and quenching J_q collisions. It should be noted, that the oxygen atoms of the ambient gas and nonthermal $O(^3P)$ atoms [equations (a) and (d)] represent physically the same species and differ only by their origin. The definitions of thermal (ambient) and nonthermal fractions of O atoms have no specific physical meaning and are introduced for the sake of convenience of the numerical analysis. *Fig. 9.2* illustrates this schematic structure.

The convergence of the physical system state to the LTE is defined by the thermal elastic collision term

$$
J_{el}^{th}(F_{th}, F_{th}) \sim n_{th}^2 < \sigma v_{th} > ,
$$

and the degree of thermal state perturbation due to superthermal O_f atoms is characterized by the collision term

$$
J_{el}^f(F_{th}, F_f) \sim n_{th}n_f < \sigma v_f > .
$$

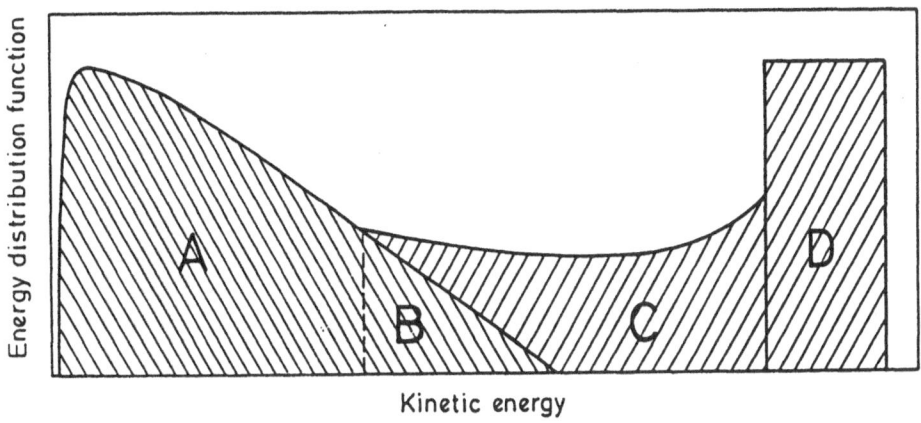

Figure 9.2. Schematic representation of the various energetic fractions of atomic oxygen: /// - thermal particles (Maxwellian); \\\ - nonthermal particles; D - primary energetic particles; C - secondary energetic particles produced by collisions between thermal and nonthermal atoms; B - hot atoms belonging to high energy tail of the Maxwellian distribution. The population designated as A is low energy part of the Maxwellian distribution.

At the lower boundary of the transition region ($J_{th} \gg J_f$), a linearized system of kinetic equations (9.10) may be used. However, in the transition region, the collision terms have comparable values and consequently, significant perturbations of the thermal state of the atmospheric gas are possible. Consequently, the kinetic system becomes essentially nonlinear, and resort to the stochastic simulation method is required.

The problems of superthermal O_f atom production, thermalization and transport in the transition region were studied under the following assumptions:

– the source terms $Q(r)$ in the system (9.10) are not time-dependent for given solar-geophysical conditions. This assumption seems acceptable for the problem under consideration because the characteristic time scales of these condition changes are longer than those for gas flow relaxation in the transition region;

– for the gas flow, the spherical symmetry approximation can be admitted, i.e. the global horizontal atmospheric circulation is not taken into account. This assumption is supported by the fact that the characteristic time scale for the vertical atmospheric profile is much less than that for the horizontal transport.

The production rate of superthermal O_f atoms in the reaction of dissociation (9.5) was calculated using the photodissociation cross-sections compiled by Conway (1988) and solar UV flux for $\lambda < 2000$ Å reported by

Tobiska (1991). The hot oxygen production rates by chemical reactions were taken from the *Field Line Interhemispheric Plasma (FLIP)* transport model (Torr *et al.*, 1990; Richards *et al.*, 1994). This model is based on the solution of the coupled time dependent energy, momentum and continuity equations for major ions beginning from 80 km in one hemisphere and extending along a magnetic field line to 80 km in the other hemisphere. The *FLIP* model represents all important ion and metastable particle densities, as well as electron and ion temperatures, photoelectron flux and flow velocities (see Torr *et al.*, 1990). The cross-section for atomic oxygen elastic collisions was taken equal to 1.7×10^{-15} cm^2, and the cross-section for collisional quenching of metastable states $O(^1D)$ and $O(^1S)$ by atomic oxygen was taken equal to 1.1×10^{-16} cm^2, in agreement with the theoretical calculations by Yee *et al.* (1990).

In order to understand the role played by chemical sources in the hot geocorona formation, we carried out the calculations for daytime (local noon) equatorial Earth's upper atmosphere and for quiet geomagnetic conditions. This means that the influence of ion precipitation in the corona formation can be neglected. Both low and high solar activity levels ($F_{10.7} = 70$ and 200) were considered to test the sensitivity of the chemical sources to the solar radiation.

In the numerical experiment [described in detail by Shematovich *et al.* (1994)], the evolution of the system of modeling particles from the initial state to steady-state was evaluated. Following the stochastic simulation method, this numerical system approximates the original kinetic system (9.10). Therefore, it is possible to use the distribution functions statistically estimated in the numerical experiment to calculate all gas macroparameters necessary to describe the atmospheric gas flow in the transition region of the Earth's atmosphere.

Steady state solutions of the Boltzmann equations were obtained for the region between 200 and 900 km. The earlier mentioned separation of atomic oxygen into thermal and nonthermal fractions was used to analyze how the initial and secondary energetic O sources effect the ambient atmospheric gas. *Figs. 9.3a, b* show the resulting energy distribution function (EDF) of the thermal and nonthermal fractions for both 'classical' and extended set of chemical reactions for high solar activity. The influence of nonthermal O atoms on the thermal population is clear as profound perturbations of the thermal EDF, which strongly deviates from the local Maxwellian distribution (*Fig. 9.3a*). These perturbations increase with altitude as the relaxation time increases with decreasing ambient density. Inclusion of the extended set of chemical sources (Gérard et al., 1995) increases the density of nonthermal O_{nth} atoms and also causes additional departure from the local equilibrium (*Fig. 9.3b*).

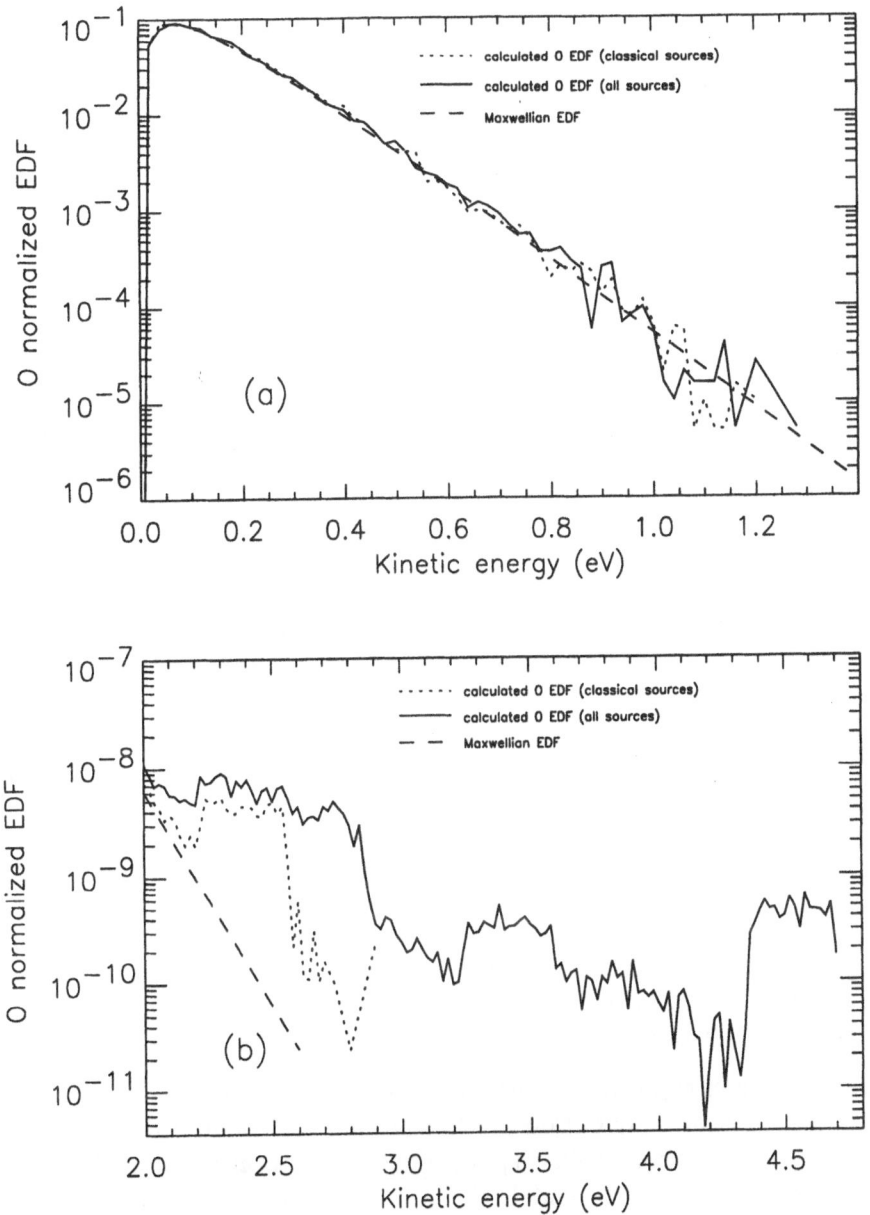

Figure 9.3.   Energy distribution function (EDF) (normalized to unity) of the oxygen atoms calculated at 355 km for high solar activity conditions. Solid and dotted lines are the calculated EDFs and dashed line is the local Maxwellian distribution. (a): low energy range of the EDF; (b): high energy range of the EDF.

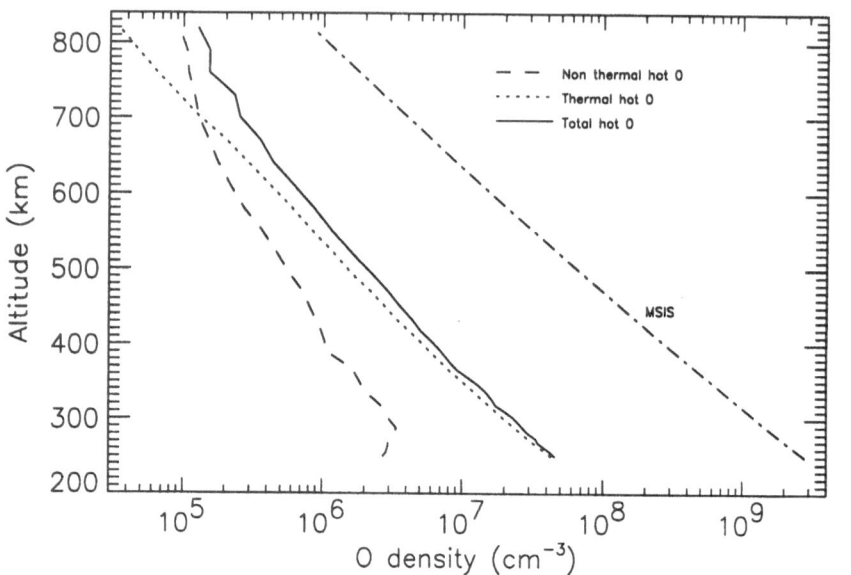

Figure 9.4. Vertical distribution of the thermal and nonthermal components and total hot oxygen density. For comparison the total atomic oxygen density from the *MSIS* model is also shown.

A detailed analysis of the distribution functions presented in *Fig. 9.3a, b* indicates that the hot O geocorona is formed by the contribution of the following populations:

- a part of the thermal population — hot atoms in the high energy tail of the Maxwellian distribution;
- nonthermal atoms including the primary source of energetic particles produced by O_2 photodissociation and exothermic chemical processes, and the secondary energetic atoms produced by collisions between the thermal and nonthermal components.

The relative contribution of these fractions depends on altitude. *Fig. 9.4* shows the vertical distribution of the thermal and nonthermal hot components, the total hot O density, as well as the total O number density according to $MSIS - 90$ model. It shows that the nonthermal O atoms become dominant above 700 km though their contribution is not negligible at lower altitudes as well. The calculated total density of hot O for the high level of solar activity is about 1.2×10^6 cm^{-3} at 550 km that is in good agreement with the results of the above discussed observations (about 1×10^6 cm^{-3} for similar level of solar activity).

Densities of the thermal and nonthermal fractions of hot O atoms with

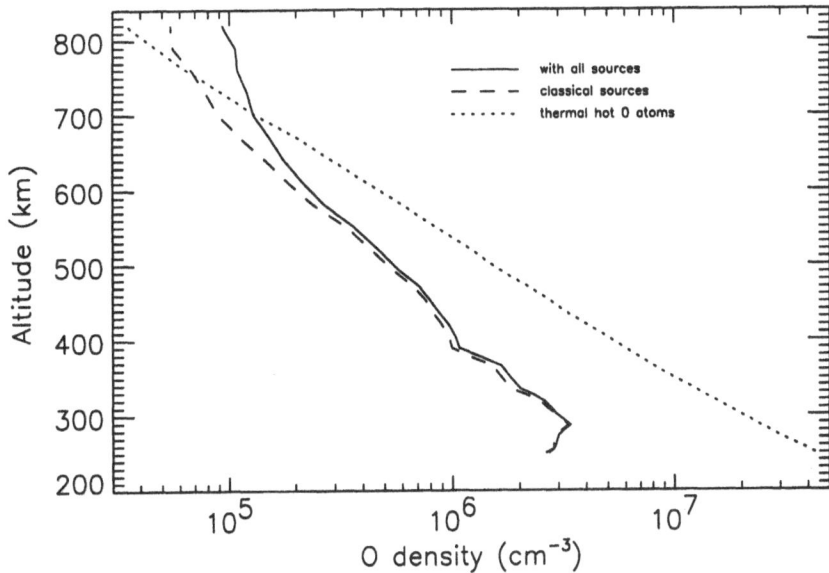

Figure 9.5. Nonthermal fractions of hot O atoms calculated with and without the new chemical sources and comparison with the hot thermal density distribution.

and without the additional chemical sources are shown in *Fig. 9.5.* It is seen that the high thermalization rate leads to comparable values of the nonthermal fraction densities in the lower exosphere and thermosphere, and that the values corresponding to all chemical sources exceed the 'classical' ones by 10–15 percent. In the exosphere above 600 km, the input of the additional sources increases and at 800 km the new chemical sources are responsible for a twofold increase of the nonthermal particle population. This energetic source leads to the dominant role of nonthermal hot O in the formation of hot oxygen geocorona at altitudes above 700 km. It should be emphasized, however, that despite the relatively small differences in the calculated densities, perturbations of the distribution function caused by the extended set of chemical reactions are more pronounced, especially for the nonthermal fraction (see *Figs. 9.3a, b*). This result implies that specific patterns of the additional hot O sources may be observed in the profile of high latitude airglow emissions. Indeed, the shape (broadening) of an airglow line is controlled to large extent by the nonequilibrium nature of the distribution function of the emitting atom or ion (Cotton *et al.*, 1993). Another possibility to evaluate the influence of hot atoms on the atmospheric processes is the study of the energy budget of the atmosphere. In the recent paper by Oliver (1997) the input of hot oxygen atoms into the

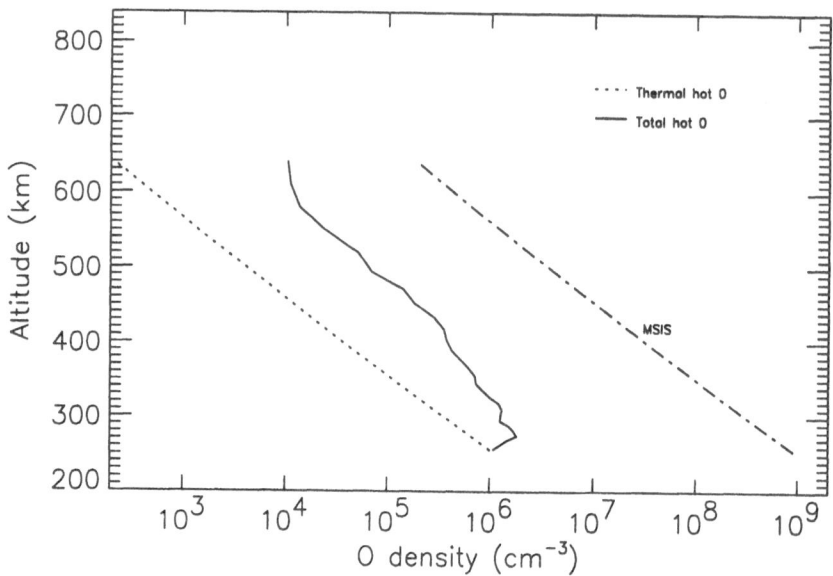

Figure 9.6. Same as *Fig. 9.4* for the low level of solar activity.

ion energy budget was analyzed, and it was shown that the hot component plays an important role.

The calculations with and without account for the additional chemical sources of hot O were also carried out for low activity conditions ($F_{10.7} = 70$) (Gérard *et al.*, 1995). The resulting vertical distribution is presented in *Fig. 9.6*. It is clear, that the calculated density of hot O atoms is significantly smaller than for the solar maximum reaching only $\sim 2 \times 10^4$ cm^{-3} at 550 km. Nevertheless, the relative contribution of the new chemical sources turns out to be more important than for solar maximum due to the hot Maxwellian fraction of O population. This result can be explained by the significant increase of the relaxation time at the lower exospheric temperature characterizing the solar minimum. The calculated range of variations from solar minimum to maximum agrees well with that derived from the *in-situ* satellite measurements.

Another important source of hot oxygen atoms (especially for high latitudes, and magnetically disturbed conditions) is the precipitation of energetic O$^+$ ions. As it was earlier mentioned, the precipitated fluxes of these ions during magnetic storms supply substantial energy in the upper atmosphere (Torr *et al.*, 1974; Kozyra *et al.*, 1982; Ishimoto *et al.*, 1992) and they may play an important role in the formation of the hot oxygen geocorona. The previous linearized models (Torr *et al.*, 1974; Kozyra *et al.*,

1982; Ishimoto et $al.$, 1992) allowed to evaluate some effects induced by O^+ precipitation (escape fluxes, heating rate of atmospheric gas, total emission rates, etc.). However, because the problem of the hot O geocorona formation is essentially nonlinear, being formed mostly by the disturbed thermal fraction of the atmospheric gas, again a more thorough study requires solving the set of nonlinear kinetic Boltzmann equations.

The first observation of the precipitation of energetic heavy ions into the atmosphere was reported by Shelley et $al.$ (1972). Further observational information on the energetic O^+ ions has appeared in a series of papers (Shelley et $al.$, 1974; Sharp et $al.$, 1974, 1976 a, b). The measurements were made at an altitude of 800 km in the energy range $0.7 \div 12$ keV. The precipitating O^+ fluxes were found to occur over a wide latitudinal range, $2 < L < 9$, and were observed in every magnetic substorm events over one-year period. However, they were also recorded at reduced intensities during quiet geomagnetic conditions. Most of the observations were made at pitch angle of $55 \div 58^o$. These measurements brought evidence of significant fluxes of fast O^+ during magnetically disturbed periods. The energy fluxes amounted to 0.4 $erg\,cm^{-2}\,s^{-1}\,sr^{-1}$ at 800 km and could occur for extended periods of time. Values more frequently measured were, however, about 0.1 $erg\,cm^{-2}\,s^{-1}\,sr^{-1}$. The energetic oxygen ions have been observed at all local times but the dayside fluxes were typically found to be factor of 5–10 less than those on the nightside (Sharp et $al.$, 1976 a).

A numerical kinetic model of the hot O geocorona formation by O^+ ion precipitation in disturbed geomagnetic conditions was developed for the nightside hemisphere by Bisikalo et $al.$ (1995). The O^+ spectrum from Shelley (1972) corresponding to $L \sim 3.4$ (i.e., at a latitude of 45°) was used. The incident O^+ flux was injected isotropically downward with an energy flux of about 0.4 $erg\,cm^{-2}\,s^{-1}$ (as in Torr et $al.$, 1982). Background data were taken from the $MSIS$-90 model (Hedin, 1991) corresponding to solar maximum ($F_{10.7} = 200$) at 3:00 local time for magnetically disturbed conditions ($Ap = 100$) and northern hemisphere in December. It is close to the conditions under which the O^+ energy spectrum was measured.

Precipitating energetic O^+ ions interact with ambient gas in the following processes:

– conversion of O^+ ions by resonance charge transfer to fast neutral atoms (as earlier, subscripts f and s refer to fast and slow atoms, respectively):

$$O_f^+ + O_s \rightarrow O_f + O_s^+ \quad ; \qquad (9.11)$$

– momentum transfer in which fast ions and atoms elastically collide with the ambient O atoms sharing kinetic energy:

$$\begin{cases} O_f^+ + O_s \rightarrow O_f^+ + O_f \\ O_f + O_s \rightarrow O_f + O_f \end{cases} ; \qquad (9.12)$$

– ionization and excitation of the ambient gas by the small fraction of fast atoms having sufficient energy:

$$O_f + O_s \rightarrow \begin{cases} O_f + O_s^+ + e_s \\ O_f + O_s^* \end{cases} . \qquad (9.13)$$

The relative importance of the above processes (9.11) – (9.13) is governed by their collision cross-sections. Virtually all the incoming energy below about 10 keV goes into reactions (9.11) and (9.12) (Rees, 1989). Because our numerical model uses the measured O^+ flux with energies up to 12 keV (Shelley et al., 1972), only reactions (9.11) and (9.12) were taken into account and reactions (9.13) were neglected.

We also adopted the charge exchange cross-section (σ_{ch}) for collisions of energetic O^+ with O measured by Rutherford and Vroom (1974) and Stebbings et al. (1964). The total cross-sections (σ_{el}) for elastic collisions of energetic O_f^+ and O_f with O are energy dependent and they are of the order of 1.7×10^{-15} cm^2 at 1 keV (Kozyra et al., 1982; Rees, 1989). The most important feature of these collisions is that the elastic differential scattering cross-sections (DSCS) are strongly forward peaked (Stebbings et al., 1964). Unfortunately, there are no measurements of DSCS for elastic O-O interactions. For this reason, the measured DSCS for elastic collisions of energetic O on O_2 (Schafer et al., 1987) were processed similar to Ishimoto et al. (1992).

The above physical system, taking into account production, elastic/inelastic relaxation, and transport processes for O atoms and ions, may be quite accurately described by the following system of kinetic equations:

$$\begin{cases} \frac{\partial}{\partial t}F_f + \mathbf{c}\frac{\partial}{\partial \mathbf{r}}F_f + \mathbf{S}\frac{\partial}{\partial \mathbf{c}}F_f = J_{ch}(F_f; F_{th}, F_i) + J_{el}(F_f, F_{th}) + \\ \qquad\qquad\qquad\qquad\qquad\qquad\qquad + J_{el}(F_f, F_i) \\[2mm] \frac{\partial}{\partial t}F_i + \mathbf{c}\frac{\partial}{\partial \mathbf{r}}F_i + \mathbf{S}\frac{\partial}{\partial \mathbf{c}}F_i = J_{ch}(F_i; F_{th}, F_f) + J_{el}(F_i, F_{th}) + \\ \qquad\qquad\qquad\qquad\qquad\qquad\qquad + J_{el}(F_i, F_f) \qquad (9.14) \\[2mm] \frac{\partial}{\partial t}F_{th} + \mathbf{c}\frac{\partial}{\partial \mathbf{r}}F_{th} + \mathbf{S}\frac{\partial}{\partial \mathbf{c}}F_{th} = J_{ch}(F_{th}; F_f, F_i) + \\ \qquad\qquad\qquad\qquad\qquad\qquad + \sum_{L=th,f,i} J_{el}(F_{th}, F_L) \end{cases}$$

where F_{th} is distribution function for thermal atoms of the ambient gas, and F_f and F_i are distribution functions for superthermal O_f atoms and

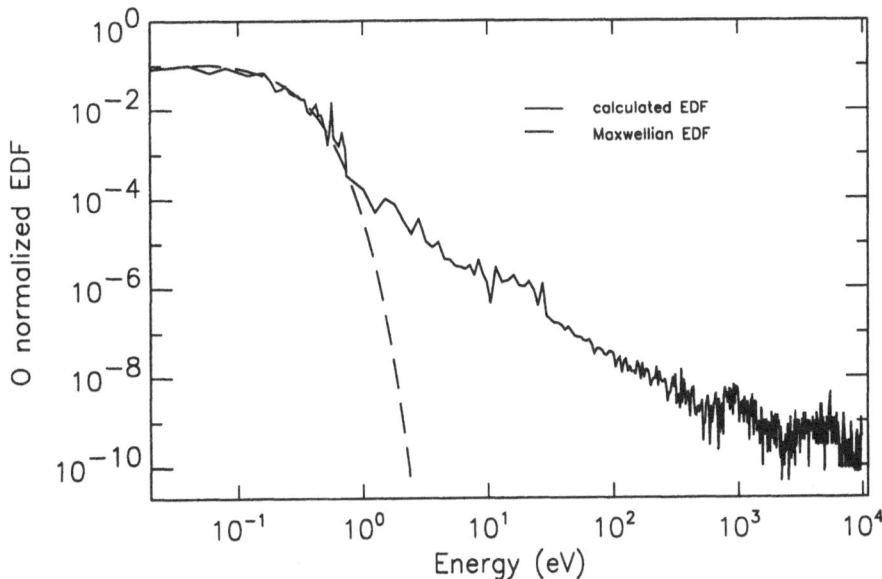

Figure 9.7. Energy distribution functions for atomic oxygen in the transition region at an altitude of 500 km. Solid line shows the calculated distribution; dashed line represents the local Maxwellian distribution function ($T = 1171$ K).

O^+ ions, respectively. In the right-hand part of the system (9.14) the charge exchange collision term

$$J_{ch} = \int \sigma_{ch}[F_f F_i - F_{th} F_i] \, |\mathbf{c}_i - \mathbf{c}_{th}| \, d\mathbf{c}_i$$

is retained, which describes the source for superthermal atoms and sink for ions and thermal atoms. The kinetic energy of fast O_f atoms and O_f^+ ions shared in elastic collisions with the ambient gas is described by the collision terms J_{el}. These terms [as in the system (9.10)] include all possible channels of elastic collisional interactions: cascade formation of secondary nonthermal O_f atoms; disturbances of thermal atoms; thermalization of nonthermal and thermal particle fractions, etc.

In the numerical experiments, the steady-state kinetic energy distribution functions for atomic oxygen were first calculated. An example of this function normalized to unity by the overall oxygen density at an altitude of 500 km is shown in *Fig. 9.7*. The dashed curve corresponds to a Maxwellian distribution function for the local temperature $T = 1171$ K. It is seen that this function is strongly in nonequilibrium as it differs from the local Maxwellian distribution, especially at high energies. It was found that these nonthermal perturbations increase with height as the relaxation

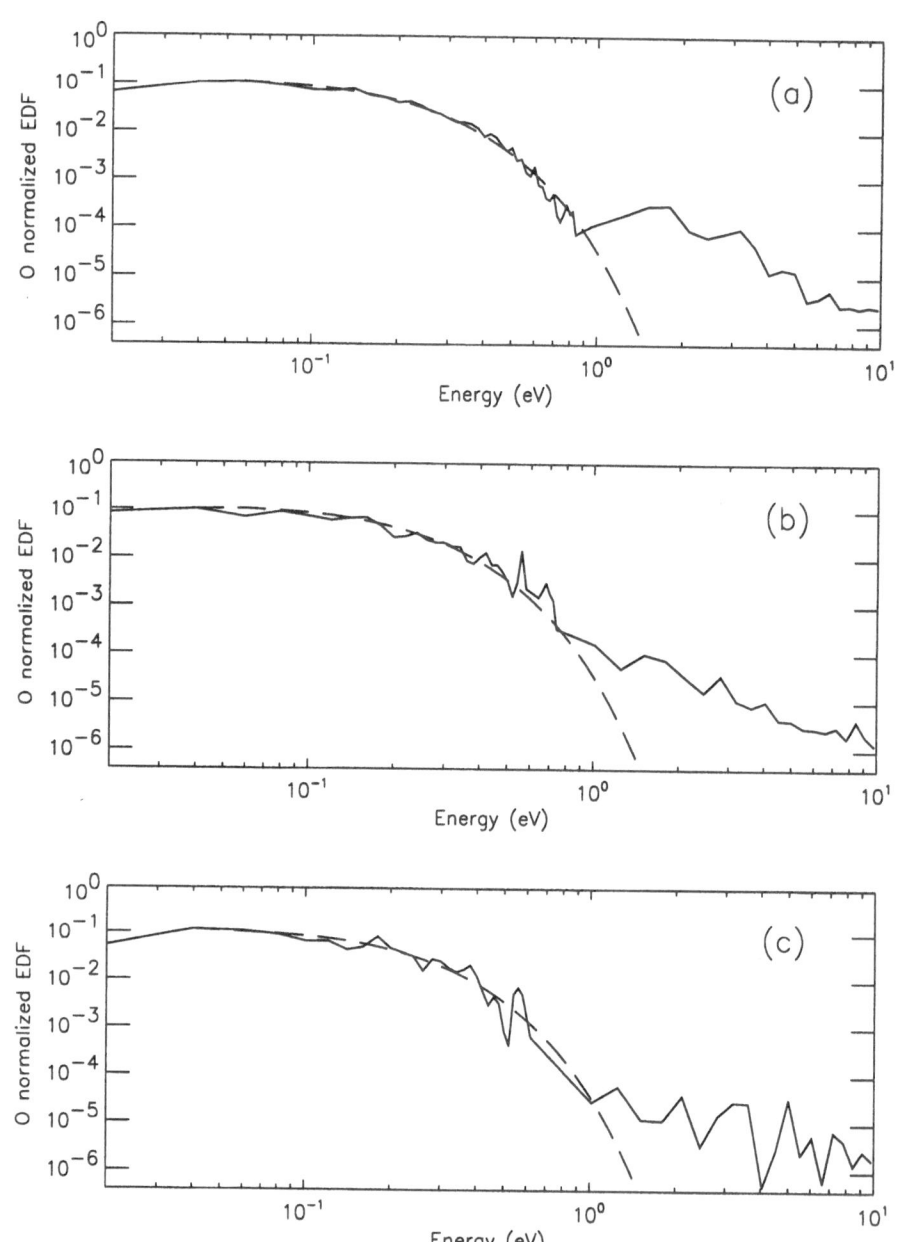

Figure 9.8. Energy distribution functions for thermal and suprathermal populations of atomic oxygen in the transition region: (a) 400 km, (b) 500 km, (c) 600 km. The solid line shows the calculated distribution; the dashed line represents the local Maxwellian distribution function.

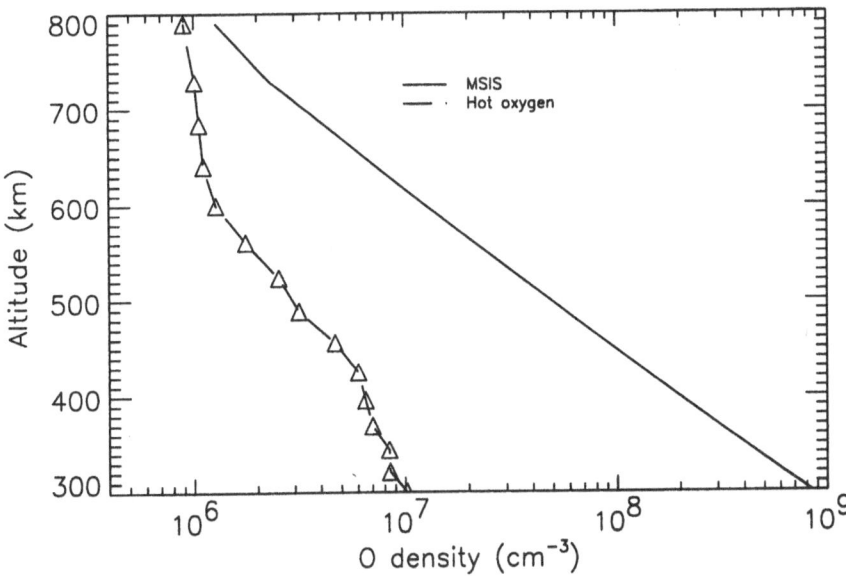

Figure 9.9. Vertical profiles of the hot and thermal fractions of atomic oxygen in the transition region. Triangles indicate the calculated values, the dashed line is a fitted curve and the solid line is the *MSIS* density profile of thermal O.

efficiency of the atmospheric gas decreases with altitude in the transition region, similar to how solar activity affects the hot O fraction we earlier discussed.

Following the procedure adopted, we can formally divide atomic oxygen into a thermal fraction, i.e. the atoms whose kinetic energy distribution may be described as Maxwellian, and a nonthermal fraction (see, *Fig. 9.2*). The latter includes: (i) superthermal atoms produced in processes of charge exchange and momentum transfer and having energies greater than the escape energy, and (ii) suprathermal atoms, i.e. atoms produced by thermalization of superthermal atoms to an energy less than the escape value, as well as by disturbances of the thermal fraction in interactions with suprathermal and superthermal atoms. This formal division of the whole population is useful for the analysis of numerical results, as the macrocharacteristics of the O^+ precipitation (such as escape flux, heating rate, etc.) are defined by the superthermal atoms, while for the hot oxygen corona formation the suprathermal atoms are mostly responsible.

An analysis of the superthermal atoms is possible in the framework of both linearized and nonlinear models. From the calculated energy distribution functions, all the macrocharacteristics characterizing the O^+ ion precipitation can easily be obtained. Our evaluation of the energy escape flux is about 12%, in agreement with the previous results of linear models (Kozyra et $al.$, 1982; Ishimoto et $al.$, 1986, 1992). Another significant characteristic of the O^+ ion precipitation is the heating rate of the neutral gas. In our calculations, we found that the maximum heating rate is 8.6×10^4 eV cm^{-3} s^{-1} at an altitude of 207 km, also in agreement with earlier studies.

However, the suprathermal atoms may be evaluated only in a nonlinear approach. The calculated energy distribution functions provide comprehensive data on atomic oxygen in the transition region (much superior to the macrocharacteristics), which are necessary for an in-depth study of the hot O geocorona formation due to O^+ ion precipitation. As it was already suggested, the hot O geocorona is formed mostly by the suprathermal component, consisting of disturbed thermal atoms and partly thermalized superthermal atoms. Although the superthermal gas fraction has a much lower density, these energetic particles significantly impact on the oxygen geocorona temperature.

Examples of steady-state kinetic energy distribution functions for thermal and suprathermal oxygen atoms normalized to unity are shown in $Fig.$ 9.8 for altitudes of 400, 500, and 600 km. The dashed curve, as in $Fig.$ 9.7, corresponds to local Maxwellian distribution function.

These functions give rise to the main macroparameters of hot O geocorona - temperature and density altitude profiles. The height profiles of density for the hot and thermal fractions of O atoms are shown in $Fig.$ 9.9. The density of the hot O atoms varies from 1.0×10^7 cm^{-3} at 300 km to 9.0×10^5 cm^{-3} at 790 km. The temperature of hot O atoms is about 4100 K for all altitudes. $Fig.$ 9.9 shows that the density scale height for the hot O is larger than for the thermal fraction (in agreement with the calculated temperatures). It is also seen that at heights above 900 km, the density of hot O becomes dominant in the upper atmosphere. These results are in a good agreement with estimates of the hot O density for high solar activity period inferred from the differences between satellite drag and mass spectrometer data based models (Hedin, 1989). According to our calculations, at about 550 km , their density is about 1.7×10^6 cm^{-3}, while Hedin's results claim for $5 \times 10^5 - 2 \times 10^6$ cm^{-3}; at 925 km the measured values are $(1 \div 4) \times 10^5$ cm^{-3}, while an extrapolation of our results gives marginally similar values of $(4 \div 5) \times 10^5$ cm^{-3}.

Summarizing the results of the study of different sources of hot oxygen geocorona formation we can conclude that chemical processes during quiet geomagnetic conditions and O^+ ion precipitation during geomagnetic storms are significant sources of nonthermal (hot) O atoms. Numerical simulations show that the distribution function of thermal oxygen becomes increasingly perturbed by collisions with the hot oxygen population at high altitudes and significantly departs from a Maxwellian distribution at all altitudes. The number density and temperature of the nonthermal oxygen atoms obtained from the respective microscopic distribution function were found to be in a good agreement with the experimental data available.

The results of the application of a nonlinear kinetic model to the study of the hot oxygen geocorona show that the collisional interaction between superthermal and thermal oxygen atoms leads to strong disturbances of the ambient atmospheric gas and, consequently, to the formation of a secondary suprathermal gas component. These suprathermal atoms produced in nonlinear processes serve as the main source for the hot O geocorona. It should be noted that this approach, involving a detailed evaluation of the nonlinear processes in the transition region of the Earth's atmosphere, may be easily extended to other planets, in order to determine the main atmospheric species that could be responsible for their coronas.

CHAPTER 10

NONTHERMAL PARTICLES
IN THE JOVIAN ATMOSPHERE

Upper atmospheres of the giant planets are characterized by a chemical composition dominated by molecular and atomic hydrogen and hydrogen-bearing compounds. Another common feature is high exospheric temperatures as compared to what could be expected at large distances from the Sun. They also possess strong magnetic fields and vastly expanding magnetospheres, especially Jupiter. The study of their thermal structures suggests the existence of a strong coupling between magnetospheres and upper atmospheres, presumably resulting from large fluxes of energetic particles precipitated from the magnetosphere along magnetic field lines or accelerated in the ionosphere by local electric fields.

Following the general description given in Section 2.2.4, we shall emphasize here the main features of the Jovian upper atmosphere with the focus on magnetosphere-ionosphere-atmosphere interactions and energetics. This approach is related to possible sources of hot H in the Jovian aurora excited by electron and protons. Their influence on the Lyman-α line profile will be evaluated.

10.1. The Jovian Upper Atmosphere and its Energetics

The temperature of the upper atmospheres of the giant planets remarkably exceeds the value deduced from thermal balance including solar EUV heating and downward conduction. For example, Jupiter's exospheric temperature at low latitudes derived from the Voyager solar and stellar occultation yield a value of 1100 ± 200 K. Such a high temperature requires the existence of additional heat sources such as dissipation of gravity waves, soft electron precipitation, or equatorward heat redistribution from very strong high latitude auroral sources. Recent measurements made by the *GALILEO* probe inside the Jovian atmosphere gave a nearly similar value of 1300 K (Seiff *et al.*, 1996) at mid-latitude. High latitude temperatures were also obtained from ground-based spectroscopy of the H_3^+ auroral infrared emission. However, although these measurements yielded values over the range 1000 – 1500 K they do not necessarily represent the exospheric temperature. Finally, the high latitude vertical temperature profile was obtained by

Trafton et $al.$ (1994) based on the H_2 quadrupole emission spectrum and $Hubble$ $Space$ $Telescope$ (HST) observations of the rotational distribution of the H_2 ultraviolet emission and with constraints placed by the Voyager IRIS infrared measurements near the homopause.

Obviously, a large energy flux precipitated into the Jovian thermosphere will generate considerable local effects and impact the energy budget of the upper atmosphere, which may also impart global effects through three-dimensional transport of heat and atmospheric species. The nature and identity of the energetic auroral particles have not been unambiguously determined, though one must keep in mind that the energetics of the atmosphere may be essentially independent of the specific details of the interaction. The most important issue is that secondary electrons are generated in any case, which contribute a large fraction of ionization and dissociation.

The local and total energy fluxes precipitated into the Jovian ionosphere–atmosphere system have been estimated from the brightness of the ultraviolet Lyman and Werner bands observed by various spacecraft. The values derived from spectral measurements made with the $International$ $Ultraviolet$ $Explorer$ (IUE) and $Voyager$ $Ultraviolet$ $Spectrometer$ (UVS) are quite consistent with those deduced from the ultraviolet images of the $Faint$ $Object$ $Camera$ (FOC) and $Wide$ $Field$ $Planetary$ $Camera$ 2 $(WFPC2)$ on board the HST. The total precipitated power in both hemispheres proved to be of the order of $10^{13} - 10^{14}$ W, that is about a factor of 100 in excess of the solar EUV flux absorbed in the planetary thermosphere. Local power fluxes have been estimated from the brightness of the HST-FOC H_2 emission. They were shown (Gérard et $al.$, 1994b) to reach values as high as 10^3 erg cm^{-1} which, at steady state, would significantly perturb hydrostatic equilibrium and generate temperatures in excess of 100,000 K. Simple energetic considerations suggest that the Jovian thermosphere is probably dominated by strong upwellings and horizontal wind fields which redistribute the energy precipitated at high latitudes.

It is therefore likely that such a large energy input into the atmosphere affects the thermal, dynamical and chemical structure of the thermosphere, possibly extended on a global scale. Some of the aeronomical and thermal consequences of these perturbations were described earlier (Waite et $al.$, 1983; Gérard et $al.$, 1994b), but no comprehensive global model was suggested so far to evaluate these effects quantitatively. One of the consequences of the interaction between energetic particles and predominantly H_2-H composition of the Jovian upper atmosphere is the direct and indirect production of fast hydrogen atoms. These atoms are produced with excess kinetic energy by electron impact on H and H_2, proton charge exchange with H_2 and H, and exothermic chemical reactions. They lose their kinetic energy in elastic and inelastic collisions with thermal particles which result in a steady state population of hot H atoms in the thermosphere, similar to the hot oxygen geocorona described in the previous chapter.

The profile of the H Ly-α line in the Jovian aurora exhibits unsual features which may provide important clues to the process of its excitation, identification of the precipitating auroral particles and thermodynamic conditions of the atmospheric gas. Clarke et al. (1989) reported IUE spectra showing Doppler-shifted emission, mainly toward the blue at 30 – 60 km/s corresponding to kinetic energy of 10 – 20 eV for fast protons or H atoms. In contrast, in the northern hemisphere red shifts (45 and 160 km/s) were detected at high latitudes. No emission was observed corresponding to energies greater than 200 eV, placing an upper limit on the contribution of protons to the Jovian aurora. Jovian auroral spectra were obtained with the Goddard High Resolution Spectrograph (GHRS) on board the HST in the 1204 – 1241 Å region with 0.57 Å spectral resolution (Clarke et al., 1994). They showed Doppler-broadened Ly-α with wings extending up to ~ 100 eV from the line center. The strength of these wings appeared to be correlated with the brightness of the auroral H_2 band emissions also present in the spectra. No measurable Doppler shift was observed in the northern hemisphere, as it had been the case in the earlier IUE observations. These earlier blue-shifted Ly-α profiles were interpreted by Clarke et al. (1989) as a possible signature of upward acceleration of ionospheric protons by electric fields inside the Jovian ionosphere analogous to plasma motions observed in the Earth ionosphere. This possibility was examined by Bhardwaj and Singhal (1993) with the use of a Monte Carlo model describing the energization and energy degradation of low energy protons in the presence of assumed parallel electric fields. They showed that acceleration of protons by parallel electric fields enhances the dayglow Ly-α intensity by creating nonthermal H atoms which resonantly scatter the solar Ly-α line. However, they also concluded that Ly-α production by accelerated ionospheric protons has a low efficiency.

10.2. Sources of Hot H in the Jovian Aurora

There was no direct in situ measurements of the energy flux of the particles precipitated into the Jovian atmosphere. Hence only indirect data based on the spectroscopic and imaging measurements may be used to place constraints on the energy and flux of auroral particles. Accumulated IUE (Skinner et al., 1984; Livengood et al., 1992) and HST (Gérard et al., 1994a,b) data indicate that the energy flux is considerably larger than in the case of the Earth. This quantity is far better constrained than the particle energy or electron and proton fluxes taken separately since the emerging brightness in the UV H_2 bands is roughly proportional to the total precipi-

TABLE 10.1. Parameters of the auroral precipitation adopted in the model runs.

Case	Precipitating particles	Energy (keV)	Energy Flux $(\mathrm{erg\,cm^{-2}\,s^{-1}})$
C1	electron	22	1
C2	electron	22	100
C3	electron	0.22	6
C4	electron +	22 +	100 +
	electron	0.22	6
C5	proton	30	10
C6	proton	3	10
C7	proton	0.3	10
C8	electron +	22 +	100 +
	electron +	0.22 +	6 +
	proton	0.3	10

tated power. Consequently, we examined two extreme (soft and hard) cases of electron precipitation and varied their energy flux from 1 $\mathrm{erg\,cm^{-2}\,s^{-1}}$ (probably typical of the weak unstructured background auroral emission) to 100 $\mathrm{erg\,cm^{-2}\,s^{-1}}$ for the auroral arc (Table 10.1). These values are considered as corresponding to fairly moderate fluxes because $HST\text{-}FOC$ observations indicate that in bright arcs the precipitated power is typically 100 $\mathrm{erg\,cm^{-2}\,s^{-1}}$, though they may reach 10 times higher values in bright events (Gérard et al., 1994b). The incident proton energy varied by 2 orders of magnitude, but was kept low as there was no observational evidence that energetic protons play a significant role. Thus the proton energy flux was set to 10 $\mathrm{erg\,cm^{-2}\,s^{-1}}$, that is 10 % of the electron contribution when the two types are energetic particles are combined.

10.2.1. ELECTRON PRECIPITATION

The main sources of hot H formation by electron impact are:

a) dissociative excitation and ionization of H_2 molecules according to the reaction chains:

$$H_2 + e \rightarrow H_2^* \rightarrow \begin{cases} H_2^*(1) \rightarrow \begin{cases} H_s(1s) + H_s(2l) + e \\ H_s^+ + H_s(1s) + e' + e \end{cases} \\ H_2^*(2) \rightarrow \begin{cases} H_f(2l) + H_f(2l) + e \\ H_f^+ + H_f(2l) + e' + e \end{cases} \end{cases} \qquad (10.1)$$

Dissociative excitation and ionization of H_2 can occur from singly $H_2^*(1)$ or doubly $H_2^*(2)$ excited states (Strathdee and Browning, 1979; Ajello et al., 1995a). The former lead to the "slow" component H_s and the latter to the "fast" component. The "slow" H atoms are formed with kinetic energies ≤ 1 eV, while the "fast" ones have an energy distribution up to 10 eV with a maximum near $4 - 6$ eV (Ajello et al., 1995b);

b) ionospheric chemical reactions induced by the electron impact as follows:

$$
\begin{array}{ll}
H_2^+ + H_2 \rightarrow H_3^+ + H_f & \text{(a)} \\
H_3^+ + e \rightarrow H_2 + H_f & \text{(b)} \\
H_3^+ + e \rightarrow H_f + H_f + H_f & \text{(c)} \\
H_2(v > 0) + H \rightarrow H_2(v' < v) + H_f & \text{(d)}
\end{array}
\qquad (10.2)
$$

The H atoms formed in these exothermic reactions have a maximum kinetic energy distribution at (a) 0.585 eV; (b) 4.5 eV; (c) 0.77 eV; and (d) $E(v = 1 - 0) = 0.5$ eV broadened by the thermal energy distribution of the interacting particles. We note that because of the low exothermicity of channel (d) the effect of vibrationally excited H_2 molecules on hot H formation was not included in our model.

Hot H atoms formed in reactions (10.1) and (10.2) have kinetic energies threshold of 10 eV and lose their energies in elastic and inelastic collisions with the ambient thermal molecular and atomic hydrogen:

$$
H_f + \begin{cases} H_2 \rightarrow H_2^* + H_{f'} \\ H \rightarrow H_{f'} + H_{f''} \end{cases}
\qquad (10.3)
$$

The processes of H_2 excitation after interaction with hot H are energy dependent and will be considered later on. The cascade sharing of the hot H atom kinetic energy in collisions (equation 10.3) leads to the formation of secondary hot H atoms with $E_{f'} < E_f$. As a result of the competition between processes of production (10.1), (10.2) and relaxation (10.3), a steady-state energy distribution of hot H is eventually reached in the Jovian upper atmosphere.

10.2.2. PROTON PRECIPITATION

The interaction of precipitating energetic protons of magnetospheric origin with molecular and atomic hydrogen of the Jovian thermosphere are described by the following additional channels (as well as the electron processes (10.1) and (10.2) from proton-induced secondary electrons):

$$
H_f^+ + H_2 \rightarrow \begin{cases} H_{f'}^+ + H_2^* & \text{(a)} \\ H_{f'}^+ + H_2^+ + e & \text{(b)} \\ H_{f'} + H_2^+ & \text{(c)} \end{cases}
\qquad (10.4)
$$

$$H_f^+ + H \rightarrow \begin{cases} H_{f'}^+ + H_f^* & \text{(a)} \\ H_{f'}^+ + H^+ + e & \text{(b)} \\ H_{f'} + H^+ & \text{.} \quad \text{(c)} \end{cases} \qquad (10.5)$$

This includes momentum and energy transfer in elastic and inelastic collisions (10.4a), (10.5a); ionization of target particles (10.4b), (10.5b); charge transfer (10.5c); and electron capture (10.4c) collisions. The excited states of H_2^* are formed in rotational states ($J = 0 - 2; 1 - 3$) and vibrational states ($v \rightarrow v' \leq 9$). For H^*, we consider excitation of the H($2l = $ s,p) state. The main channels of proton energy loss are charge transfer (10.5c) and electron capture (10.4c) reactions in which the hot H_f atoms produced have the kinetic energy of the precipitating protons.

The energetic H atoms produced by proton impact further interact with the main constituents of the Jovian upper atmosphere, transfering their momentum and kinetic energy by elastic and inelastic collisions, ionization and stripping processes:

$$H_f + H_2 \rightarrow \begin{cases} H_{f'} + H_2^* & \text{(a)} \\ H_{f'} + H_2^+ + e & \text{(b)} \\ H_{f'}^+ + H_2 + e & \text{(c)} \end{cases} \qquad (10.6)$$

$$H_f + H \rightarrow \begin{cases} H_{f'} + H_{f''}^* & \text{(a)} \\ H_{f'} + H^+ + e & \text{(b)} \\ H_{f'}^+ + H + e & \text{.} \quad \text{(c)} \end{cases} \qquad (10.7)$$

The inelastic channels considered for H_f-H_2 collisions are rotational ($j = 0 - 1, 1 - 3$) and vibrational ($v = 0 - 1$) excitations, Ly-α, Hα, and Hβ excitation (10.6a), ionization (10.6b) and stripping (10.6c) processes. Excited states of H formed in (10.7a) are the same as in the reaction (10.5a). Formally, in the interactions (10.6) and (10.7) the negative H_f^- ions may be formed in electron capture reactions (Gealy and Van Zyl, 1987):

$$H_f + H(H_2) \rightarrow H_f^- + H_s^+(H_{2,s}^+) \quad .$$

These negative ions are efficiently lost in electron loss (or charge transfer) reactions:

$$H_f^- + H_s(H_{2,s}) \rightarrow H_f + H_s^-(H_{2,s}^-) \quad .$$

Therefore, these fast negative hydrogen ions are immediately converted into fast H atoms.

The secondary fast $H_{f'}$ atoms and $H_{f'}^+$ ions produced by momentum transfer (10.6a, b; 10.7a, b) and stripping (10.6c, 10.7c) reactions recycle the reaction set (10.4)-(10.7). This means that interaction of the precipitating protons with the main neutral constituents of the Jovian atmosphere must be considered as a cascade process producing a growing set of fast secondary $H_{f'}$ atoms. Finally, this cascade process of elastic and inelastic interactions results in the formation of the hot H atom population in the auroral Jovian upper atmosphere.

The cross-sections for the fast proton and H atom interactions with H and H_2 are characterized by a strong and complex energy dependence (Janev et al., 1987; Phelps, 1990). This fact leads to variations of the branching ratios of the different channels of the reaction set (10.4) – (10.7) as a function of the impact energy. In this model, we use recent available data for total and differential cross-sections as described below. Further details are given by Bisikalo et al. (1996).

10.3. Model Description

The physical system taking into account the production, elastic and inelastic relaxation and transport processes for the hydrogen atoms and ions in the transition region of the Jovian thermosphere may be adequately represented by the system of nonlinear Boltzmann-type kinetic equations earlier described in detail. The stochastic method is applied to solve the system and investigate the energy degradation of the H^+-H beam and the formation of hot hydrogen atoms due to proton and electron precipitation in the Jovian aurora. In the numerical simulations that follows, the evolution of the system of modeling particles due to collisional processes (10.1) – (10.7) and particle transport is calculated from the initial state to the steady state.

In order to minimize boundary effects, the lower boundary was set at an altitude below the homopause (about 350 km) and the upper boundary was set at 6500 km where the atmospheric gas flow is practically collisionless. The region of the Jovian atmosphere under study was divided into 24 radial cells. The altitude dependent cell size was chosen from the condition that it must be equal to or smaller than the free path length near the lower boundary of each cell.

The auroral atmospheric model (Fig. 10.1) was developed using the recent HST GHRS H_2 vibrational temperature determination in the auroral zone (Trafton et al., 1994). It utilizes the one-dimensional chemical diffusion approach (Waite et al. 1983) updated with regard to the Jupiter/Ulysses data. Vertical profiles of the neutral species H, CH_4, C_2H_2, C_2H_6, and CH_3 and the major ions H^+ and H_3^+ were calculated using an eddy diffusion coefficient of 2×10^6 $cm^2 s^{-1}$ beginning at the CH_4 homopause level (pressure ~ 5 μbars) upward. The resulting hydrocarbon and hydrogen densities proved to be consistent with the work of Gladstone et al. (1996). In turn, the neutral temperature structure was constrained near the homopause by

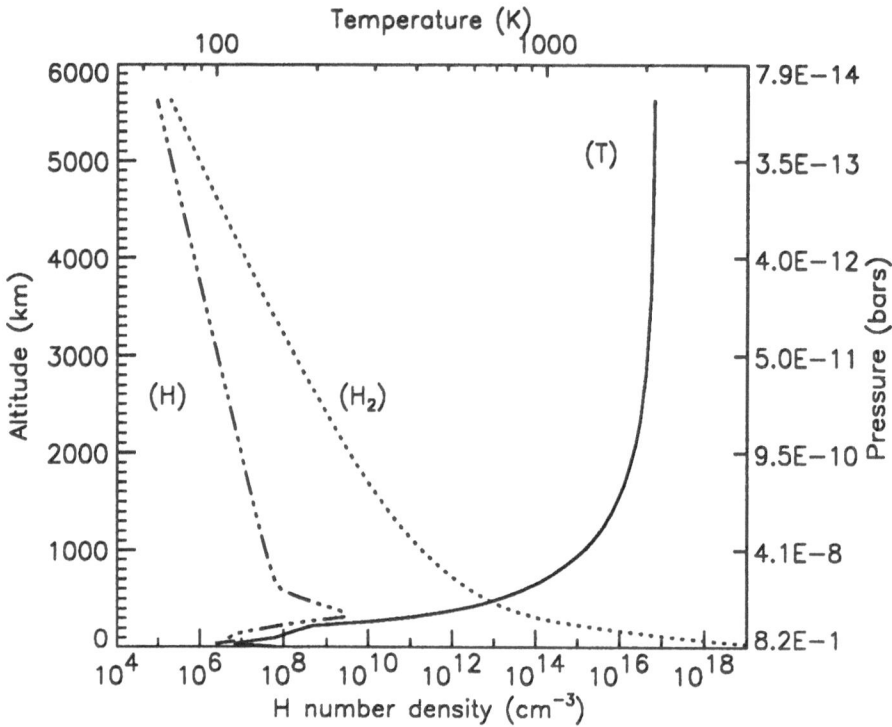

Figure 10.1. Model of the Jovian neutral upper atmosphere used in this study.

the value resulting from Drossart *et al.'s* (1993) analysis, which exhibited a steep thermal gradient of 2 K/km starting at the pressure level of 55 microbars and reaching the temperature of 220 K at the 20 μbar level. The height profile also matched the H_2 rovibrational temperatures at the peak of the auroral emissions (550 K at 20 μbar) obtained by Trafton *et al.* (1994), being also constrained by the H_2 quadrupole temperatures of 740(+490, −200) K between 1 and 0.01 μbar pressure levels (Kim *et al.*, 1990). In a similar height range, though with presumably lower pressures, H_3^+ temperature limits of 800 to 1200 K (Drossart *et al.*, 1989) were used. These combined observational constraints proved to be best fitted by an empirical Bates atmospheric profile having an exospheric temperature of 2070 K at the pressure level of 0.1 nbar.

Different models were used to determine the effects of electron and proton precipitation. In particular, an updated version of the Waite *et al.* (1983) two-stream electron transport model was used to evaluate the electron degradation patterns.

Interaction of proton-hydrogen beams with the Jovian atmosphere was calculated using the same method as that described in Chapter 9 to investigate the precipitation of energetic O^+ ions into the high latitude Earth's thermosphere. It is also similar to the approach used in that study to describe the energy distribution function of fast H atoms. Because observational information on the energy spectrum of precipitated auroral protons is not available, monoenergetic proton fluxes at different energies were taken. It was futher assumed that the incident proton flux is injected isotropically into the downward hemisphere, carrying an energy flux of 10 $erg\,cm^{-2}\,s^{-1}$. Although this value is relatively small compared to the energy delivered by electron precipitation it seems admittable, as there is no observational evidence that proton precipitation is dominant on Jupiter.

10.4. Fast H Atom Distribution

The kinetic energy distribution functions (EDF, in particles/cm^3 eV) calculated for different electron impact conditions (runs C1 – C3) at 3630 km are shown in *Fig. 10.2*. These functions are normalized to the local hydrogen density. The dashed curve corresponds to a Maxwellian distribution function for the local temperature ($T = 2070$ K). It is evident that the real steady state kinetic energy distribution function departs from equilibrium and significantly diverges from the local Maxwellian distribution. It was found in the calculations performed that these nonthermal perturbations increase with the energy flux. It is interesting to note that the presence of soft electrons (run C3) leads to substantial increase of the EDF perturbations. This feature may be caused by two factors: 1) the hot H production rates due to soft electrons are significantly larger; and 2) the peak altitude of the energy deposition increases in the case of soft electron precipitation. The decrease of relaxation properties of the atmospheric gas with altitude in the transition region leads to more pronounced EDF perturbations at high altitudes.

The calculated hot H EDF (run C4), representing the total input of combined soft and hard precipitating electron fluxes, is shown in *Fig. 10.3*. The comparison of *Figs. 10.2* and *10.3* shows that the main contribution to hot H formation is provided by precipitation of soft electrons. These electrons appear to be responsible for the presence of hot H atoms with energies less than 10 eV, as can be expected when considering the energy outputs in reactions (10.1) and (10.2). This means that it is hardly possible to explain the observed broadening of the Ly-α profiles (up to $\sim$ 100 eV, Clarke *et al.*, 1994) by hot H atoms produced by electron impact only.

Detailed analysis of the distribution functions presented in *Figs. 10.2* and *10.3* implies that the hot H population fraction in the Jovian aurora is formed by the contribution of the following populations:

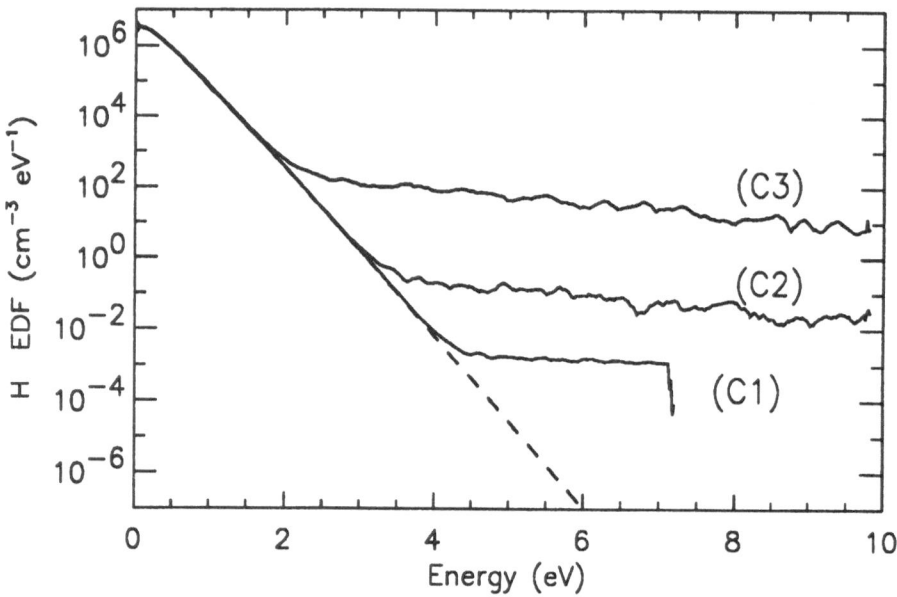

Figure 10.2. Kinetic energy distribution of the hot hydrogen atoms in the Jovian aurora calculated for different electron energies and fluxes at 3630 km: (C1) case 1 of Table 10.1; (C2) case 2; (C3) case 3. The dashed curve is the Maxwellian distribution at $T = 2070$ K.

- thermal hot atoms part of the high energy tail of the Maxwellian distribution (we formally define hot particles as those with kinetic energies higher than three times their thermal energy);
- nonthermal atoms including: (a) primary energetic source particles produced in dissociative excitation, ionization, and chemical processes (10.1), (10.2); (b) secondary energetic atoms produced in collisions between the thermal and nonthermal gas components.

The relative magnitude of the thermal and nonthermal H fractions depends on altitude. *Fig. 10.4* shows the vertical distribution of the thermal and nonthermal hot species, the total hot H density, and the total H density in the adopted model of the Jovian atmosphere. It shows that contribution of the nonthermal H atoms dominates above 4000 km, though their input becomes significant above 2500 km. The peak density of nonthermal hot H is located near 3600 km. It is also seen from *Fig. 10.4* that the inclusion of the nonthermal H population leads to increase of the scale height in comparison with the thermal hydrogen scale height.

As it was discussed earlier, electron precipitation cannot explain the formation of hot H atoms with energies higher than 10 eV, in contrast to the observed broadening of the auroral Ly-α profile. Since the Bhardwaj

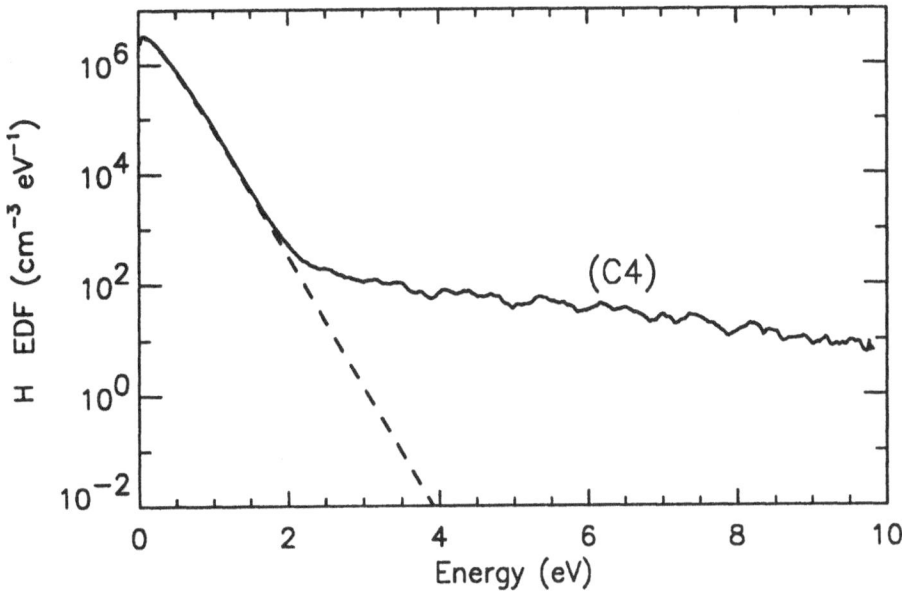

Figure 10.3. Same as *Fig. 10.2* for the combination of soft and hard electron precipitation of case 4 (solid line). The dashed line represents the local Maxwellian distribution at the ambient gas temperature.

and Singhal (1993) study indicated that protons acceleration by parallel electric fields cannot be involved as an efficient production mechanism, we consider proton precipitation as an alternative source of high energetic H atoms.

The calculated EDFs for different proton flux conditions (runs C5-C7) at 3630 km are shown in *Fig. 10.5*. Again, the dashed curve corresponds to the Maxwellian distribution function at the local temperature $T = 2070$ K. It is clear that, ulike electron precipitation, proton precipitation results in the formation of a high energy EDF tail. The EDFs are shown up to 150 eV because these high values correspond to the energies of hot H required by the observed Ly-α profile in the Jovian aurora. As in the case of electron precipitation, incorporation of soft protons into the model leads to an increase of the EDF perturbations.

It is also worthwhile to consider the fate of precipitating protons injected isotropically into one hemisphere and propagating downward. Following the schemes (10.4) and (10.5), the incident high energy protons either become forward scattered in the interaction processes accompanied by momentum transfer, or form fast forward scattering H atoms in charge exchange reactions. In these cascade processes, the initial protons or high energy H atoms

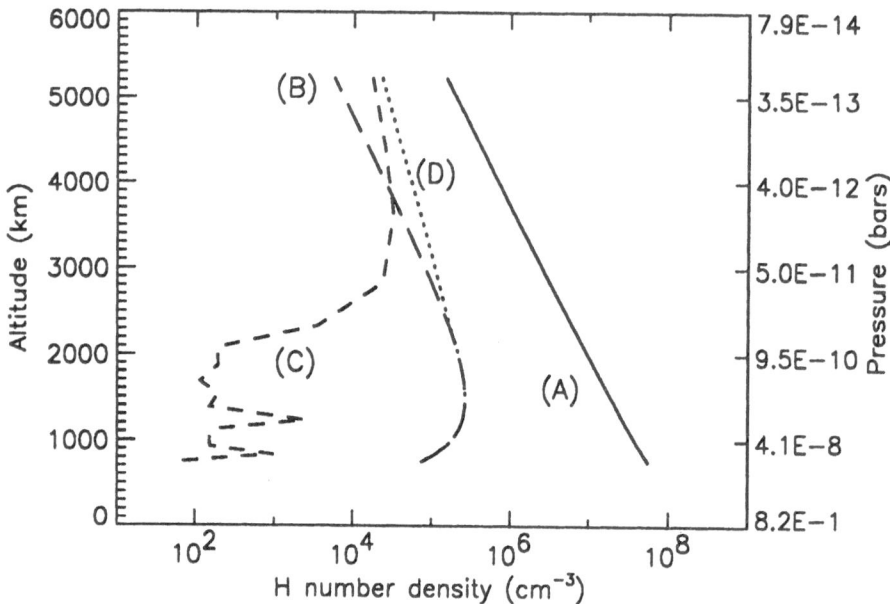

Figure 10.4. Altitude distribution of the different components of the hot H population: (A) total H; (B) hot H atoms in the Maxwellian tail; (C) superthermal atoms; (D) total hot H. The electron precipitation is identical to that shown in *Fig. 10.3*.

generate a growing amount of hot H atoms of secondary origin which, in turn, also participate in elastic and inelastic collisions with the ambient gas. Momentum transfer in these collisions leads to an increasing isotropy of the secondary hot H flux. The calculated EDF at 3630 km for downward and upward moving particles are shown in *Figs. 10.6* and *10.7*, respectively. It is seen that up to 150 eV, the EDFs in both directions are practically identical, i.e., the downward and upward fluxes are equal in this energy range.

To simplify the analysis of the EDF patterns it is convenient to examine the cumulative distribution function

$$n_H(E) = \int_E^\infty f_H(E') \, dE' \quad ,$$

where $f_H(E')$ is the calculated EDF for hot thermal and nonthermal H atoms. This function defines the number of particles having energy greater than a given energy E. Examples of these cumulative distribution functions at the altitude of the hot H peak density for combined electron and proton precipitations are shown in *Fig. 10.8*. These functions may be useful to

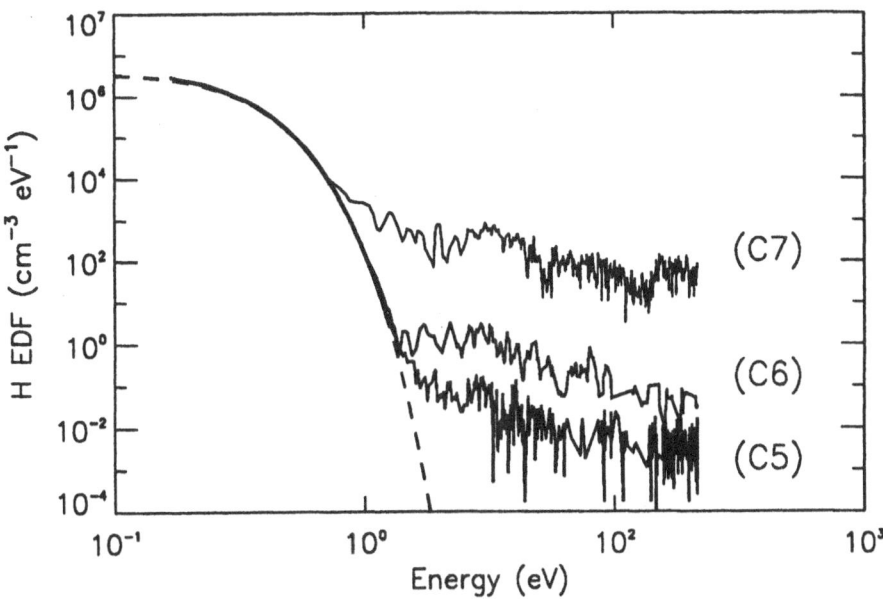

Figure 10.5.   Energy distribution function of hot H due to proton precipitation at 3630 km: (C5) case 5; (C6) case 6; (C7) case 7. The dashed line represents the local Maxwellian distribution.

interpret the observed spectra, especially the wing broadening, as they give the density at each energy interval of the EDF. The set of cumulative distribution functions at different heights may be also used to obtain the energy dependent column densities of hot H atoms.

In *Fig. 10.9* the height profiles of nonthermal H atoms produced by electron (run C4) and proton (runs C5 – C7) precipitation are compared. It is seen that only in the case of soft protons densities of nonthermal H do become higher than in the case of electron precipitation. It should be noted in this regard that, although the densities of hot H produced by protons and electrons are comparable, hot atoms produced by protons have higher energies and are responsible for the formation of the far wings of the EDF.

Fig. 10.10 presents the height profiles of the total hot H atom density calculated in run C8 when the electron (22 keV + 0.22 keV) and proton (0.3 keV) contributions are combined. A crude estimation of the earlier mentioned contribution of a 10% proton energy flux (as in the case C8) may be based on the lack of detection of red Doppler-shifted Ly-α emission observed with the *IUE* (Clarke *et al.*, 1989). For a total H_2 emission of 76 kR, an upper limit of 100 R/0.3 Å due to protons with energy in excess

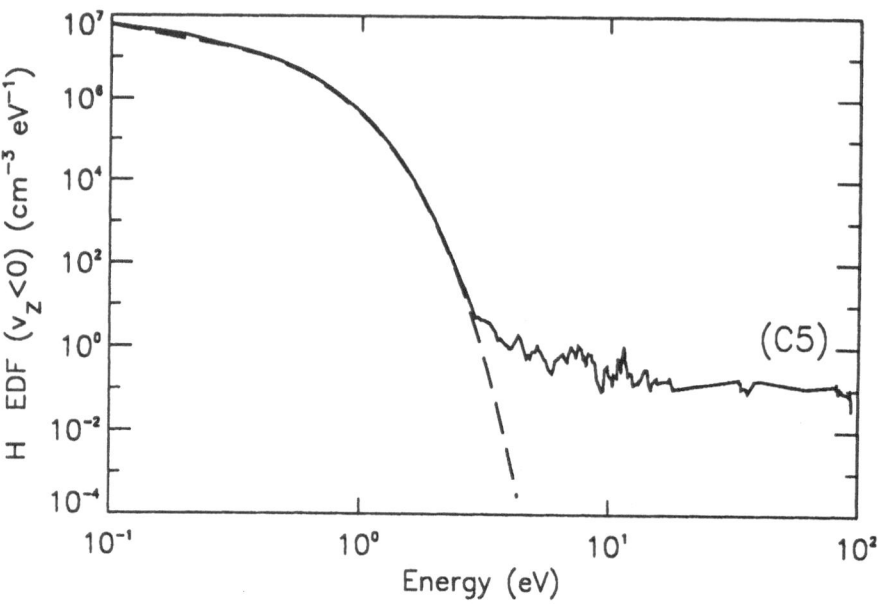

Figure 10.6. Energy distribution function of the downward moving hot H atoms for case 5. The dashed line is the Maxwellian EDF.

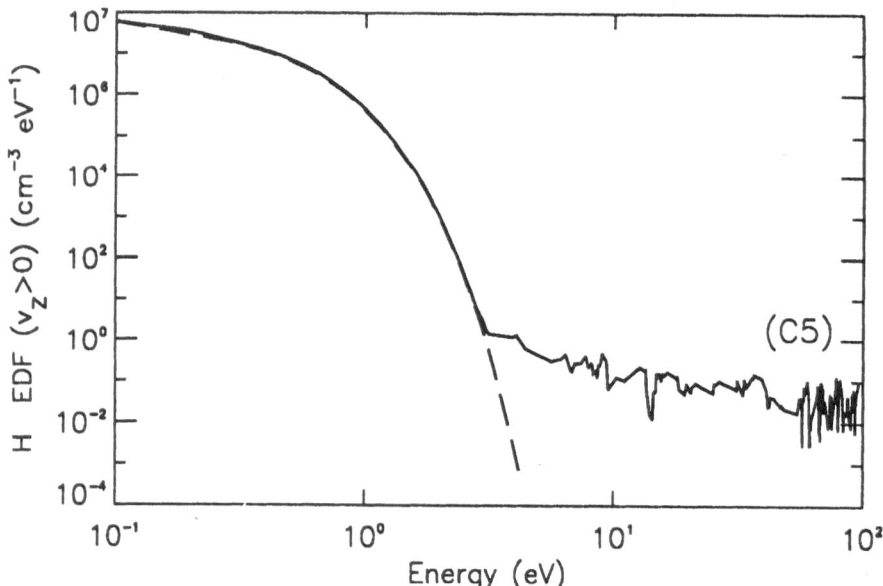

Figure 10.7. Energy distribution function of the upward moving hot H atoms for case 5. The dashed line is the Maxwellian EDF.

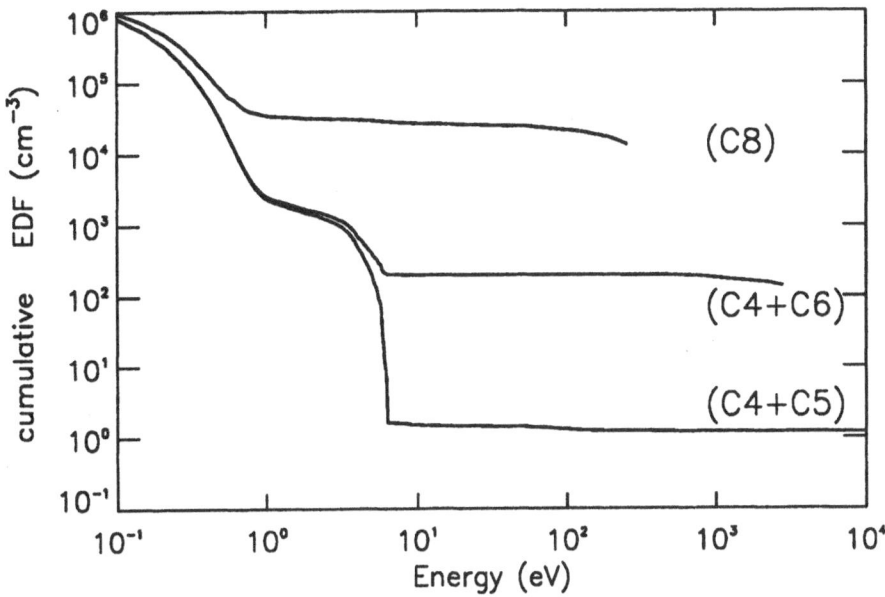

Figure 10.8. Cumulative distribution function of the hot H atoms calculated at 3630 km: (C4+C5) combined cases 4 and 5 of Table 10.1; (C4+C6) combined cases 4 and 6 of Table 10.1; (C8) case 8 of Table 10.1.

of 130 eV was set from the observation. Assuming an effective width of ~ 5 Å for a hypothetical shifted and broadened line, this limit would correspond to about 1.5 kR. Rego *et al.* (1994) estimated the column production rate of Ly-α by direct 300 keV proton impact precipitated in the Jovian atmosphere as equal to 1.5 kR, for a proton aurora with total H_2 emission of 12 kR. Consequently, scaling considerations show that a 10% proton flux would lead to ~ 1 kR of shifted Ly-α emission for 76 kR total H_2 emission, what is quite close to the above referred limit of the IUE observations. We thus conclude that the 10% proton flux assumed for the case C8 is close to the upper limit set by spectroscopy of the Jovian aurora. The vertical distribution of the thermal hot components, the hot H density of electron origin (run C4) and the total H density are also shown in this figure. It is evident that in this combined case the density of hot H at high altitudes becomes comparable to the density of ambient H. The peak density of hot H is located at an altitude of about 3600 km. The column density of hot H in run 8 is about $1 - 2 \times 10^{14}$ cm^{-2} for altitudes up to 6500 km.

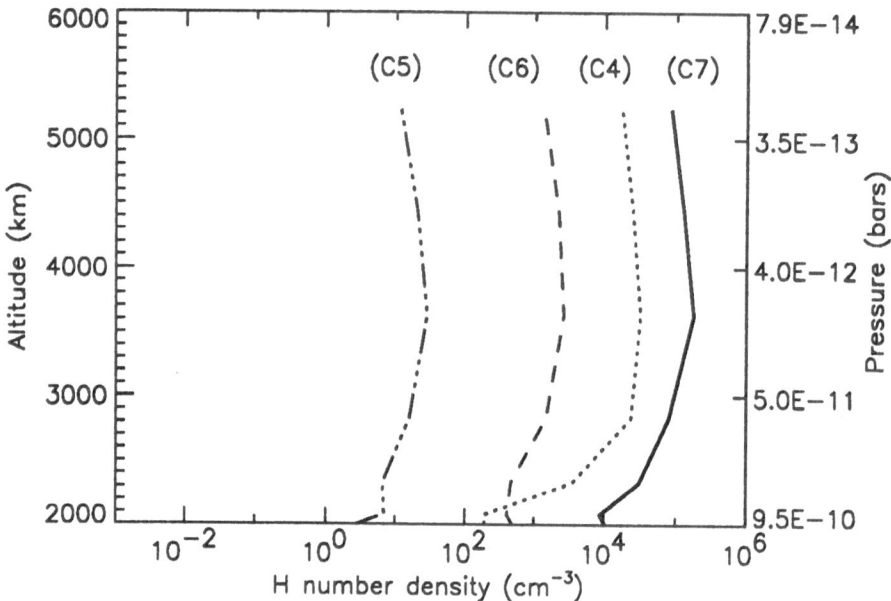

Figure 10.9. Altitude profiles of the superthermal H atoms produced by electron and proton precipitation: (C4) case 4, (C5) case 5, (C6) case 6 and (C7) case 7.

10.5. Ly-α Profile in the Presence of hot Hydrogen

The EDFs of fast H atoms calculated at all altitudes for the case 8 of Table 10.1 were used in a radiation transfer code to calculate the profile of the emerging Ly-α emission for an emission angle of 60° ($\mu = 0.5$). Calculations were performed with and without hot H atoms. The radiative transfer code used to simulate the auroral Ly-α line profiles is a modified version of the angle-averaged partial redistribution code described by Gladstone (1988). This code uses a Feautrier-type matrix solution of the radiative transfer equation for an optically thick resonance line transition in a plane-parallel atmosphere having variable properties (e.g., temperature, composition). The Ly-α line parameters used in the study are as listed in Table 2 of Gladstone (1988). For this study, the Jovian atmosphere from 250 km ($P \sim 2$ μbars) to 10,000 km above the ammonia cloud tops was discretized into 99 levels. Line intensities were represented with 4 streams (2 up and 2 down) and 36 wavelengths (from 0 to 0.74 Å from the line center).

The Ly-α flux at 1 AU is assumed to be 3.2×10^{11} ph/cm^2 s with a line profile consistent with that given by Lemaire *et al.* (1978). The symmetric profile due to various sources is given in *Fig. 10.11* for the case of no hot

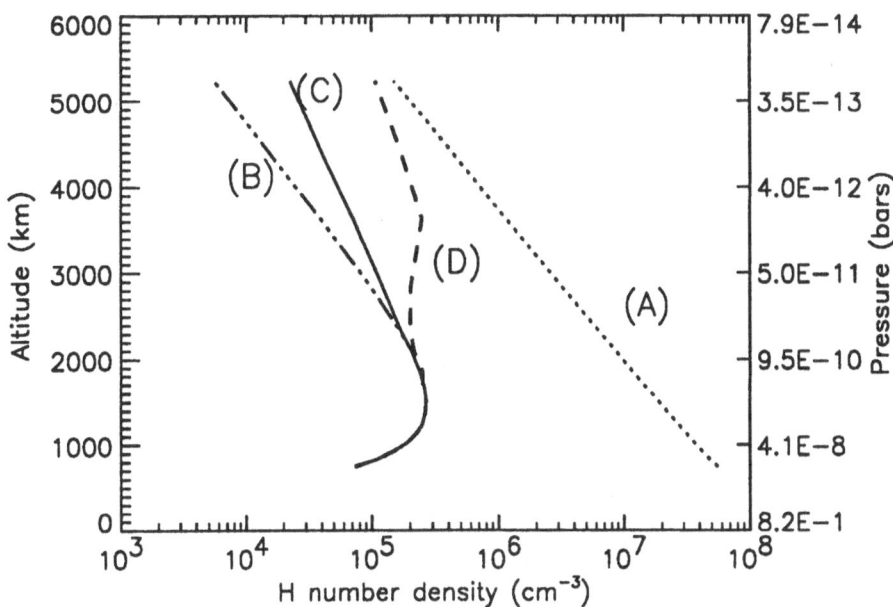

Figure 10.10. Altitude profiles of different components of hot H formed by electron and proton precipitation: (A) total H density; (B) hot H atom in the Maxwellian tail, (C) hot H atoms population due to electron precipitation (case 4); (D) hot H population due to combined electron and proton precipitation (case 8).

H contribution. Scattered solar light contribution is quite small. For the $e + H_2 \rightarrow H^* + H + e$ dissociative excitation, it is assumed that the nascent line profile of slow H fragments is characterized by a temperature of 3000 K associated with an energy of 0.27 eV, typical of the slow H atoms released in the process (Ajello *et al.*, 1995a). The fast $e + H_2$ fragments in the (2s, 2p) states are assumed to have a nascent line profile corresponding to 50,000 K, which is about 4.3 eV, in agreement with the Ajello *et al.*'s measurements as well. The effect of the presence of the fast H atom population is shown in *Fig. 10.12*. For inclusion into the radiative transfer code, the calculated hot H distribution function was approximated by the sum of three Maxwellian functions whose density and temperature were fitted to those calculated for the case 8 of Table 10.1. In this case, all 4 contributions are considerably broadened by scattering due to the non thermal H atoms.

Finally, the energy distribution function of the nonthermal hydrogen atoms in the Jovian aurora was calculated as function of altitude for various energies and fluxes of incident electrons and protons. The calculated spectrum extended from 0 eV up to the energy of primary particles. It was shown that the EDF perturbation depends on the energy of primary beam

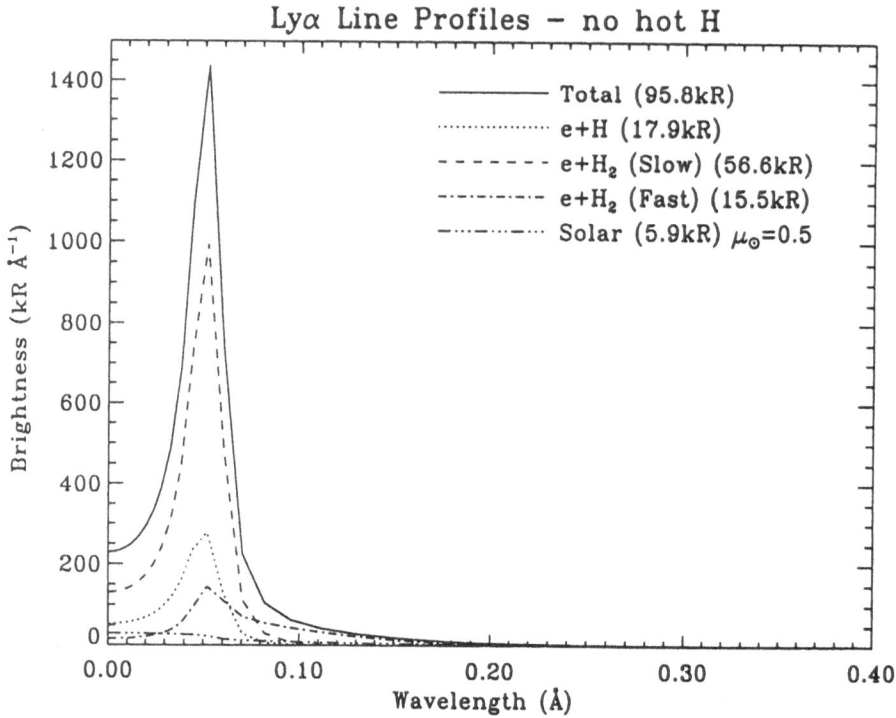

$Figure\ 10.11.$ Model Ly-α line profiles (a symmetric line profile with deviation from 0 to 0.74 Å from line center is assumed) for the auroral line emergent at 60° from the vertical ($\mu = 0.5$), with no hot H contribution to the scattering. The total emission shown is the sum of contributions from direct excitation of H by electron impact, dissociative excitation of H_2 by electron impact (with consideration of both fast and slow H fragments) and resonantly scattered solar Ly-α ($\mu_\odot = 0.5$).

and is more pronounced in the thermosphere for soft than for hard particles. The model revealed that only protons form an extended non thermal ($E > 10$ eV) energy tail, while at intermediate energies ($E < 10$ eV) both protons and electrons contribute significantly to the formation of this nonthermal population. No anisotropy was found in the EDF up to 150 eV for both electron and proton precipitation. The point should be stressed, however, that this conclusion applies only to the H ground state. Excitation and emission from (2p, 2s) excited states were not considered in this study. One may assume that direct excitation of H(2p, 2d) by energetic protons would produce anisotropy of the atom distribution.

We may conclude that radiative transfer calculations in the presence of the nonthermal H population calculated for $\sim$ 100 erg/cm^2 s flux of

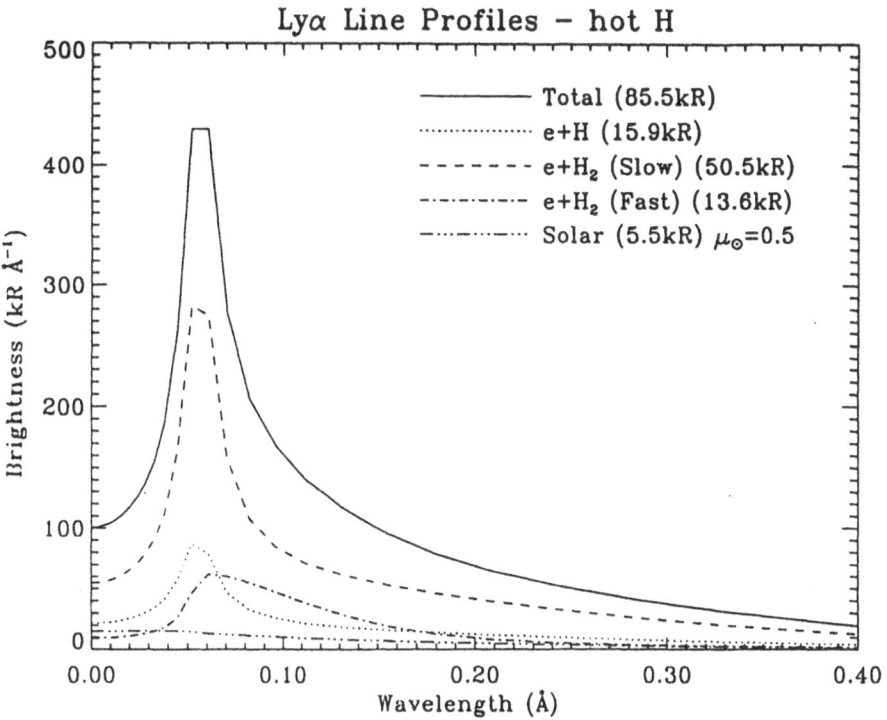

Figure 10.12. Same as Fig. 10.11, with the effects of scattering by hot H atoms, consistent with the distribution of case 8 of Table 10.1, included in the radiative transfer calculations.

electron and 10 erg/cm^{2}s of proton show a significant effect on the line profile indeed. In particular, the line wings become considerably broader than for thermal hydrogen only. Such line broadening was observed on *HST* with the *GHRS* spectrometer.

CHAPTER 11

TRANSITION REGIONS IN COMETARY ATMOSPHERES

The study of the physics and chemistry of a cometary atmosphere encompasses numerous complicated processes of its formation and evolution. From a mathematical point of view, the most adequate description of the problems involved is possible based on the methods of the nonequilibrium kinetic theory of rarefied gases. This is due to the strong gradient change of gas parameters in the inner coma and the predominantly nonequilibrium nature of the collisional physical and chemical processes induced by the impact of solar UV radiation on the cometary gas (Marov and Shematovich, 1985; 1987; Bisikalo and Shematovich, 1987; Bisikalo et al., 1989).

We shall first briefly describe the observational data available and some theoretical constraints on the chemical structure of the cometary nucleus and inner coma and then focus on the appropriate kinetic models development using stochastic simulation approach for their evaluation (Marov and Shematovich, 1985; Bisikalo and Shematovich, 1987).

11.1. Chemical Composition of the Cometary Nucleus and Coma

11.1.1. GENERAL CHARACTERISTICS AND COMPOSING ELEMENTS

Absorption of solar radiation by a cometary nucleus results in heating and sublimation of its volatile constituents, which gives rise to the formation of the cometary atmosphere. The model of cometary nucleus chemical composition is defined on the basis of the analysis of spectral observations of cometary near-nuclear regions in the ultraviolet, optical, infrared, and radio wavelengths. These observations were carried out from both ground-based and space observatories (see, e.g., Feldman, 1990; Grewing et al. (Eds.), 1988; Huebner (Ed.), 1990).

Insufficiently reliable knowledge of the structure and composition of the nucleus greatly complicates mathematical modeling of comets, in particular the processes of chemical kinetics in coma. Although baseline models are usually confined to the analysis of pure H_2O nucleus, an accurate description of phenomena in the coma requires more complete information about its composition because it defines the abundances of both parent and daughter products of sublimation, with allowance for photolysis and subsequent chemical transformations. Thus the correctness of models de-

239

pends on the validity of the assumptions about the chemical composition and surface patterns of the nucleus and the processes occurring in the subsurface and boundary zones.

Whipple's model (Whipple, 1950; 1984), which has won the most recognition, starts from the assumption that a cometary nucleus contains mainly water ice, while other fractions of the ice component (CH_4, NH_3, and CO_2) appear as macroscopic ingredients of its composition. Attempts have been made to modify this model by assuming that compounds were already formed at the stage of formation of the nucleus: clathrate-hydrate inclusions of the (CH_4, NH_3, CO_2) nH_2O type (Delsemme and Swings, 1952; Delsemme, 1976). This hypothesis encounters grave difficulties, however, since comet spectra contain no signs of methane or ammonia, which should be liberated more easily than CH_4 and NH_3 condensates as the clathrates easily decay with increasing temperature. Moreover, the mechanism of condensation of individual molecules (without the formation of a hydrate envelope) at the stage of growth of the cometary nucleus seems more likely.

The idea that cometary ices contain strongly bound ion-molecular complexes or clusters, rather than weakly bound groups of molecules (including clathrate-hydrates), seems preferable at first glance (Shul'man, 1982). They form a structure in which one or several neutral molecules are attached to an ion by a monopole-dipole interaction, i.e., a structure of the (H^+, HNO_3^-, N_2^+, ·) nH_2O type. It is important to emphasize that an energy of ~ 10 eV is released in the recombination of a pair of such ions (Stakhanov, 1979), and since the energy required to sublime a molecule of water ice is ~ 0.5 eV, this is enough to sublime ~ 20 H_2O molecules. In other words, even a small admixture of cluster ions (several percent) could serve as a fairly efficient, heterogeneous, internal energy source for a cometary nucleus. From these standpoints, attempts have been made to explain, in particular, certain anomalous features in the observed behavior of comets, such as the perihelion brightness asymmetry (up to the complete disappearance of a comet after perihelion passage), brightness variations over characteristic times of $\sim 10^3$ sec, departures from the r^{-4} dependence in the luminosity variation, and signs of dust in faint comets at heliocentric distances $r \geq 2.5$ AU (Shul'man, 1982). The latter could be caused, in principle, by ion-molecular clusters (capable of simulating optical scattering by dust particles) that appear during explosive decomposition of a frozen cluster into a plasma.

A simpler explanation should, nevertheless, be sought for the experimental data presently available. Indeed, our present understanding rests on an H_2O ice model of a nucleus with inclusions of highly volatile ices. With increasing insolation, the latent heat of the phase transition due to the restructuring of amorphous ice into a cubic structure is released, starting with depths of several dozen meters and at a temperature above 125 K

TABLE 11.1. Elements and compounds detected in comets.

	Groups of Chemical Compounds	Region of Observations
I	C, C_2, C_3, CH, CN, CS, CH_3CN, CO, HCN, NH, NH_2, O, OH, H_2O, CH_3OH, H_2CO, H_2S, Na, S, Si, H_2, CO_2, NO	In the cometary atmosphere at different distances from the Sun
II	Si, Ca, CO, Cr, Cu, Fe, V, K, Mg	Near the Sun
III	C^+, CH^+, CO^+, CO_2^+, CN^+, N_2^+, OH^+, H_2O^+, and silicate particles	

Original data from Feldman (1990); Hoban *et al.* (1993); and Whipple (1984) with supplements reflecting recent results.

(Bar-Nun and Kleinfeld, 1989; Prialnik and Bur-Nun, 1990). Sublimation of highly volatile ices in the mixture would lead to the formation of cavities (pockets) in which pressure rises, which, together with thermal deformations, is capable to produce cracks in the surface crust or its partial ejection and the uncovering of individual 'fresh regions'. A gradual change in the activity of the nucleus results, causing the formation of an atmosphere and tail, the observed brightness anomalies, and broad and narrow gas-dust ejections — the jets. Such a scenario seems fairly well founded, and it is important to understand to what extent it is reinforced by the available data on chemical composition.

Unfortunately, only phenomena associated with the liberation of cometary gas and dust, rather than those occurring in the nucleus itself, are yet accessible to observation. About 40 chemical components in the form of molecules, atoms, radicals, and ions, listed in Table 11.1, have been discovered on the basis of such observations. They consist of both parent components and dissociation fragments of those and other (unidentified) original compounds (Mendis, 1988). The results of groundbased spectrometry, supplemented by spectroscopic and mass-spectrometric measurements in the coma of comet Halley returned from spacecraft, have provided the main information for numerous comets.

The spectral properties of comets are fairly uniform, on the whole, and differ little from one to another. This pertains to both long- and short-period comets, regardless of the degree of activity they display, which probably indicates the closeness of their chemical compositions and a common origin. At the same time, individual features are also observed in the spectra (see, e.g., Cochran, 1987; Cochran *et al.*, 1992; A'Hearn and Millis, 1980; A'Hearn, 1988; Newburn and Spinrad, 1984; 1985; 1988). This may depend both on local irregularities in the material of the cometary nucleus and its age (and hence the peculiarities of the torn-off surface crust) and on the degree of dust release (the dust-to-gas ratio), which varies greatly for different comets and affects the characteristics of space around the comet.

Table 11.1 includes three groups of compounds and atoms present in the coma and tail and also dependent of distance from the Sun. The main gaseous components form group I. Only near the Sun do heavy metals (group II) and ionized components (group III) appear in the makeup of the cometary atmosphere. These daughter products indicate that the parent molecules of the nucleus itself are primarily H_2O, CO_2, hydrogen cyanide (HCN), and methyl cyanide (CH_3CN). Such typical representatives of ices at the periphery of the Solar system as ammonia and methane (which, along with H_2O obviously serve as the original products of the components of group I and of complexes of chemical reactions with their participation and must, with a high probability, appear in the makeup of cometary nuclei) have not yet been, however, detected directly in comets.

As one may see, comets are composed of four main elements: H, C, N, and O. In comparing the observed abundances with the solar (cosmic) abundances of elements, lower relative abundances of hydrogen (by about three orders of magnitude) and inert gases are found in comets. In connection with the H deficit, the redox ratio H/O for comets can be expected to be 1.8, which places important limits on the ratios of components and features of the chemical composition of the nucleus.

In some comets sulfur and sulfur-bearing compounds are presented. They were detected in the S_2 form in comets Iras-Araki-Alcock, Chernis, and Halley and as hydrogen sulfide (H_2S) and the radical CS in comets Austin and Levy (A'Hearn and Feldman, 1985; Krishna Swamy and Wallis, 1987; Crovisier et al., 1991). This radical is probably also responsible for the identification of SI 1815, 1475, and 1425 Å lines in the spectra of comet Wilson (Roettger et al., 1989), belonging to the common parent molecule CS_2, although this latter result is not sufficiently reliable. The short-lived sulfur dimer S_2, which dissociates rapidly in the coma, is also found in the ultraviolet spectrum of comet IRAS-Araki-Alcock; this dissociation probably explains the fact that only an upper limit for its detection has been established for other comets (A'Hearn et al., 1983; Budzien and Feldman, 1992).

As for SO and SO_2, the presence of which is postulated in a number of coma models to be the result of the interaction of S and OH (which should lead to the formation of sulfur monoxide SO with subsequent oxidation to SO_2), they have not been reliably identified in cometary spectra. For now we can evidently speak only of an upper limit of detection for these daughter products, the admixture ratios of which are considerably lower, however, than for other sulfur-bearing compounds (Kim and A'Hearn, 1991).

11.1.2. CARBON AND ORGANICS

It is typical for the coma to contain carbon atoms and radicals, as indicated by the presence of strong C_2, C_3, and CN bands in their ultraviolet spectra. In those spectra the aforementioned four main elements of group I and some simple molecules composed of them have strong resonant transitions, which are preserved even at distances of more than ~ 3 AU (Feldman, 1990; Feldman and Budzien, 1989). This wavelength range is just as effective for detecting and estimating the abundances of other components listed in Table 11.1.

Direct determinations of CO_2 in the coma have yielded only an upper limit for its abundance, so indirect estimates must be used. They can be obtained, for example, by measuring the characteristic CO_2 emission features (the B-X and A-X bands) in the ultraviolet wavelength, although mechanisms of excitation of those bands have been little studied. As for CO, most of these molecules in the coma are in the ground vibrational and rotational states, so the far-IR and radio wavelengths are more informative for estimating their abundance and its variations. Nevertheless, the ultraviolet is also effective, especially in studying chemical transformations in the coma. Bands of the Cameron system are typical of CO in this range. The corresponding measurements of CO_2 and CO emission features, made with high resolution for comets Hartley 2 and Shoemaker-Levy 9 using the *HST*, have yielded $\sim 4 - 9\%$ for the CO_2/H_2O ratio and ~ 2 % forthe CO/H_2O ratio (Weaver *et al.*, 1993).

Such carbon-bearing compounds as methanol (CH_3OH) and formaldehyde (H_2CO) that remain in the nonvolatile state only at temperatures below ~ 100 K have been detected, in turn, in the millimeter and submillimeter ranges (0.3 - 3 mm) in the spectra of comets Austin, Levy, and Swift-Tuttle (Crovisier *et al.*, 1991; Bockelee-Morvan *et al.*, 1993a, b). This means that these original compounds are part of the ice component formed at low temperatures at the periphery of the Solar system.

Methanol, which turned out to be an important component of cometary atmospheres, has also been identified in IR spectra of comets Wilson, Swift-Tuttle, Austin, Halley, Brorsen-Metcalf, and Levy based on the measured fluorescent emission at 3.4 μm (n_2 and n_9 transitions) and 3.52 μm (n_3 transition). Its abundance relative to H_2O ranged from 1 % to 6 %, with the highest values characterizing the new comets Wilson and Austin, as well as comet Swift-Tuttle, for which it reached more than 8 % (Hoban *et al.*, 1993; Mumma *et al.*, 1993). It is therefore tempting to associate the presence of CH_3OH with a cosmogenic source for its origin. The presence of organic indicators in the 3.2-3.6 μm emission spectrum (at 3.36 μm, in particular, associated with the stretching mode of the C-H molecule) has also been reported to present in comets Okazaki-Levi-Rudenko and Bradfield. In four comets, moreover, evidence has been found for the presence of unsaturated aromatic molecules (Bockelee-Morvan et $al.$, 1993a, b), which correlates well with such complex hydrocarbons of the C_nH_m type as naphthalene, phenatrine, pyrene, anthrol, etc. At the same time, a residual feature, centered at 3.43 μm, is revealed in all of these comets; it correlates well with water vapor and thereby indicates gas transport of organic matter, which is present in comparable amounts from comet to comet.

The presence of carbon-containing compounds and various organic compounds indicates that the observed radicals are not formed by chemical reactions in the coma, but appear to be primordial products of their decomposition (Ulrich and Conklin, 1975; Huebner et $al.$, 1982). From an analysis of neutral and ionized components in the coma, with allowance for probable photolysis and subsequent chemical transformations, it was attempted to reconstruct acceptable ratios of anticipated parent molecules (in the form of frozen gases) in the cometary nucleus (Huebner, 1985). The results of such an analysis are of undoubted interest, since they broaden our ideas about the possible complexes of chemical compounds forming the nucleus.

Also noteworthy are data from experiments intended to clarify the photochemical and chemical evolution due to irradiation of the mixture of ices formed from the presumed primordial compounds from the interstellar medium or at the periphery of the solar system. In the experiments (Allamandola et $al.$, 1988) H_2O, CH_3OH, NH_3, and CO in the proper proportions and at 10 K were taken as the original compounds, and they were subjected to UV irradiation from a gas-discharge lamp. One may assume that the products of ultraviolet photolysis of such ice analogs could have originally gone into the makeup of cometary ices. It was found that methanol plays the main role in these transformations. It turned out, however, that on its basis one can expect that more complex organic compounds, involving nitrile (isonitrile) and carbonyl groups, would also be formed in nuclei.

The hypothesis that methane clathrates ($CH_4 \times nH_2O$) are an important component of cometary nuclei (or of particles of the interstellar medium) in combination with hydrocarbons (CH_4, C_2H_6, or C_2H_2), water, and ammonia was used, in turn, as a starting point by Khare et $al.$ (1989). That mixture was irradiated using a plasma discharge at 77 K, followed by the measurements of the infrared spectra of the resulting products. A whole series of organic compounds, including alkanes, aldehydes, alcohol, and substituted aromatic functional groups, were identified. In particular, among the aldehydes formaldehyde (H_2CO) and its polyoxymethylene polymer ($H_2CO)_n$, ions of which may be present in the coma of comet Halley (Huebner et $al.$, 1989; Snyder et $al.$, 1989a, b; Moller and Jackson, 1990), merit special

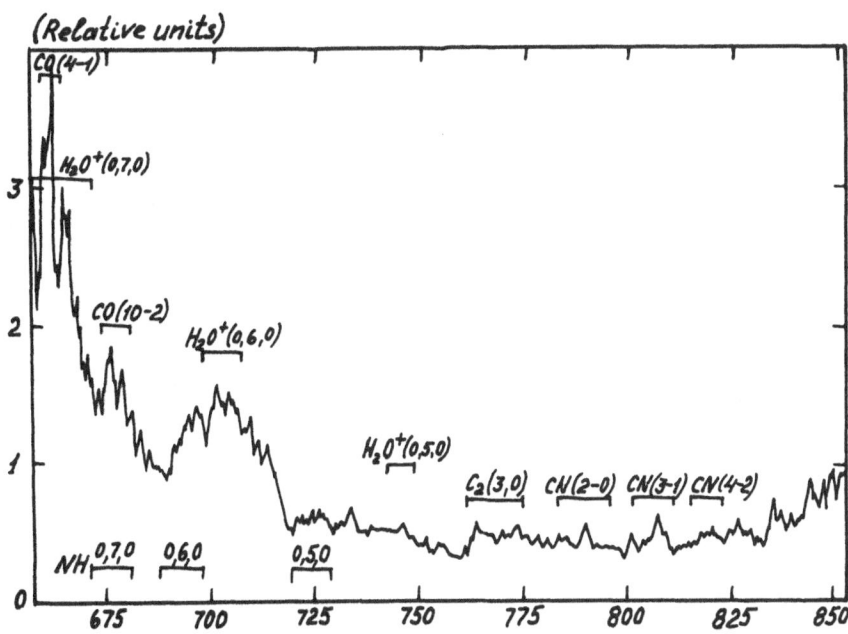

Figure 11.1.   Example of a spectrum of Halley's comet obtained from ground-based observations in the 650 – 850 nm range (according to Esipov *et al.*, 1994).

attention. A number of interesting results on transforming the composition of cometary ices have been also obtained in laboratory experiments of Ibadinov and Aliev (1989).

11.1.3. WATER AND PRODUCTS OF ITS DISSOCIATION

There is no doubt about the predominance of H_2O molecules in cometary nuclei. Water is detected directly in the near infrared, along with other parent molecules in the coma. An example of the total spectrum of comet Halley from ground-based observations in the 650 – 850 nm wavelength is shown in *Fig. 11.1*. Besides the characteristic H_2O bands, C_2, CO, CN, and NH_2 molecules and radicals are reliably identified in it.

The most direct method of determining water content, by observing the fluorescence caused by the vibrational transition at 2.7 μm, has been implemented successfully in experiments on board spacecraft (Moroz *et al.*, 1987; Weaver *et al.*, 1987). Convincing data on H_2O content have also been obtained in the experiments using Fourier spectrometer on the *Kuiper airborne observatory* in observations of comets Halley and Wilson. The high resolution ($\lambda/\Delta\lambda$ up to 3×10^5) attained in the profiles of H_2O spectral lines made it possible (as in the case of OH and HCN in the radio wavelength) to obtain both the spatial distribution and the velocity field of neutral gas in the atmospheres of those comets (Larson $et\ al.$, 1986; 1987; 1990; 1991).

In confirmation of the idea that the nucleus has a heterogeneous surface structure, it has been shown that the actual picture differs considerably from the idealized model of spherically symmetric gas outflow at a constant velocity. The positions and widths of the observed H_2O spectral lines indicate that anisotropy of the streams extends out to radial distances beyond 10^4 km, while retention of an approximately Gaussian line shape indicates a velocity spread due to collisional processes. Those processes in the inner coma (sometimes called the Greenstein effect) also cause asymmetry in the observed emission line profiles. A particularly strong asymmetry, superimposed onto the blue shift, is probably due to regions of strong outgassing associated with jets. The expansion velocity of the H_2O coma is consistent, on the whole, with the theoretical $\sim r^{-2}$ dependence, although for comet Wilson, for example, high outflow velocities have been observed that differ from the simple kinematic model. Examples of maps of H_2O intensities in the coma of comet Halley on the pre- and postperihelion segments of its orbit, obtained by Larson $et\ al.$ (1991), are shown in $Fig.\ 11.2.$

An attempt at direct identification of H_2O bands in the microwave spectrum of comet Bradfield was made by Jackson $et\ al.$ (1976) (see also Jackson, 1982), and later on in the $1 \leq \lambda \leq 6$ cm emission continuum and individual emission lines of that wavelength by Irvine $et\ al.$ (1984). No direct evidence on the presence of water in the coma were obtained, however, whereas the abundances of OH, CN, CS, HCN, CH_3CN, and some other components were estimated. A spectral resolution comparable to that obtained for H_2O in the IR range and an acceptable signal-to-noise ratio under the conditions of escaping gas streams was achieved for some radicals (OH, HCN). Nevertheless, since the radio spectral and kinematic parameters of H_2O are strongly influenced by numerous processes in coma, the appropriate model must be used to reconstruct the spatial distribution of water from the OH/H_2O ratio.

The products of water dissociation, primarily hydroxyl (OH), are reliably detected in the vacuum ultraviolet (3085 Å), which corresponds to strong HI Ly-α emission, providing clear evidence of hydrogen envelopes of comets, extending to millions of kilometers. From the emission intensity in the resonant fluorescence spectrum of OH, due to the rapid transition from the $A^2\Sigma^+$ photodissociative state with highly excited rotational levels to the $X^2\Pi$ state, in turn, one can estimate the distribution of water content at distances of some 100 km from the nucleus (Bertaux, 1986; Budzien and Feldman, 1991). Observations of the spatial distribution of H and OH agree well, on the whole, with the model of photodissociation of parent wa-

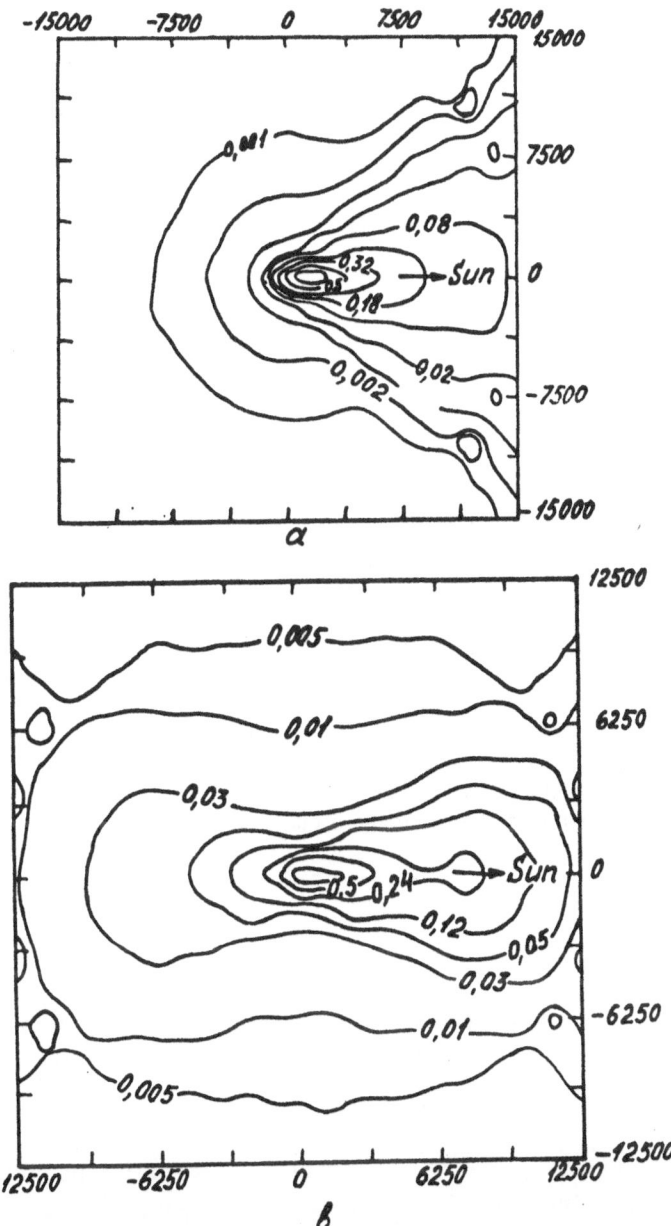

Figure 11.2.   Isolines of H_2O density in the coma of Halley's comet, obtained from an analysis of spectral line profiles in the ν_3 2.65 μm band (measurements using the Fourier spectrometer on the Kuiper airborne observatory, compared with synthetic profiles of the respective lines calculated on the basis of a kinematic model): a) segment of the orbit before perihelion; b) after perihelion. Along the coordinate axes: distance from the center in kilometers (corresponding to a round spectrometer field of view with a radius 1.5×10^4 km). The densities on the isolines are in relative units, corresponding to the relative intensity of the spectral line profile (according to Larson *et al.*, 1986).

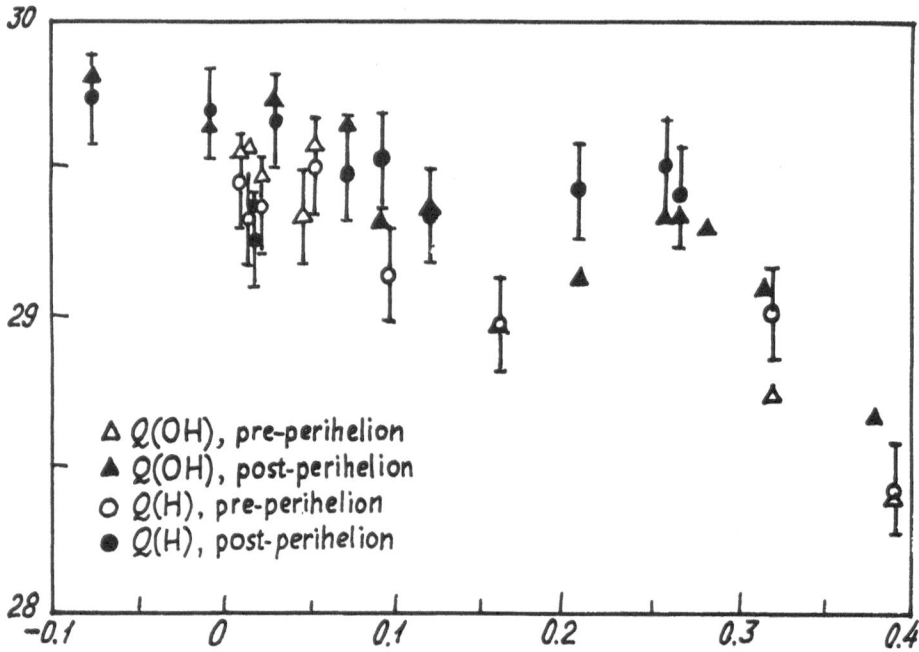

Legend within figure:
△ Q(OH), pre-perihelion
▲ Q(OH), post-perihelion
○ Q(H), pre-perihelion
● Q(H), post-perihelion

Figure 11.3. The water production efficiency (H - circles; OH - triangles) as a function of heliocentric distance based on measurements by the International Ultraviolet Explorer satellite. Solid and open circles and triangles correspond to post perihelion and pre-perihelion segments of the cometary orbit, respectively, (according to Combi and Feldman, 1992).

ter molecules. The water production efficiency as a function of heliocentric distance based on measurements by the *IUE* satellite is shown in *Fig. 11.3* (Combi and Feldman, 1992).

Another efficient tracer of water content and spatial distribution in the coma in this spectral range is excited atomic oxygen [OI] in the ^{1}D (6300 Å) and ^{1}S (5577 Å) states. The two emission features were observed in comet Halley in a 10:1 ratio, consistent with theoretical estimates at a distance $\sim$ 10^4 km from the nucleus (Festou and Feldman, 1981; Smyth and Schempp, 1989). A certain symmetry in H_2O topology in the coma of comet Halley was found from observations of [OI] 6300 Å emission, in particular an excess of up to 10% toward the Sun (Combi and McCrosky, 1991), which agrees well with IR measurements (Larson *et al.*, 1986; 1989).

11.1.4. NITROGEN-CARBON RADICALS AND IONS

It is important to emphasize again that no significant qualitative changes are found in the spectra of comets in due course of their evolution; in other words, the continuum-to-emission ratios (corresponding, to first order, to the dust-to-gas ratio), which are usually low at heliocentric distances $R < 1$ AU and high at $R > 1$ AU, are maintained, on the whole. The intensities of coma emission features, excited not only in resonant scattering but also by fluorescence (Donn $et\ al.$, 1984), increase considerably as the Sun is approached, however, both due to the increasing rate of sublimation of the nucleus and because of the increase in the number of solar photons and the efficiency of photochemical processes. All of the main features of molecular spectra at ultraviolet and visible wavelengths, including the Swann bands for the C_2 molecule, the CO, CS, CN, CO^+, CN^+, and OH bands (Krishna Swamy, 1984), and H_2O^+ bands (Disanti $et\ al.$, 1990; Scherb $et\ al.$, 1990), can be explained, in particular, in terms of the resonant fluorescence mechanism.

This is also related to the characteristic lines, detected in a number of comets, of atomic carbon (CI, CII) and nitrogen (NI) in the region between 100 and 200 nm, excited due to dissociation of the parent molecules and radicals C_2, C_3, CH, CH^+, CO, and CN by solar ultraviolet radiation (Singh $et\ al.$, 1991). This makes it possible to estimate the original abundances of those components and the efficiencies of photochemical processes in the coma at different distances from the nucleus and different heliocentric distances, including the possible contribution to those processes from solar wind electrons, as has been suggested, for example, for the strong CO^+ emission detected at ~ 5.8 AU in the coma of comet Schwassmann-Wachmann (Cochran $et\ al.$, 1991).

Along with water dissociation products (OH and [OI]), lines belonging to the NH_2 (amine) radical, which (like the imidogen radical NH) is a product of ammonia photolysis, have been identified in spectra of comet Halley at 6300 Å (Arpigny $et\ al.$, 1987; Magee-Sauer $et\ al.$, 1989). Indirect evidence of the presence of NH_2 has also been found at 6334 Å (Combi and McCrosky, 1991). This yielded an estimate in the $0.1 - 0.4\%$ range for the NH_3 abundance in the coma of comet Halley relative to H_2O, depending on the fluorescence efficiency factor used for the NH_2 (0,8,0) band. At the same time, an estimate of $NH_3/H_2O \simeq 0.5 - 1\%$ was obtained based on NH emission in the $A^3\Pi$-$X^3\Sigma$ (0,0) band at 3360 Å, measured by the $Astron$ satellite, though a possible influence of photodissociation on this estimate was neglegted in the data analysis.

A value almost two orders of magnitude lower (0.02 %) was obtained for comet Giacobini-Zinner, however, which probably indicates approximately tenfold ammonia depletion of its nucleus (Konno and Wyckoff, 1989). The analysis of UV spectra obtained with the *HST* for comets Hartley 2 and Shoemaker-Levy (Arpigny *et al.*, 1993) promises to yield improved estimates of NH_3 abundance in comets.

From emission intensity, in turn, one can independently estimate the rates of gas release, primarily for water from the (0,0), (1,0), and (1,1) hydroxyl bands. An investigation of the spatial distribution of ions in the coma also enables one to determine the size of the region within which collisions dominate over the effects of sweeping away by the solar wind. For comet Halley, for example, at the time of the encounter with the *Vega 1* spacecraft, its boundary for H_2O^+ ions located at a radial distance of $\sim$ 25,000 km. Certain features of the structure and kinematics of the thermal plasma around the comet can be thereby revealed. Outside this region, ions in the outer coma (mainly H^+ and O^+) are responsible for effects of loading of the solar wind, the development of bow shock, and the formation of plasma tail.

Unfortunately, relationships with the respective parent molecules can be established for far from all of the detected radicals. Besides OH (a product of H_2O) and NH_2 (a product of NH_3), here we can also include CN, CH, and CH_2 (products of CH_3CN) with a certain probability. At the same time, we do not fully understand the appearance in the coma of such typical molecules as CO, C_2, and C_3, for which dust (ice) particles enriched with carbon might be responsible, or the N_2^+ ion, which is one of the main components of cometary plasma tails (the upper limit of the relative abundance of neutral nitrogen is ≤ 0.1 according to mass-spectrometric measurements on the *Giotto* spacecraft (Eberhardt *et al.*, 1987). Analysis of chemical processes in the cometary atmosphere places some constraints on the probable sources of these and other radicals. The most important of these processes are the reactions due to the direct impact of solar UV photons and solar wind electrons on products of sublimation of the nucleus: photodissociation, photoionization, photodissociative ionization, ionization by electron impact, etc. Here one must also include the reverse reactions that accompany these processes: recombination, quenching of excited states, as well as ion-exchange and charge-transfer reactions. The following sections of this chapter are specifically focused on these processes, involving the aeronomy of mainly hydrogen-oxygen cometary atmosphere.

Due to the large number of impurities in an actual cometary nucleus and its heterogeneous composition discussed earlier, the processes of chemistry and photochemistry occurring in a cometary atmosphere are undoubtedly more complicated and diverse. Enormous difficulties are therefore involved in modeling the entire complex of reactions, even for a limited set of identified parent components. The most probable main parent molecules in the inner coma of a comet, their dissociation products, and a set of chemical reactions were summarized by Gombosi *et al.* (1986) incorporating the studies of Huebner (1985) and Mendis *et al.* (1985).

It is evident that the aeronomic processes these authors listed can be extremely extensive and that they differ greatly in probabilities (rates) of occurrence. In some cases they also depend on whether the intermediate compounds can be formed in the coma. When the activity of the photolysis products is high, these compounds can be fairly complex, but exist for only a limited time, being themselves subject to the energetic photon and electron impacts. This is related, in particular, to the problem of formation of C_3 and C_2, for which CO_2 and CH_4 could be responsible, though the chains of respective processes lacks a quantitative agreement between the calculated and observed C_2 and C_3 abundances.

With this regard it is worth to note that the chemical processes in coma have an unsteady nature, in contrast to the situation with interstellar clouds, where temperatures are lower, and many reactions occurring in an expanding gas of variable density (and hence with a changing optical depth to solar photons) are therefore slowed down (Marov and Shematovich, 1987). This analogy is justified because the compounds making up the comets may be identical to those detected in interstellar space, although we admit that some similarities (e.g. C_2 and C_3 molecules) could be just a coincidence.

11.1.5. COMPOSITION OF DUST PARTICLES AND GAS-DUST RELATIONSHIP

The spatial distributions of a number of chemical components in the coma of comet Halley can be judged from the data of multipositional spectrophotometry, obtained by *Giotto* spacecraft with a spatial resolution of about 103 km. The maps of emission intensities for a number of transitions with characteristic lines, such as CN (3883 Å), C_3 (4040 Å), C_2 (5100 Å), NH_2 (6107 Å), CH (4322 Å), [OI] (6300 Å), CN (4216 Å), were compiled, which indicates the complicated structure of the coma at distances of some 10^5 km from the nucleus. An especially large asymmetry was observed in the CN, C_2, C_3, O, and H_2O^+ profiles. It may be related to the presence of jets seen in the images of comet Halley obtained in the same period through the *International Halley Watch (IHW)* program of ground-based observations (A'Hearn *et al.*, 1986; Cosmovici *et al.*, 1988; 1993; Esipov *et al.*, 1994). There is no doubt about the 'cyanic' and 'carbonic' jets. As far as the [OI] emission is concerned, its source could be OH photodissociation in $L\alpha$, rather than photodissociation of CO_2, in other words, an asymmetry of the [OI] profiles may be related to hydroxyl. The mechanism of formation of the H_2O^+ jets is not yet entirely clear.

It cannot be also ruled out that for certain features of the intensity patterns recorded in coma, some molecules are responsible that have a disperse source. Their origin is associated with dust particles that have retained volatiles in their composition and release them after being ejected from the nucleus, in the inner coma (Wallis et $al.$, 1987; Lichtenegger and Komle, 1991). This assumption has been related, in particular, to the considerably lower degree of outgassing from the comet Brorsen-Metcalf during its passage in 1989 compared to the comet Halley, while their orbital parameters were similar: the ratios of H_2O, NH_2, C_2, and CN abundances were 15%, 41%, 75%, and 70%, respectively. Similarly, a lower IR emission in the continuum near 3 μm was found for the comet Brorsen-Metcalf, and there were no spectroscopic signs of silicate dust (Disanti and Fink, 1991). The small excess of the color temperature and the satisfactory conformity of the emission in the 3.5 - 20 μm range to that of an absolutely black body also indicated a low abundance of subliming dust particles.

As more data becomes available, the evidence come up that dust not only affects the spectral properties of the coma, but can itself contribute to the kinetics of the processes (through heterogeneous chemistry). The question of the chemical composition, size spectrum, and dust abundance is therefore of enormous importance for understanding both the nature of the coma evolution and genesis of the cometary material. In this regard, the emission features in the near infrared at 3.2 - 3.6 μm earlier discussed, which have been identified with methanol and formaldehyde and suggest the presence of other organic compounds, merit special attention.

The possible reasons for the appearance of emission near 3.4 μm have been found and analyzed in detail in the spectra of comets Halley and Wilson, while there were no traces of this feature in comets West, IRAS-Araki-Alcock, and Encke (Combi et $al.$, 1986; Wickramasinghe and Allen, 1986; Danks et $al.$, 1987; Allen and Wickramasinghe, 1987; Hanner et $al.$, 1985; R. Gehrz et $al.$, 1989). It seems very likely that the nature of this emission may be due both to molecular components (the characteristic C-H bond in hydrocarbons) and organic dust particles. In this case, as a comparison with the data of a laboratory spectra reconstruction for certain classes of organic residues showed, it is possible that such particles are similar in their properties to the material of interstellar dust particles (Greenberg, 1982; Hanner, 1986; Chyba et $al.$, 1989; Greenberg et $al.$, 1989) bringing support to the idea about intergalaxy connections. Additional arguments were brought in some resemblence found between the composition of solid fragments of cometary matter and particles of the interstellar medium.

One must be cautious, however, treating all these data as favoring this challenging idea. Firstly, one must keep in mind that the IR spectra of the dusty material of comets and interstellar dust (at $\lambda = 3.4$ μm) have marked differences, which may reflect either their different original chemical compositions or subsequent metamorphic processes that these primordial particles underwent in cometary nuclei. Secondly, there are unequal abundances in cometary ices and in the icy mantles of interplanetary dust particles of such important molecular compounds as CH_4 and CH_3OH. Thirdly, the spectra of silicates in comets and in molecular clouds also differ (Tokunaga and Booke, 1990). Finally, it is rather evident that the gas mixture ratio due to a given dust component of the coma may differ strongly from the true ratio of volatiles in the nucleus owing to different temperatures and the presence of phase transitions.

The relatively large particles collected on Earth and in space (Brownlee particles), the origin of which has been associated with comets, are accessible to laboratory analysis. Their composition differs little from that of carbolnaceous chondrites of class C. At the same time, very fine dust particles with a mass $\sim 10^{-16}$ g, highly (by about 30 %) enriched with carbon (C), hydrogen (H), oxygen (O), and nitrogen (N) and hence abbreviated as CHON, have been detected in space experiments on the *Vega* and *Giotto* craft. This reinforced the idea that nuclei contain a considerable amount of organic matter, about an order of magnitude higher than, for example, the CHON abundance in the Murchison meteorite (Jeisberger *et al.*, 1989). This simultaneously gave reason to assume that nitrogen (possibly serving as the source of N_2^+ in a cometary tail) and other gases are carried off with these particles.

The aforementioned hydrocarbon emission at 3.4 μm, as well as the observed rise in C_2 and C_3 emission in jets at radial distances between 10^4 and 3×10^4 km in the coma of comet Halley, may also be related to CHON particles (Clairemidi *et al.*, 1990). It is possible that carbon-containing molecules liberated from CHON particles were responsible for this effect. This supports the idea (Whipple, 1989a, b, c) that CHON particles may be an important source for the appearance of certain additional gases in the coma. They probably are also responsible for a considerable extent in the mass spectra of heavy ions recorded by the cluster ion analyzer (PICCA) on the *Giotto* spacecraft, being the organic remnants of the decomposition of complex compounds, or the potential source of the formation of cluster molecules by means of numerous chemical reactions in the inner coma (Wegman *et al.*, 1987; Mitchel *et al.*, 1992).

The discovery of CHON carbonate particles starkly poses the question of their origin in connection with the general problem of the formation of comets and meteorites, as well as their relative potential contribution to enriching the inner regions of the Solar system, and Earth in particular, with organic matter. In fact, particles smaller than ~ 100 μm are able to reach Earth's surface without being destroyed in the atmosphere, and they may be the main source of exogenic material containing complex organic compounds synthesized under the conditions of the cosmic medium, up to amino acids or their direct precursors, such as aldehydes or nitrites - the foundations of prebiogenic chemistry (Anders, 1989). In particular, the considerable content of amino acids of extraterrestrial origin discovered at Stevens Klint (Denmark) in rocks belonging to the time of the Cretaceous-Tertiary boundary has been associated with such a source (Zahnle and Grinspoon, 1990). The corresponding particles could have been produced by the disintegration of a large comet and the showers of meteor dust associated with it.

It has also been hypothesized that meteoroids that consist mainly of high-molecular-weight organic compounds, similar to CHON particles, produce meteors at anomalously high altitudes in Earth's atmosphere, and that polymerization of organic compounds occurs in the surface layer of the icy nucleus of a comet under the action of cosmic radiation (Lebedinets, 1990; 1991). From the difference in coefficients of luminosity between chondritic and organic meteorites, it has been concluded that in the nucleus of comet Halley, as well as in the interplanetary medium, about half of the dust material consists of organic matter, whereas in the nucleus of comet Giacobini-Zinner and in the Draconid meteor showers there is two orders of magnitude even more organic than chondritic dust, i.e., the fraction of organic dust may reach 99 %. The existence of the terrestrial dust cloud is now being associated with a supply of friable, easily vaporized organic dust from "minicomets" (Frank et al., 1986).

Now we shall briefly discuss the data available about dust particles of other composition, which can be addressed as the main source of knowledge about the nucleus composition and serve as a clue for finding the relationship between the minerals and volatiles liberated from the matrices they compose.

In addition to CHON in the coma of Halley's comet, many other dust particles with a high abundance of silicon (Si) and other elements (Mg, Al, Fe, Ag, Ca, O, C) have been detected. The presence of lines of these elements was first discovered in the spectrum of comet Ikeya-Seki 1965 8 (Preston, 1967). It was attempted to determine the mass ratios of the elements and compare them with the analogous ratios for the Sun, carbonaceous chondrites, and Earth's crust. The ratios of Cr, Mn, Ni, Mg, Si, Ti, V, and Co to Fe in the spectra associated with comet's nucleus were found close to those for the Sun and for chondrites, though an unexpected enrichment of nucleus with Cu and a depletion of Al, K, and Li atoms was also revealed.

Spectroscopic measurements in the IR wavelength also made it possible to identify silicate dust in the comas of comets Halley and Wilson based on the characteristic feature at 10 μm, starting with R $\leq$ 1.3 AU (Hanner, 1986). The corresponding reconstruction of data from the dust-impact instruments (PUMA on *Vegas* and PSA on *Giotto*) is consistent with the possibility that some dust particles have a fairly unusual composition. Indeed, while the ratio of Mg to Si abundance remains fairly constant, the iron abundance (Fe/Si) varies widely. The Mg seems to be mainly in silicates, whereas the wide range of variation of Fe indicates that it may be present not only in metallic form but in an extensive class of minerals, such as magnetite, sulfides, and silicates with different Fe abundances (Lawler *et al.*, 1989). None of these detected particles seem to consist of a single mineral (with the possible exception of FeS), but more likely they are collections of considerably smaller grains and possibly contain a large amount of glass.

The ratio Fe/(Fe+Mg) is quite consistent with the broad, smooth distributions found for the submicron dust particles collected in Earth's stratosphere, in contrast to the narrow distributions typical for matrices of carbonaceous chondrites where hydrated minerals dominate. This suggests that coma dust particles (except for CHON particles) contain magnesium-rich, high-temperature silicates and that anhydrous minerals predominate in them.

11.2. Nonequilibrium Processes in the Inner Coma

In the introductory Chapter 2 we have already noticed that the increase of the solar incident radiation as a comet approaches the Sun is responsible for the intense outgasing of volatiles from its nucleus and coma formation. The structure of the coma is inhomogeneous and contains both gas and dust particles, which form, in particular, jet-like configurations. Similarly, as shown in the previous section, the chemical composition of the coma is extremely complicated as well. This means that the modeling a cometary gas-dust atmosphere involves many problems, such as energy exchange and dynamics of the flow, photolysis and chemical reactions between the parent and daughter molecules, as well as a significant contribution of the dust particles to the balances of mass, momentum, and energy. Both methods of the molecular-kinetic theory of gases and the heterogeneous mechanics of a continuous medium are required for an elaborative treatment of this problem. However, there are many constraints to model jointly the processes of chemical kinetics and gas-dust flow at the same level of accuracy. Hence, the models that have been developed until recently are still limited, being specifically focused on only a part of the phenomena involved.

11.2.1. GAS DYNAMIC MODELING OF THE INNER COMA

Attempts to model gas-dust mixture flow near the nucleus, with a limited set of chemical processes in coma, have been undertaken in many studies (see, e.g., Probstein, 1969; Shul'man, 1972; 1987; Marconi and Mendis, 1982a, b; 1983; Krasnobaev, 1983; Gombosi et al., 1986). It has been shown that the interaction between phases causes a number of qualitatively new effects that markedly influence the dynamics and energetics of the carrier and disperse phases. In particular, this disrupts the flow's adiabaticity, and it may become transonic. The spatial distributions of the main parameters of the gas suspension have been determined under the assumption of spherical symmetry.

In a more detailed theoretical analysis of the problem, a closed system of equations of heterogeneous mechanics (of the hydrodynamic type in Lagrangian coordinates) describing the evolution of the two-phase coma, with properly formulated boundary conditions, was obtained by Kolesnichenko and Marov, (1985) and Marov et al., (1993). The conditions of heat transfer were analyzed for a dust particle in steady, free-molecular stream. In the numerical implementation of this model (Kolesnichenko and Skorov, 1987) methods developed to solve conjugate and nonsteady problems of the mechanics of chemically active media were used, as well as a program package constructed on the basis of fully conservative schemes for gas dynamic equations (Grishin and Fomin, 1984; Samarsky, 1971; Samarskii and Popov, 1981).

Calculations carried out for the case of a slightly dusty coma in the approximation of spherically symmetric outflow made it possible to obtain radial profiles of hydrodynamic parameters in the inner coma using quasi-three-dimensional thermal model of the nucleus like that of Halley comet. As an illustration, *Fig. 11.4* shows the curves of the temperature dependence in the equatorial plane of the nucleus, its axis of intrinsic rotation being perpendicular to the orbital plane while the comet is 0.8 AU from the Sun. It is seen that, due to horizontal temperature gradients taken into account in the quasi-three-dimensional model, the high-temperature region on the surface of the nucleus in the vicinity of the subsolar point (shifted about 26° from longitude $\Phi = 0$) is considerably "smeared out", while the temperature variation in regions near the terminators ($\Phi = 90$ and 270°) becomes smoother. Moreover, considerable asymmetry is observed in the temperature distribution on the night and day sides of the nucleus relative to the direction to the Sun and the positions of the temperature extreme are clearly displayed.

The radial distributions of the main parameters of the carrier and disperse phases calculated on this basis made possible a qualitative estimate

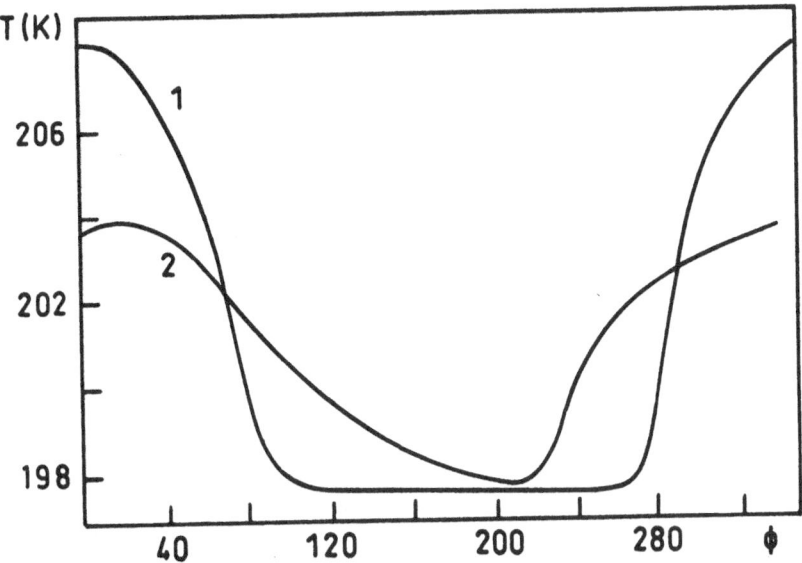

Figure 11.4.   Surface temperature T_s of a cometary nucleus as a function of longitude Φ at the heliocentric distance $R = 0.8$ AU: 1) one-dimensional model; 2) quasi-three-dimensional model.

of the contribution of dissipative processes that disrupt the adiabaticity of flow and also lead to qualitative changes in the parameters of the gas stream. The profiles of temperature T, density ρ, and velocity V as functions of the cometocentric distance r (at distances up to ~ 1000 km from the nucleus) are shown in *Fig. 11.5*. They correspond to the original model of chemically neutral, dust-free coma in which the departure from adiabatic regime is caused only by the thermal conductivity of the carrier phase. Here the velocity (corresponding to gas production rate $Q \simeq 10^{29}$ mol/s) reaches some constant value fairly rapidly, the density varies in proportion to r^{-2}, and the temperature is proportional to $r^{-2/3}$. With allowance for the chemical activity of the carrier phase (primarily due to photodissociation of H_2O and the subsequent complex of chemical reactions), a corresponding power source was added to the energy equation. This resulted in an increase of hydrodynamic velocity and a more monotonic variation of stream density, but a temperature minimum (~ 20 K) appeared at about 100 km from the nucleus (*Fig. 11.6*).

Allowance for dust (for the degree of dustiness $\eta = 0.1$) qualitatively did not change the ρ, V, and T profiles, what is indicative of an absence of significant gas heating in the inner coma due to heat transfer from hot dust at the relatively low r. In other words, this channel of heat supply is not

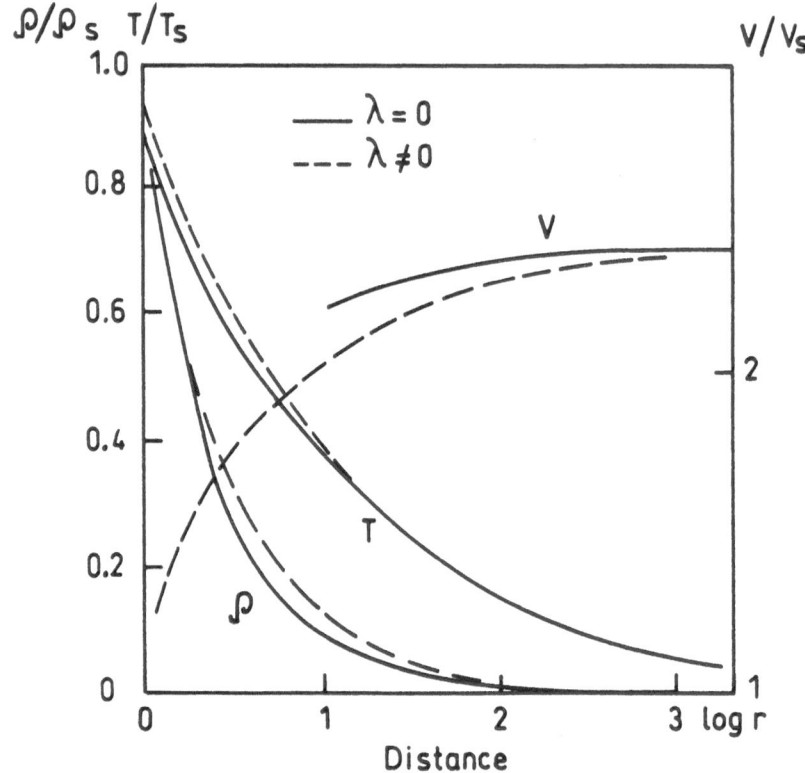

Figure 11.5. Radial profiles of temperature T, density ρ, and velocity V of the gaseous carrier phase (normalized to the respective values T_s, ρ_s, and V_s at the surface of the nucleus) as functions of cometocentric distance r, according to the model of a chemically neutral, non-dusty inner coma of Halley's comet.

as strong as the chemical energy sources at considerable heliocentric distances R, since the heat transfer intensity is proportional to r^{-4}, while the efficiency of the photodissociation processes (at a fixed R) is proportional to r^{-2}. A comparison of the parameters of the disperse and carrier phases for this model is shown in *Figs. 11.7* and *11.8*.

Fig. 11.7 illustrates the behavior of gas (carrier phase 1) and dust particles (phase 2) in the immediate vicinity of the nucleus (within several dozen meters) for different values of the initial velocity V_{2s} of grain separation from the surface (1 – 5 m/s). It is evident that the smaller V_{2s}, the more significant is the energy supply to the interaction between phases. In other words, deceleration of the gaseous component (decrease in the velocity V_1) and its heating occur more efficiently. Since the velocity of gas outflow equals or slightly exceeds the speed of sound ($M = 1.05$), a transition to the subsonic range with the formation of bow shock is possible for values $V_{2s} \leq 5$ m/s. It should result in a change in the optical charac-

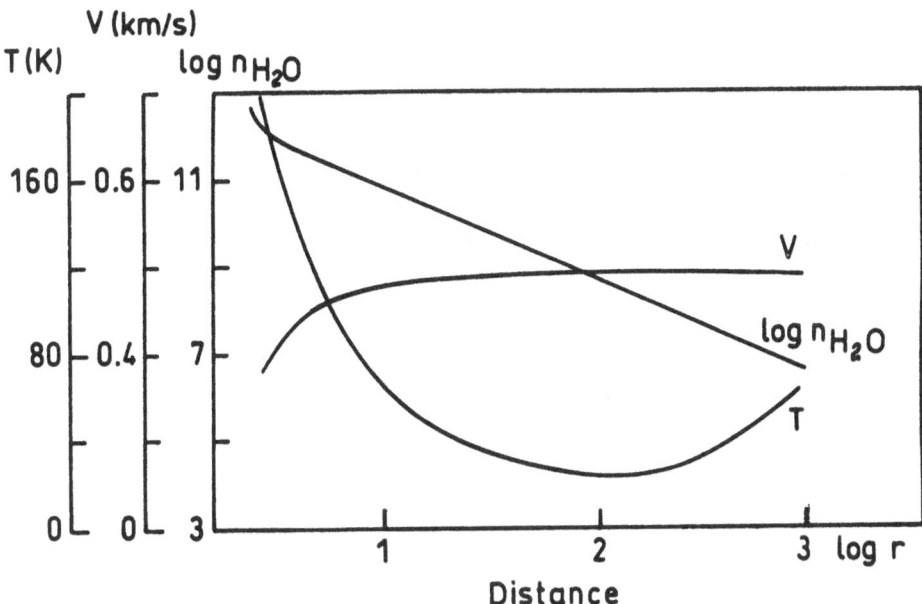

Figure 11.6. Radial profiles of temperature T, number density n, and velocity V of the carrier phase in absolute values as function of cometocentric distance r, according to the model of a chemically active, slightly dusty ($\eta = 0.1$) gas in the inner coma of Halley's comet.

teristics of the shock layer and the formation of optically bright structures (having spatial sizes smaller than the telescopic resolution, however). Starting with $V_{2s} \sim 8$ m/s, the outflow regime remains supersonic in the entire region. Calculations also showed the existence of the relationship between the velocity V_2 of dust particles (up to which nonmonotonicity is displayed in the variation of V_1) and the degree of dustiness η, which turned out to be almost directly proportional. The density distribution of the carrier phase in this case differs markedly from the inverse-square law.

Fig. 11.8a, b shows the results of modeling the behavior of the main parameters of heterogeneous stream at distances up to several hundred kilometers from the nucleus. The shapes of the V_1 and T_1 curves in *Fig. 11.8a* for the cases of chemically neutral and chemically active gas mixture supports the conclusions drawn earlier in the analysis of *Fig. 11.6*. Indeed, as distance from the nucleus increases, the role of the gas interaction with the slower and hotter disperse phase grows, while at smaller distances the influence of photolysis of water vapor remains dominant. A decrease in the rate of photochemical heating of the inner coma leads to some rise in the mass-averaged velocity V_1 of the carrier phase. As for the dust, its temperature varies insignificantly (*Fig. 11.8b*). Some variation of T_2 closer to the

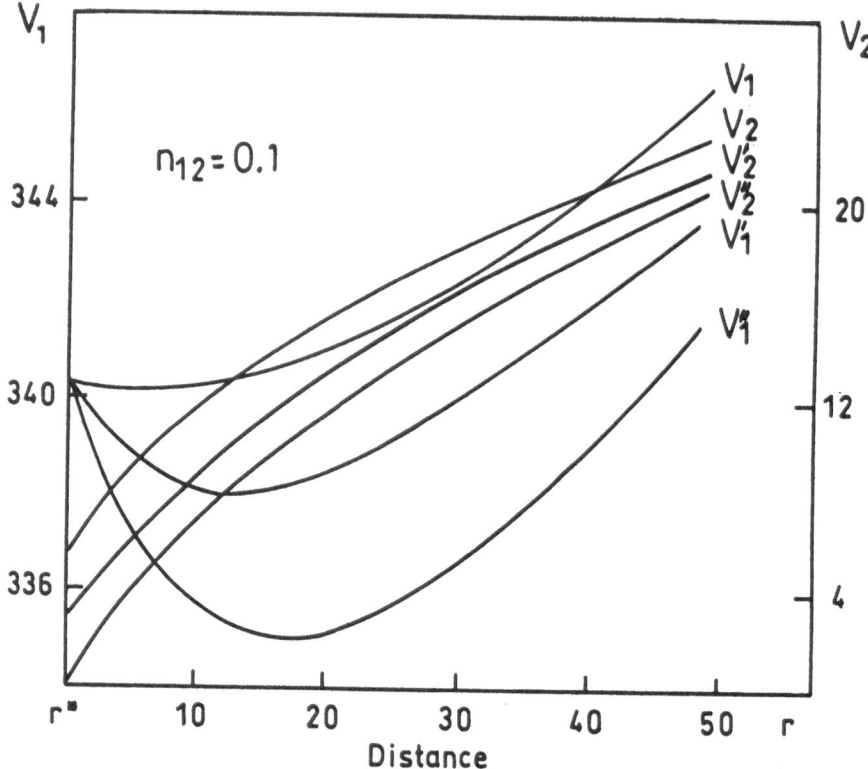

Figure 11.7. Profiles of V_1 and V_2 near the surface of the nucleus ($M \geq 1$) for different grain separation velocities V_{2s}: 5 m/s (V_1, V_2), 3 m/s ($V_{1'}$, $V_{2'}$), and 1 m/s ($V_{1''}$, $V_{2''}$).

nucleus is explained the effects of radiation and the interaction between phases in this region of the coma. Efficient acceleration of dust particles occurs in this zone, after which V_2 and T_2 reach horizontal asymptotes. These estimates of the velocity of the gas-dust flow agree well with the observational data.

It must be emphasized, however, that the use of one-velocity approximation to describe the carrier phase in the gas dynamic model under consideration becomes rather coarse when studying the behavior of the daughter components. Strictly speaking, such an approach does not apply, in particular, for describing the "fast" hydrogen and hydroxyl formed by water photolysis. Therefore, a more complete investigation of the evolution of the multicomponent nonequilibrium gas of the inner coma in the solar radiation field is needed, that it is possible to accomplish most effectively on the basis of a detailed analysis of the kinetics of photolytic and elastic collisions, with the involvement of nonequilibrium processes. The results of such modeling are discussed in the next section.

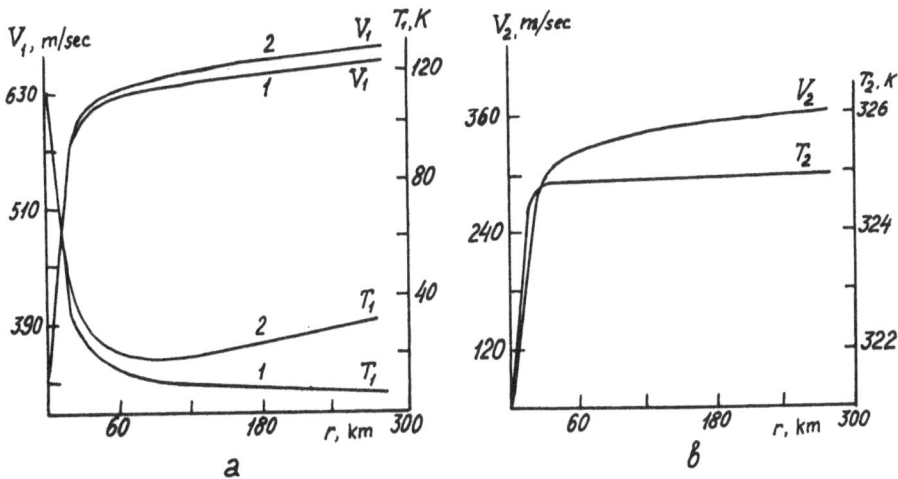

Figure 11.8. Radial profiles of the macroscopic parameters of the gaseous (a) and disperse (b) phases as functions of r for the model of a slightly dusty ($\eta = 0.1$) inner coma. T_1 and V_1 are the gas temperature and velocity and T_2 and V_2 are the dust temperature and velocity, respectively: a) variation of T_1 and V_1 for the cases of neutral gas (1) and chemically active gas mixture (2); b) corresponding results of calculations for the V_2 and T_2 profiles.

11.2.2. KINETIC MODELING OF THE NEAR-SURFACE LAYER

It was common practice in previous works to use simplified boundary conditions for the parameters of gas-dust flow on the surface of the cometary nucleus: gas temperature and pressure were assumed to be equal the equilibrium values for ice. As was shown in Shul'man (1972), this approach gives a correct order of magnitude, but it is contradictory from the viewpoint of the kinetic theory of evaporation and should be improved. The exact definition of the aforementioned parameters is only one of the problems that should be solved on the basis of modern numerical models of the near-nuclear cometary region. Other problems that require detailed consideration of the processes near the interphase boundary are the determination of the velocity distribution of subliming molecules and the calculation of the molecular flux back to the surface.

Evaluation of the flow in the boundary layer cannot be performed in the framework of continuous mechanics; instead methods of the kinetic theory must be applied to treat the problem. The Boltzmann equation, with consideration for collisions and possible reactions, needs to be solved for each component of the carrier phase. The condition at the external boundary should correspond to the equilibrium velocity distribution of the molecules, while the condition at the interphase boundary can be obtained from the kinetics of interaction between gas and surface. It is a very dificult problem which is further complicated when the dispersed phase is included. Obviously, through some assumptions the statement of the problem must be resonably simplified.

First, the chemical transformations may be neglected, since the lifetimes of basic molecules in the cometary atmosphere, which are determined by the rates of chemical reactions, are considerably greater than the time of motion in the near-surface (Knudsen) layer. Second, the fraction of molecules subjected to collisions with dust particles in the boundary layer is within a fraction of percent of their total number. This can be deduced from the fact that, for the majority of comets, the ratio of dust to gas mass production η is less than or equal to unity, and the typical size of dust particle is less then a micron. Hence, as a first approximation, one can neglect the perturbations of the carrier phase produced by the dust particles and assume that the dispersed phase is accelerated only by the gas flow. Therefore the original problem is subdivided into two separate problems: variation of dynamic characteristics of the carrier phase without chemical reactions and dust suspension in the atmosphere; and variation of momentum and energy of the dispersed particles due to interaction with molecules whose velocity distribution function is known.

In solving the first problem two basic approaches are generally used: either a numerical modeling of the relaxation of the velocity distribution function, or an analytic approach to define the relation between gas parameters at the internal and external boundaries of the Knudsen layer.

In the analytic approach, the Knudsen layer is considered as a gas dynamic discontinuity. From the laws of conservation of mass, momentum and energy fluxes, with the involvement of certain model assumptions on the kinetics of evaporation (sublimation), one can relate the values of density n_0, and temperature T_0 at the external boundary of the layer with their values n_s, and T_s at the internal boundary (the surface). This approach was developed, for example, by Anisimov et al. (1970). The rate of sublimation was determined by the Hertz-Knudsen formula. In the region of discontinuity the nonequilibrium distribution function was approximated by the sum of given distribution functions at the boundaries of intermediate regions with certain attributed weights. These weights depend on the distance from the surface. Proceeding from three laws of conservation and from the equation of ideal gas state, and also introducing an additional assumption that the local Mach number number (M_0) is equal to unity, the following algebraic relations were obtained: $n_0/n_s=0.31$, $T_0/T_s=0.67$. For this case an upper limit of the ratio of molecular flux back to the surface to direct flux of molecules turned out to be approximately 0.18. In an analogous approach Crifo (1986), based on the assumption that Mach number at external boundary is a free parameter of the model, the following relations were obtained for the case $M_0 = 1$, $n_0/n_s=0.25$, $T_0/T_s=0.85$. The molecular backward flux for this latter model was about 0.25.

The sublimation process has an intrinsically nonequilibrium nature because subliming particles flow out into semi-space (Ytrehus, 1977) and consequently, to determine accurately the structure of near-surface layer it is necessary to solve the set of kinetic equations at the boundary of the phase separation.

Let us briefly consider the results of the numerical kinetic simulation of the subliming gas flow in a thin layer of the cometary atmosphere immedeately adjacent to the nucleus' surface (Bisikalo et al., 1989). Radial profiles of water vapor macroparameters in the Knudsen layer of a cometary nucleus are shown in $Fig.\ 11.9$. Since the pattern of these profiles does not exhibit any noticeable changes depending on the surface temperature in the range $T_s = 180$ to 200 K, only the results for $T_s = 200$ K are taken as an example. The radial distance from the nucleus surface is given in the mean free path lengths λ. The value of λ for the typical gas number density $n = 2.13 \times 10^{13}$ cm^{-3} and for the gas-kinetic collision cross-section $\sigma = 1 \times 10^{-15}$ cm^2 is equal to about 50 cm. In $Fig.\ 11.9a$, the radial profile of the kinetic gas temperature T and its components $T_{\parallel}$ and $T_{\perp}$ corresponding to the radial and normal chaotic gas motions are shown. It is seen that near the nucleus surface (at the distance $\lambda/2$) the difference between these temperatures is high: $T_{\parallel} = 90.8$ K, $T_{\perp} = 182.6$ K, and $T = 152$ K. This result shows evidence of considerable deviation from the thermal equilibrium state for the translational degrees of freedom of the subliming particles. Collisions lead to a near-equilibrium state characterized by a single kinetic temperature $T = 113$ K at the distance 9.5λ from the nuclear surface. It is reasonable to refer to this distance as the outer boundary of the Knudsen layer and to adopt the gas parameters at this distance as the boundary values for the subsequent gas dynamical treatment of the problem. The radial profile of mass velocity of H_2O is shown in $Fig\ 11.9b$. The velocity changes in the Knudsen layer lie within the range from 146 to 305 m/s. Values of the number density also exhibit significant changes: from 2.13×10^{13} cm^{-3} to 1.02×10^{13} cm^{-3} ($Fig.\ 11.9c$).

Similar calculations were carried out for all temperatures in the range of 180 to 200 K. The near-equilibrium state in this temperature interval is accomplished at (10-16) λ from the nucleus surface with the following gas dynamical boundary values: $n = 0.34n_s$; $T = 0.56T_s$; and Mach number $M = 1.15$. These results define more accurately the parameters of the near-surface gas flow, as compared to the above quoted results of the analytic assessments.

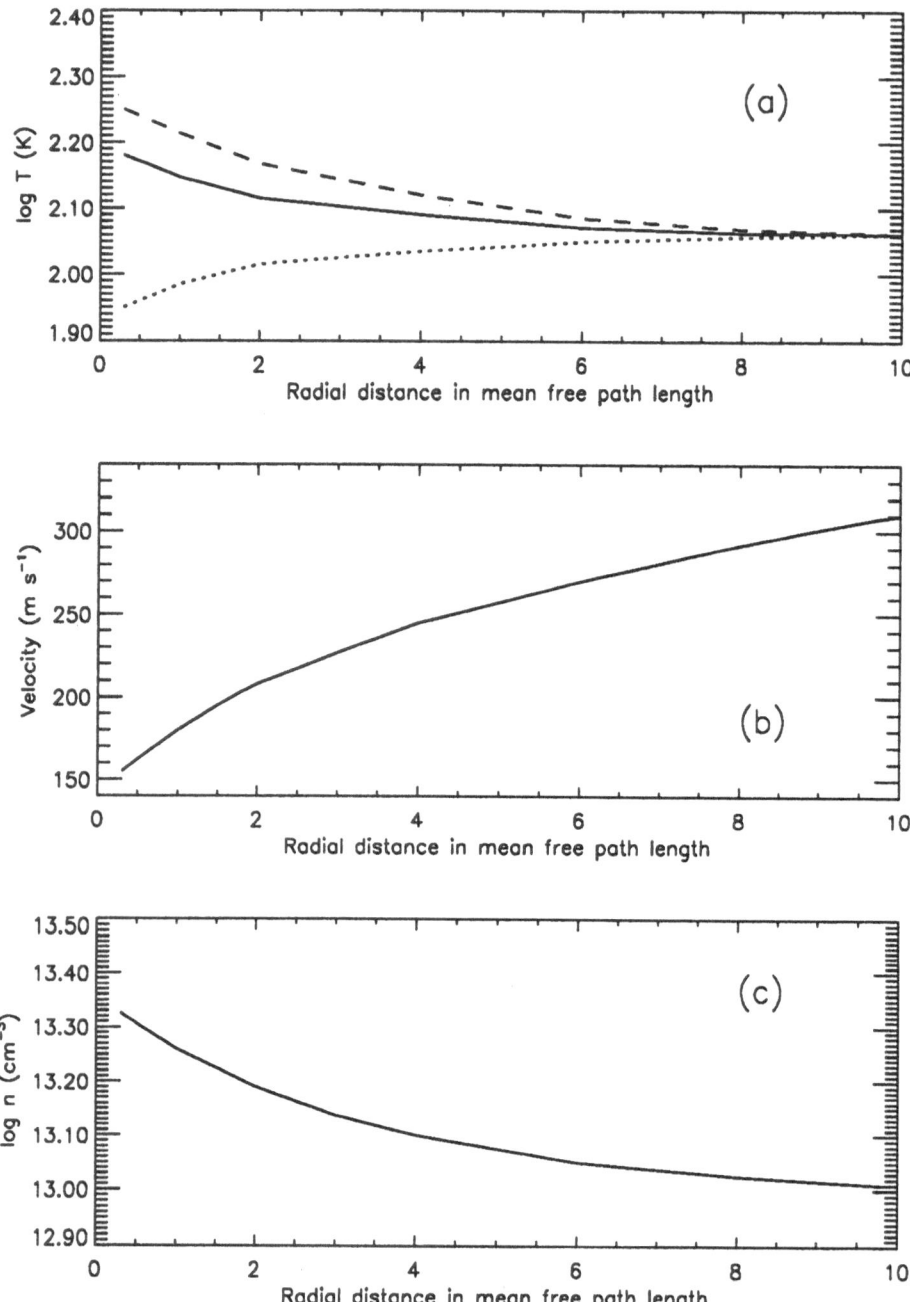

Figure 11.9. Radial profiles of gas macroparameters in the near-surface layer: (a) - kinetic temperature (T - solid line; $T_\parallel$ - dotted line; $T_\perp$ - dashed line); (b) - mean-mass velocity; (c) - number density; distances from the surface are given in the mean free path length ($\lambda = 50$ cm).

11.3. Hot H Atoms in the Inner Coma

The complexity of kinetics and dynamics of the cometary gas outflow in the inner coma is caused by the presence of two nonequilibrium state regions of the gas flow:

– the Knudsen layer, where collisional relaxation of semi-spherical Maxwellian velocity distribution of subliming particles is accomplished. As we have seen, the gas parameters at the other boundary of this layer can be properly defined only through kinetic consideration (Bisikalo *et al.*, 1989);

– the transition (from collisional to free-molecular flow regime) region, where the frequency of elastic collisions becomes insufficient for the maintenance of the local thermal balance.

Following the results of gas dynamic modeling of the inner coma, which revealed the most pronounced effect of the chemical energy sources (see Section 11.2.1), we consider in more detail the nature of these sources based on the kinetic approach. We consider the photochemical reactions, accompanied by the formation of translationally excited (hot) particles, which cause the nonequilibrium character of the cometary gas outflow.

The interaction between these energetically active particles - for example, the high-energy hydrogen atoms formed in the process of H_2O photodissociation - and the dominant components of the coma results in disturbances of the cometary gas thermal state. Products of the photolysis have velocity distribution functions strongly deviated from the equilibrium Maxwellian functions due to permanent injection of new particles and insufficient efficiency of elastic energy exchange. Indeed, the mean coefficient of energy transfer in H-H_2O collisions is about 0.1, i.e. ~ 20 collisions are necessary for the total thermalization of a high-energy hydrogen atom.

In the general case, the photochemistry and dynamics of rarefied gas outflow in the inner coma subjected to the incident solar UV radiation can be described by the set of stationary kinetic equations:

$$\begin{cases} c\frac{\partial}{\partial r}F_\alpha = Q_\alpha + \sum\limits_{\alpha_1,r} J_r^{s\,\alpha\alpha_1}(F_\alpha, F_{\alpha_1}) \\ F_{\bar\alpha}\mid_{r=r_0} = F_\alpha^{(0)}(\mathbf{c}, t) \\ F_{\hat\alpha}\mid_{r=r_0} = 0\,, \quad \alpha = 0, 1, \ldots, M^a \end{cases} \qquad (11.1)$$

Here $\bar\alpha$ and $\hat\alpha$ are primary and secondary components of the cometary gas respectively, and $Q_\alpha(\mathbf{c}, \mathbf{r})$ is the local α particle production rate in photolytic and electron impact reactions. Primary molecules subliming from the nucleus surface ($r = r_0$), are distributed by velocities in a solid angle 2π in accordance with the Maxwellian function with the nucleus surface temperature T_s.

In order to solve the set of equations (11.1), a space-nonuniform numerical stochastic model of cometary gas outflow (Marov and Shematovich, 1985; Bisikalo and Shematovich, 1988) was developed. Based on this approach, some results of numerical kinetic simulations of photochemical and dynamic processes in the inner coma were obtained (Bisikalo and Shematovich, 1987, 1988; Bisikalo et al., 1989; Marov and Shematovich, 1987; Marov et al., 1990; 1996).

For a typical comet of moderate brightness (like the Halley comet) at a distance of 1 AU from the Sun it was adopted that nucleus consists of pure water ice; its radius is about 5.5 km; surface temperature is about 200 K; and the production rate of water molecules is $\sim 10^{29}$ s^{-1}. It was also assumed that the gas outflow from the comet nucleus is steady-state and spherically symmetric. An excitation of rotational and vibrational levels of OH was not taken into account, i.e., it was postulated that the energy excess due to dissociation transfers to the kinetic energy of the dissociation products. In addition, the gas cooling due to radiation in IR lines of water molecules was not taken into account because the inner coma in these lines is optically thick (Bisikalo and Strel'nitskij, 1985). Finally, the presence of a dust component was neglected, because of the insignificant influence of mineral dust on the dynamics and energetics of cometary gas (Bockelee-Morvan and Crovisier, 1987; see also the results of Section 11.2.1). All molecules of the cometary gas components (H_2O, OH, H) were considered as rigid spheres with respective cross-sections and their collisional interaction was reduced to an elastic exchange of kinetic energy.

A simple scheme of the coma photochemistry was suggested, in which only the main photolytic reactions — photodissociation of primary component H_2O and photolysis of secondary radical OH were included.

Dissociation of the water molecule in the ground state $\tilde{X}^1A_1$ by UV radiation occurs through two main channels:

$$H_2O + h\nu \rightarrow H_2O(\tilde{A}^1B_1, \tilde{B}^1A_1) \rightarrow \begin{cases} H(^2S) + OH(\tilde{X}^2\Pi_i, A^2\Sigma^+) \\ O(^3P, {}^1D) + H_2(\tilde{X}^1\Sigma_g^+) \end{cases} \quad (11.2)$$

with the binding energies $E(\text{O-H}_2) = 5$ eV and $E(\text{H-OH}) = 5.11$ eV. It implemented via the following singlet electronic states of the water molecule: a) the dissociative continuum $\tilde{A}^1B_1$ near the energy $E(\tilde{A}^1B_1) = 7.40$ eV with the channel branching ratio $\sim 0.99 : 0.01$; b) the wide dissociative continuum $\tilde{B}^1A_1$ near the energy $E(\tilde{B}^1A_1) = 9.67$ eV with the channel branching ratio $\sim 0.9 : 0.1$ (Crovisier, 1989).

The OH radical is a secondary component of the inner coma playing an important role in its photochemistry. Photolysis of OH can be presented by the following scheme (Van Dishoeck and Dalgarno, 1984):

$$OH + h\nu \rightarrow H(^2S) + O(^3P, {}^1D) \ , \quad (11.3)$$

and the binding energy is equal to $E(\text{O-H}) = 4.40$ eV.

The photolytic reactions (11.2) and (11.3) are accompanied by the formation of translationally excited molecules and atoms H, OH, O, and H_2 bringing the excess energy of the absorbed UV photon with the account of pertinent photodissociation channels thresholds. For photodissociation of H_2O the mean excess energy is about ~ 1.9 eV, so that superthermal H atoms are preferably formed (Crovisier, 1989).

The velocity distribution functions of all the gas components were obtained in the numerical simulation of the kinetic system (11.1). For the photolysis products, significant deviations of the calculated distributions from the equilibrium Maxwellian functions were found. It should be noted that these deviations are displayed in the presence of superthermal tails of distribution, as well as in symmetry disturbances of the calculated distributions. The latter effects (the difference of distributions in the forward and backward directions) become appreciable for H atoms at the distances $\geq 10^2$ km from nucleus, and for OH radicals - for distances $\geq 4 \times 10^2$ km.

Using the calculated distribution functions, the radial profiles of the macroparameters of the main gas in the inner coma were evaluated. These results are shown in *Fig. 11.10*. The densities of H_2O, OH and H are presented in *Fig. 11.10a*, and the mean-mass velocities in *Fig. 11.10b*. It can be seen that the velocity of H exceeds the velocities of heavy particles H_2O and OH at a distance $r \geq 4$ km from the nucleus due to contribution of the hot particles formed in the process of photolysis. This velocity divergence increases with the distance (at $r \simeq 10^3$ km the velocity of H is about 3030 m/s and that of water vapor is only ~ 625 m/s). In *Fig. 11.10c* the radial profiles of kinetic temperature and their components in the radial and orthogonal directions are presented. It can be seen that at the distances $r \leq 130$ km the gas cooling due to the outflow is the dominant process, while at $r > 130$ km the gas heating due to energy exchange with hot H and OH particles prevails (at $r \simeq 130$ km, $T_{min} \simeq 3.5$ K; at $r = 10^3$ km, $T \simeq 14.3$ K). It is interesting to note that even at the beginning of the outflow (at $r \geq 9$ km) the main component is characterized by deviations from the thermal balance. These deviations, in particular, are observed in the divergence of the kinetic temperature components $T_{\parallel}$ and $T_{\perp}$. The detailed kinetic consideration of photolytic and elastic collisions shows that the efficiency of energy exchange between the translationally excited photolysis products and the main gas component is smaller than is generally used in gas dynamical models, and consequently, the temperature of H_2O is lower as well.

The comparison of the results of kinetic simulation with those obtained in gas dynamical one- and multifluid models (Marconi and Mendis, 1982a, b, 1983; Huebner and Keady, 1983; Huebner (Ed.), 1990; see also our results

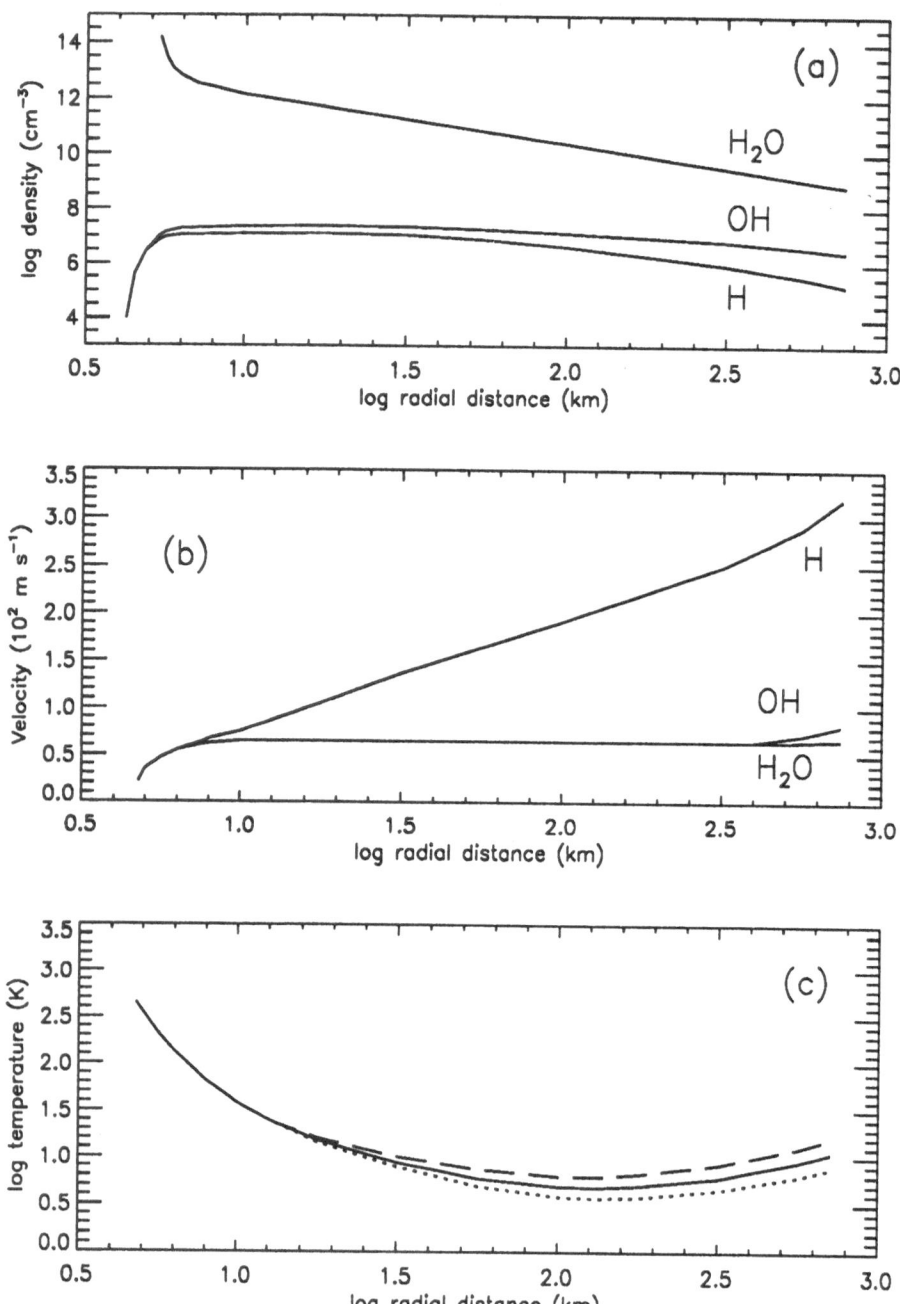

Figure 11.10.   Radial profiles of gas macroparameters in the cometary coma: (a) – number density; (b) – mean-mass velocity; (c) – kinetic temperature of water vapor (T – solid line; $T_{\parallel}$ – dashed line; $T_{\perp}$ - dotted line).

in Section 11.2.1) shows qualitative difference in the gas macroparameter calculations. In particular, in gas dynamical models, H and OH are characterized by the same radial density profiles, while in the kinetic model H has lower densities (see *Fig. 11.10a, b*).

The nonequilibrium character of the cometary gas outflow clearly indicates the importance of the kinetic approach, specifically the simulation of the translationally excited (hot) particles in a cometary atmosphere (see also Bocklee-Morvan and Crovisier, 1987; Ip 1989, 1990 b; Hodges 1990). Kinetic modeling is the most efficient approach to correctly include photolysis and thermalization processes which determine the contribution of the hot particles in the gas energetics and in the formation of the hydrogen corona (Smyth *et al.*, 1991).

CONCLUSIONS

In this monograph we summarized the results of our study of the kinetic approach to the modeling of rarefied gas envelopes of the celestial bodies. The focus was given to the problem of non-equilibrium processes occurring in the planetary upper atmosphere including transition regions and corona formation and cometary coma. We specifically considered the basic mathematical ideas underlying numerical models for stochastic simulation of atmospheric kinetics and the development of algorithms for the evaluation of these models.

A nonequilibrium state of the atmospheric multicomponent gas as an open thermodynamical system is triggered and maintained by the incident solar EUV and corpuscular radiation, with the involvement of successive collisional relaxation and numerous chemical reactions responsible for the system evolution. The original photolytic and energetic electron impact processes of interaction and the physical and chemical evolutionary processes can be described by a set of nonlinear Boltzmann type equations with the source term. Instead of a direct solution of these equations facing many obstacles, a special technique of stochastic simulation and Monte-Carlo algorithms were used for the numerical computer evaluation of such a system. This powerful technique allowed to study in detail the nonequilibrium state of the rarefied atmospheric gas at both microscopic (molecular) and macroscopic (continuum) levels and, in particular, to estimate the input of nonthermal particles into the composition, energetics, and dynamics of the planetary upper atmospheres. It also allowed to study specifics of transition regions in the atmosphere of the Earth, as well as some peculiarities of nonequilibrium processes pertinent to the Jovian atmosphere and cometary coma.

The main objectives were to reveal specific effects of the processes of non-equilibrium kinetics in terms of their domain and/or significant contribution to the macroscopic state of the atmospheric system. These included:

– the detailed quantitative study of the kinetics of photolysis and energetic electron impacts involving estimates of the dissociation and ionization production rates and formation of molecules and atoms in different states of excitation;

- the multichannel processes of thermal dissipation in the Earth's thermosphere and the evaluation of the processes of collisional relaxation of the "hot" particles of photochemical origin (superthermal O and N) involving their perturbing effects on the atmospheric gas and the formation of nonthermal atomic oxygen of secondary origin;
- the photochemistry of odd-nitrogen and its influence on the composition of the lower thermosphere of the Earth;
- the kinetics of the thermal dissipation of light atmospheric species from the Earth's exosphere and the formation of "hot" oxygen corona, relevant to the formation of coronas of nonthermal particles on other inner planets;
- the hot hydrogen sources, their distribution and auroral hydrogen emission in Jupiter's thermosphere;
- the kinetic effects and peculiarities of the subliming gas flow in the near-surface (Knudsen) layer of the cometary nucleus;
- the photochemistry and dynamics of the rarefied gas in the inner cometary coma.

These results made it possible to get more insight into rather delicate aeronomy processes occuring in the planetary and cometary atmospheres. The kinetic approach and algorithms for stochastic simulation of the rarefied atmospheric gas contribute to the solution of some important aeronomical problems related to the development of global atmospheric models involving elements of prognosis of the state of the near-planetary space. They also open vast opportunities for a more rigorous analysis of experimental data provided both by ground-based observations and returned by space missions.

REFERENCES

REFERENCES

Aamodt, R.E., Case, K.H. (1962) Density in a simple model of the exosphere, *Phys. Fluids*, **5**, 1019.

A'Hearn, M.F. (1988) Comets. A decade of UV astronomy using IUE, *ESA Spec. Publ.*, No. 281, 47.

A'Hearn, M.F., and Millis, R.L. (1980) Abundance correlations among comets, *Astron. J.*, **85**, 1528.

A'Hearn, M.F. and Feldman, P.D. (1985) S_2: a clue to the origin of cometary ice?, in: *Ices in Solar System*. Proceedings of NATO Advanced Resenrch Workshop, D. Reidel, Dordrecht, 463.

A'Hearn, M.F., Feldman, P.D., and Schleicher, D.G. (1983) The discovery of S_2 in comet IRAS-Araki-Alcock 1983d, *Astrophys. J.*, **274**, L99.

A'Hearn, M.F., Hoban, S., Birch, P.V., *et al.* (1986) Cyanogell jets in comet Halley, *Nature*, **324**, 649.

Ajello, J.M., Ahmed S.M., Kanik I., and Multari R. (1995a) Kinetic energy distribution of H(2p) from dissociative excitation of H_2, *Phys. Rev. Lett.*, **75** , 3261.

Ajello, J.M., Kanik I., Ahmed S.M., and Clarke J.T. (1995b) Line profile of H Lyman α from dissociative excitation of H_2 with application to Jupiter, *J. Geophys. Res.,,* **100**, 26411.

Akasofu, S.-I., and Chapman, S. (1972) *Solar terrestrial physics*, Oxford University Press, London.

Alder, B.J., and Wainwright, T.E. (1957) Studies in molecular dynamics, *J. Chem. Phys.*, **27**, 1208.

Alekseev, B.I. (1982) *Mathematical kinetics of reactive gases*, Nauka, Moscow, (in russian).

Allamandola, L.J., Sandford, S.A., and Valero, G.L. (1988) Photochemical and thermal evolution of interstellar/precometary ice analogs, *Icarus*, **76**, 225.

Allen, D.A., and Wickramasinghe, D.T. (1987) Discovery of organic grains in comet Wilson, *Nature*, **399**, 615.

Allen, M.P., and Tildesley, D.J. (1987) *Computer simulation of liquids*, Oxford University Press.

Amiot, C. (1982) The infrared emission spectrum of NO': analysis of the $\Delta v = 3$ sequence up to $v = 22$, *J. Mol. Spectrosc.,* **94**, 150.

Anders, E. (1989) Pre-biotic organic matter from comets and asteroids, *Nature*, **342**, 255.

Anisimov, S.I., Imas Ya.A., Romanov G.S., and Hodyako Yu.V. (1970) *Influence of high power radiation on metals*. Nauka, Moscow.

Aristov, V.V., and Tcheremissine, F.G. (1980) The conservative splitting method for solving the Boltzmann equation, *USSR Comput. Math. Math. Phys.*, **20**, 208.

Armstrong, P.S, Lipson, S.J., Dodd, J.A., Lowell, J.R., Blumberg, W.A.M., and Nadile, R.M. (1994) Highly rotationally excited NO(v,J) in the thermosphere from CIRRIS 1A limb radiance measurements, *Geophys. Res. Lett.*, **21**, 2425.

Arpigny, C., Magain, P., Manfroid, J., *et al.* (1987) Resolution of the OI + NH_2 blend in comet P/Halley, *Astron. Astrophys.*, **187**, 485.

273

Arpigny, C., A'Hearn, M.F.,Weaver, H.A, and Feldman, P.D.(1993) Observations of CH, NH, and NH$_2$ in comets P/Hartley 2 (1991 XV) and Shoemaker-Levy (1991 a1) using the Hubble Space Telescope, in: IAU Symp. No. 160: *Asteroids, Comets, Meteors*, Book of Abstracts, Belgirate (Novara), Italy, 17.

Atreya, S.K. (1986) *Atmospheres and ionospheres of the outer planets and their satellites*. Springer-Verlag, Berlin.

Babovsky, H. (1989) A convergence proof for Nanbu's Boltzmann simulation scheme, *European J. of Mechanics B/Fluids*, **8**, 41.

Banks, P.M., and Kockarts, G. (1973) *Aeronomy*. Academic Press, New York.

Bar-Nun, A. and Kleinfeld, I. (1989) On the temperature and gas composition in the region of comet formation, *Icarus*, **80**.

Barghouthi, I.A., Barakat, A.R., and Schunk, R.W. (1993) Monte-Carlo study of the transition region in the polar wind: an improved collision model, *J. Geophys. Res.*, **98**, 17583.

Barlier, F., Berger, C., Falin, J.L., and Thuillier, G. (1978) A thermospheric model based on satellite drag data, *Ann. Geophys*, **34**, 9.

Barth, C.A. (1964) Rocket measurement of the nitric oxide dayglow, *J. Geophys. Res.*, **69**, 3301.

Barth, C.A. (1990) Reference models for thermospheric NO, *Adv. Space Res.*, **10** , 103.

Barth, C.A., Farmer, C.B., Siskind, D.E., and Perich, J.P. (1996) ATMOS observations of nitric oxide in the thermosphere and lower mesosphere, *J. Geophys. Res.*, **101**, 12489.

Barth, C.A., Tobiska, W.K., Siskind, D.E., and Cleary D.D. (1988) Solar–terrestrial coupling: low–latitude thermospheric nitric oxide, *Geophys. Res. Lett.*, **15**, 92.

Bates, D.R. (1989a) Rapid deactivation of N(^{2}D) by O(^{3}P) and 630 nm emission, *Planet. Space Sci.*, **37**, 1145.

Bates, D.R. (1989b) Theoretical considerations regarding some inelastic collision processes of interest in aeronomy: deactivation and charge transfer, *Planet. Space Sci.*, **37**, 363.

Bauer, S.J. (1973) *Physics of planetary ionospheres*. Springer-Verlag, Berlin.

Belotserkovskii, O.M., and Yanitskii, V.E. (1975) Statistical particle-in-cell method for solving rarefied gas dynamics problems, *Zhurn. Vych. Mat. Mat. Fiz.*, **15**, 1195, (in russian).

Bernstein, R.B., Dalgarno, A., Massey, H., and Percival I.C. (1963) Thermal scattering of atoms by homonuclear diatomic molecules, *Proc. Roy. Soc. Lond. A*, **274**, 427.

Bertaux, J.L. (1976) Observations of hydrogen in the upper atmosphere, *J. Atmos. Terr. Phys.*, **38**, 821.

Bertaux, J.L.(1986) The UV bright spot of water vapor in comets, *Astron. Astrophys.*, **160**, L7.

Bertaux, J.L., Blamont, J.E., Lepine, V.M., Kurt, V.G., Romanova, N.N., and Smirnov, A.S. (1981) Venera 11 and Venera 12 observations of EUV emissions from the upper atmosphere of Venus, *Planet. Space Sci.*, **29**, 149.

Bhardwaj A., and Singhal R.P. (1993) Optically thin H Lyman alpha production on outer planets : low-energy proton acceleration in parallel electric fields and neutral H atom precipitation form ring current, *J. Geophys. Res.*, **98**, 9473.

Bhatnagar, P.L., Gross, E.P., and Krook, M. (1954) Model for collision processes in gases, I. Small amplitude processes in charged and neutral one-component systems, *Phys. Rev.*, **94**, 511.

Bird, G.A. (1970) Direct simulation and the Boltzmann equation, *Phys. Fluids*, **13**, 2676.

Bird, G.A. (1976) *Molecular gas dynamics*. Clarendon Press, Oxford.

Bird, G.A. (1994) *Molecular gas dynamics and the direct simulation of gas flow*. Clarendon Press, Oxford.

Bisikalo, D.V., and Shematovich V.I. (1987) Kinetic model of cometary inner coma, *Astron. Tsirc.*, **1507**, 1, (in russian).

Bisikalo, D.V., and Shematovich, V.I. (1988) Numerical modeling of nonequilibrium rarefied gas flow in inner coma of comet, *Astron. Herald,* **21**, 41, (in russian).

Bisikalo, D.V., and Shematovich, V.I. (1989) Variation of the thermal dissipation rate in the Earth's upper atmosphere, in: *Our changing atmosphere,* Univ. de Liège, 267.

Bisikalo, D.V., and Strel'nitskij V.S. (1985) Ice halo of comet and temperature of inner coma, *Letters to Astron. J.,* **11**, 475, (in russian).

Bisikalo, D.V., Marov M.Ya., Shematovich V.I., and Strelnitsky V.S. (1989) The flow of the subliming gas in the near-nuclear (Knudsen) layer of the cometary coma, *Adv. Space Res.,* **9**, 53.

Bisikalo, D.V., Shematovich, V.I., and Gérard, J.C. (1995) 'Kinetic model of the formation of the hot oxygen geocorona. II. Influence of O^+ ion precipitation, *J. Geophys. Res.,* **100**, 3715.

Bisikalo, D.V., Shematovich, V.I., Gérard, J.C., Gladstone, R. and Waite J.H. (1996) The distribution of hot hydrogen atoms produced by electron and proton precipitation in the Jovian aurora, *J. Geophys. Res.,* **101**, 21157.

Blochintsev, D.A. (1961) *The fundamentals of quantum mechanics.* High School Publ., Moscow, (in russian).

Blum, P., Harris, I., and Priester, W. (1972) The physics of the neutral upper atmosphere, in: *CIRA-72,* Academic Verlag, Berlin, 221.

Bockelee-Morvan D., and Crovisier J. (1987) The role of water in the thermal balance of the coma, in: *Proceedings of Symposium on the Diversity and Similarity of Comets,* ESA SP-278, Paris, 235.

Bockelee-Morvan, D., Brooke, D.T.,and Crovisier, J. (1993a) The origin of the 3.2 - 3.6 micron emission features in comets: gas and dust, in: IAU Symp. No. 160: *Asteroids, Comets, Meteors,* Book of Abstracts, Belgirate (Novara), Italy, 34.

Bockelee-Morvan, D., Padman, R., Davies, J.K., and Crovisier, J. (1993b) Observations of submillimetre lines of CH3OH, HCN and H2CO in comet P/Swift-Tuttle with the James Clerk Maxwell Telescope, in: IAU Symp. No. 160: *Asteroids, Comets, Meteors,* Book of Abstracts, Belgirate (Novara), Italy, 35.

Brinkmann, R.T. (1971) More comments on the validity of Jeans escape rate, *Planet. Space Sci.,* **19**, 791.

Budzien, S.A., and Feldman, P.D. (1992) Upper limits to the S2 abundance in several comets observed with the International Ultraviolet Explorer, *Icarus,* **99**, 143.

Budzien, S.A., and Feldman, P.D. (1991) OH prompt emission in comet IRAS-Araki-Alcock (1983 VII), *Icarus,* **90**, 308.

Buonsanto, M.J., Solomon, S.C., and Tobiska, W.K. (1992) Comparison of measured and modeled solar EUV flux and its effect on the E-F1 region ionosphere, *J. Geophys. Res.,* **97**, 10513.

Cercignani, C. (1969) *Mathematical methods in kinetic theory.* Plenum Press, New York.

Cercignani, C. (1988) *The Boltzmann equation and its applications.* Springer, New York.

Chamberlain, J.W.(1978) *Theory of planetary atmospheres.* Academic Press, New York.

Chamberlain, J.W. and Hunten, D. (1987) *Theory of planetary atmospheres. An introduction to their physics and chemistry.* Second Edition, Academic Press, New York.

Chamberlain, J.W., and Smith, G.R. (1971) Comments on the rate of evaporation of a non-maxwellian atmosphere, *Planet. Space Sci.,* **19**, 675.

Chapman, S. (1916-17) The kinetic theory of simple and composite gases. Viscosity, thermal conduction, and diffusion, *Proc. Royal Soc.,* **A93**, 1.

Chapman, S., and Cowling, T.G. (1952) *The mathematical theory of non-uniform gases* (2nd edn). Cambridge University Press.

Chyba, C.F., Sagan, C., and Mumma, M.J. (1989) The heliocentric evolution of cometary infrared spectra: Results from an organic grain model, *Icarus,* **79**, 362.

Clancy, R.I., Rusch, D.W., and Muhleman, D.O. (1992) A microwave measurement of high levels of thermospheric nitric oxide, *Geophys. Res. Lett.,* **19**, 261.

Clairemidi, J., Moreels, G., and Krasnopolsky, V.A. (1990) Gaseous CN, C_2, and C_2 jets in the inner coma of comet P/Halley observed from the Vega 2 spacecraft, *Icarus,* **86**, 115.

Clarke J.T., Trauger J., and Waite J.H. (1989) Doppler shifted H Ly α emission from Jupiter's aurora, *Geophys. Res. Lett.*, **16**, 587.

Clarke J.T., Ben Jaffel L., Vidal-Madjar A., Gladstone G.R., Waite J.H., Prangé R., Gérard J.C., Ajello J., and James G. (1994) Hubble Space Telescope Goddard High-resolution spectrograph H_2 rotational spectra of Jupiter's aurora, *Ap. J.*, **430**, L73.

Cleary, D.D. (1986) Daytime high-latitude rocket observations of the NO γ, δ and ϵ bands, *J. Geophys. Res.*, **91**, 11,337.

Cochran, A.L. (1987) Another look at abundance correlations among comets, *Astron. J.*, **93**, 231.

Cochran, A.L., Barker, E.S., Ramseyer, T.F., and Storrs, A.D. (1992) The McDonald Observatory faint comet survey: gas production in 17 comets, *Icarus*, **98**, 151.

Cochran, A.L., Cochran, W.D., Barker, E.S., and Storrs, A.D. (1991) The development of the CO^+ coma of comet P/Schwassmann-Wachmann 1, *Icarus*, **92**, 179.

Combes, M., Moroz, V.I., Crifo, J.F., it et al. (1986) Infrared sounding of comet Halley from Vega 1, *Nature*, **321**, 266.

Combi, M.R., and Feldman, P.D.(1992) IUE observations of H Lyman-α in comet P/Giacobini-Zinner, *Icarus*, **91**, 260.

Combi, M.R., and McCrosky, R.E.(1991) High resolution spectra of the 6300 Å region of comet P/Halley, *Icarus*, **91**, 270.

Conway, R.A. (1988) Photoabsorption and photoionisation cross section of O, O_2 and N_2 for photoelectron production calculations : a compilation of recent laboratory measurements, Report 6155, Naval Research Laboratory, Washington.

Cosmovici, C.B., Schwarz, G., Ip, W.-H., and Mack, P.(1988) Gas and dust jets in the inner coma of comet Halley, *Nature*, **332**, 705.

Cosmovici, C.B., Schwarz, G., Ip, W.-H., and Fink, V.(1993) H_2O jets in the coma of comet Halley? in: IAU Symp. No. 160, *Asteroids, Comets, Meteors*, Book of Abstracts, Belgirate (Novara), Italy, 75.

Cotton, D.M., Gladstone, G.R., and Chakrabarti, S. (1993) Sounding rocket observation of a hot atomic oxygen geocorona, *J. Geophys. Res.*, **98**, 21651.

Cravens, T.E., and Killeen, T.L. (1988) Longitudinally asymmetric transport of nitric oxide in the E-region, *Planet. Space Sci.*, **36**, 11.

Cravens, T.E., and Stewart, A.I. (1978) Global morphology of nitric oxide in the lower E-region, *J. Geophys. Res.*, **83**, 2446.

Crifo, J.F. (1986) Comets as large dirty snowballs sublimating in interplanetary space: A review of gasdynamics models, in: *Rarefied Gas Dynamics*, Springer Verlag, **2**, 229.

Crovisier, J. (1989) The photodissociation of water in cometary atmospheres, *Astron. Astrophys.*, **213**, 459.

Crovisier, J., Despois, D., Bockelee-Morvan, D., *et al.* (1991) Microwave observations of hydrogen sulfide and searches for other sulfur compounds in comets Austin (1989 ci) and Levy (1990 c), *Icarus*, **93**, 246.

Danks, A.C., Encrenaz, T., Bouchet, P., et al. (1987) The spectrum of comet P/Halley from 3.0 to 4.0 μm, *Astron. Astrophys.*, **184**, 329.

Davenport, J.E., Slanger, T.G. and Black, G. (1976) The quenching of $N(^2D)$ by $O(^3P)$, *J. Geophys. Res.*, **81** , 12.

Delsemme, A.H., and Swings, P.(1952) Hydrates de gaz dans les noyaux cométaires et les grains interstellaires, *Ann. Astrophys.*, **15**, 1.

Delsemme, A.H.(1976) Chemical nature of the cometary snows, *Mem. Soc. R. Sci. Liège*, *Ser. 6*, **9**, 135.

DeMore, W.B. Sander, S.P., Golden, D.M. Molina, M.T., Hampson, R.F., Kurylo, M.J., Howard, C.J. and Ravishankara, A.R. (1990) Chemical kinetics and photochemical data for use in stratospheric modeling, Evaluation number 9; JPL publication 90-1, Pasadena, CA.

Deshpande, S.M. (1978) An unbiased and consistent Monte Carlo game simulating of the Boltzmann equation, Indian Institute of Science, Report 78FM4.

Disanti, M.I., Fink, U., and Shultz, A.B. (1990) Spatial distribution of H_2O^+ in comet P/Halley, *Icarus*, **86**, 152.

Disanti, M.I., and Fink, U. (1991) Composition comparison between comets P/Halley and P/Brorsen-Metcalf, *Icarus*, **91**, 105.

Donn, B., Wdowiak, T., Nuth, J., and Chappelle, E. (1984) Laboratory experiments suggesting a fluorescent source for the Red Rectangle visual emission, *Bull. Am. Astron. Soc.*, **16**, 462.

Donnelly, R.F. and Pope, J.H. (1973) The 1-3000 Å solar flux for a moderate level of solar activity for use in modeling the ionosphere and upper atmosphere, Tech. Rep. ERL 276, Natl. Oceanic and Atmos. Admin., Silver Spring, Md.

Drossart, P., Maillard J.P., Caldwell J., Kim S.J., Watson K.G., Majewski W.A., Tennyson J., Waite Jr. J.H., and Wagener R. (1989) Detection of H_3^+ on Jupiter, *Nature*, **340**, 539.

Drossart, P., Bézard B., Atreya S.K., Bishop J., Waite J.H., and Boice D. (1993) Thermal profiles in the auroral regions of Jupiter, *J. Geophys. Res.*, **98**, 18803.

Du, M.L. and Dalgarno, A. (1990) The radiative association of N and O atoms, *J. Geophys. Res.*, **95**, 12265.

Duff, J.W., Bien, F., and Paulsen D.E. (1994) Classical dynamics of the $N(^4S) + O_2(X\ ^3\Sigma_g^-) \rightarrow NO(X\ ^2\Pi) + O(^3P)$ reaction, *Geophys. Res. Lett.*, **21**, 2043.

Eberhardt, P., Krankovsky, D., Schulte, W.,et al. (1987) The CO and N_2 abundance in comet P/Halley, *Astron. Astrophys.*, **187**, 481.

Engebretson, M.J. and Mauersberger, K. (1983) The response of thermospheric atomic nitrogen to magnetic storms, *J. Geophys. Res.*, **88**, 6331.

Engebretson, M.J., Mauersberger, K. and Potter, W.E. (1977a) Extension of atomic nitrogen measurements into the lower thermosphere, *J. Geophys. Res.*, **82**, 3291.

Engebretson, M.J., Mauersberger, K., Kayser, D.C. Potter, W.E. and Nier, A.O. (1977b) Empirical model of atomic nitrogen in the upper atmosphere, *J. Geophys. Res.*, **82**, 461.

Enskog, D. (1917) *Kinetische theorie der Vorgänge in mässig verdünnten Gasen*. I. Allgemeiner Teil, Almqvist and Wiksell, Uppsala.

Eparvier, F.J., and Barth C.A. (1992) Self absorption theory applied to rocket measurements of the nitric oxide (1, 0) γ band in the daytime thermosphere, *J. Geophys. Res.*, **97**, 13723.

Esipov, V. F., Korsun, P. P., Mamadov, O., and Parusimov V. G. (1994) Relative spectrophotometry of comet Halley in the near-IR range, *Astron. Herald (Solar System Research)*, **28**, 37.

Fahr, H.J. (1976) Reduced hydrogen temperatures in the transition region between thermosphere and exosphere, *Ann. Geophys.*, **32**, 277.

Fahr, H.J., and Nass, H.U. (1978) Concerning the structure of the transition layer between the terrestrial atmosphere and the exosphere, *Ann. Geophys.*, **34**, 219.

Fahr, H.J., and Shizgal, B. (1983) Modern exospheric theories and their observational relevance, *Rev. Geophys.*, **21**, 75.

Feldman, P.D.(1990) Ultraviolet spectroscopy of comets, in: *Comets in the Post Halley Era* (R. L. Newburn ed.).

Feldman, P.D., and Budzien, S.A.(1989) Upper limits on C3 abundances in several comets observed with IUE, *Bull. Am. Astron. Soc.*, **21**, 937.

Feldman, P.D., and Takacs, P.Z. (1974) Nitric oxide gamma and delta band emission at twilight, *Geophys. Res. Lett.*, **1**, 169.

Feldman, P.D., Fournier, K.D., Grinin, V.P., and Zvereva, A.M. (1993) The abundance of ammonia in comet P/Halley derived from ultraviolet spectrophotometry, *Astrophys. J.*, **104**, 348.

Fell, C. , Steinfeld, J.I. and Miller, S. (1990) Quenching of $N(^2D)$ by $O(^3P)$, *J. Chem. Phys.*, **92**, 4768.

Feller, W. (1970) *An introduction to probability theory and its applications* (third ed.). Wiley, New York.

Fennelly, J., and Torr, D.G. (1992) Photoionization and photoabsorption cross sections of O, N_2, O_2 and N for aeronomic calculations, *Atomic Data Nucl. Data Tables,* **51**, 321.

Ferziger, J., and Kaper, H. (1972), *Mathematical theory of transport processes in gases.* North-Holland, Amsterdam.

Fesen, C.G., Gérard, J.C., and Rusch, D.W. (1989) Rapid deactivation of $N(^2D)$ by O : Impact on thermospheric and mesospheric odd nitrogen, *J. Geophys. Res.,* **94**, 5419.

Fesen, C.G. , Gérard, J.C., and Rusch, D.W. (1990) The latitudinal gradient of NO peak density, *J. Geophys. Res.,* **95**, 19053.

Festou, M.C., and Feldman, P.D. (1981) The forbidden oxygen lines in comets, *Astron. Astrophys.,* **103**, 154.

Fox, J.L. (1993) On the escape of oxygen and hydrogen from Mars, *Geophys. Res. Lett.,* **20**, 1847.

Fox, J.L., and Dalgarno, A. (1980) The production of nitrogen atoms on Mars and their escape, *Planet. Space Sci.,* **28**, 41.

Frank, L.A., Sigwarth, C.B., and Craven, J.D. (1986) On the influx of small comets into Earth's upper atmosphere. II. Interpretation, *Geophys. Res. Lett.,* **13**, 307.

Frederick, J.E. and Rusch, D.W. (1977) On the chemistry of metastable atomic nitrogen in the F region deduced from simultaneous satellite measurements of the 5200-Å airglow and atmospheric composition, *J. Geophys. Res.,* **82**, 3508.

Fuller-Rowell, T.J. (1993) Modeling the solar cycle change in nitric oxide in the thermosphere and upper mesosphere, *J. Geophys. Res.,* **98**, 1559.

Gardner, J.L., and Samson, J.A.R. (1975) Photoion and photoelectron spectroscopy of CO and N_2, *J. Chem. Phys.,* **62**, 1447.

Garstang, R.H. (1956) Transition probabilities in auroral lines, in: *The Airglow and the Aurora* (Edited by Armstrong, E.B. and Dalgarno, A.), 324, Pergamon, New York.

Gealy M.W., and Van Zyl B. (1987) Cross sections for electron capture and loss. I. H^+ and H^- impact on H and H_2, *Phys. Rev.,* **36** , 3091.

Gérard, J.C. and Barth C.A. (1977) High Latitude Nitric Oxide in the Lower Thermosphere, *J. Geophys. Res.,* **82** , 674.

Gérard, J.C., and Roble, R.G. (1986) The role of nitric oxide on the zonally averaged structure of the thermosphere: Solstice conditions for solar cycle minimum, *Planet. Space Sci.,* **34**, 131.

Gérard, J.C., and Roble, R.G. (1988) The role of nitric oxide on the zonally averaged structure of the thermosphere: Solstice conditions for solar cycle maximum, *Planet. Space Sci.,* **36**, 271.

Gérard, J.C., and Taieb, C. (1986) The E region electron density diurnal asymmetry at Saint-Santin: observations and role of nitric oxide, *J. Atmos. Terr. Phys.,* **48**, 471.

Gérard J.C., Dols V., Prangé R., and Paresce F. (1994a) The morphology of the north Jovian ultraviolet aurora observed with the Hubble Space Telescope: a full longitudinal mapping, *Planet. Space Sci.,* **42**, 905.

Gérard J.C., Grodent D., Prangé R., Waite J.H., Gladstone G.R., Dols V., Paresce F., Storrs A., Ben Jaffel L. and Franke K.A. (1994b) A remarkable auroral event on Jupiter observed in ultraviolet with the Hubble Space Telescope, *Science,* **266**, 1675.

Gérard, J.C., Fesen, C.G., and Rusch, D.W. (1990) Solar cycle variation of nitric oxide at solstice, *J. Geophys. Res.,* **95**, 12235.

Gérard, J.C., Richards, P.G., Shematovich, V.I., and Bisikalo, D.V. (1995) The importance of new chemical sources for the hot oxygen geocorona, *Geophys. Res. Lett.,* **22**, 279.

Gérard, J.C., Roble, R.G., Rusch, D.W. and Stewart, A.I.F. (1984) The global distribution of thermospheric odd-nitrogen for solstice conditions during solar cycle minimum, *J. Geophys. Res.,* **89**, 1725.

Gérard, J.C., Shematovich, V.I., and Bisikalo, D.V. (1991) Non thermal nitrogen atoms in the Earth's thermosphere 2. A source of ntric oxide, *Geophys. Res. Lett.,* **18**, 1695.

Gérard, J.C., Shematovich, V.I., and Bisikalo, D.V. (1993) Effect of hot $N(^4S)$ atoms on the NO solar cycle variation in the lower thermosphere, *J. Geophys. Res.*, **98**, 11581.

Gérard, J.C., Shematovich, V.I., and Bisikalo, D.V. (1995) The role of fast N(4S) atoms and photoelectrons on the distribution of NO in the thermosphere, in: *The Upper Mesosphere and Lower Thermosphere: A Review of Experiment and Theory, Geophysical Monograph 87*, American Geophysical Union, Washington D.C., 235.

Gérard, J.C., Shematovich, V.I., Bisikalo, D.V., and Duff, J.W. (1996) An updated model of the hot nitrogen atom kinetics and thermospheric nitric oxide, *J. Geophys. Res.*, **102**, 285.

Gehrz, R.D., Ney, E.P., Rozental, E., and Tokunaga, A.T. (1989) Infrared photometry and spectrometry of comet P/Encke 1987, *Icarus*, **80**, 280.

Gigulev, V.N. (1971) Investigation of Bogolubov equation chain for strongly correlative systems, *Theor. Math. Phys.*, **7**, 106, (in russian).

Gladstone, R.G. (1988) UV resonance line dayglow emission on Earth and Jupiter, *J. Geophys. Res.*, **93**, 14623.

Gladstone, R.G, Allen M. and Yung Y.L. (1996) Hydrocarbon photochemisty in the upper atmosphere of Jupiter, *Icarus*, **1191**, 1.

Godefroid, M.,and Froese-Ficher, C. (1984) MCHF-BP fine-structure splittings and transition rates for the ground configuration in the nitrogen sequence, *J. Phys. B. Atom. Molec. Phys.*, **17**, 681.

Gombosi, T.I., Nagy, A.F., and Cravens, T.F.(1986) Dust and neutral gas modeling of the inner atmospheres of comets, *Rev. Geophys. Space Phys.*, **24**, 667.

Gordiets B.F., Kulikov Yu.N., Markov M.N., and Marov M.Ya. (1982) Numerical modeling of the thermospheric heat budget, *J. Geophys. Res.*, **87**, 4504.

Grad, H. (1949) On the kinetic theory of rarefied gases, *Commun. Pure Aappl. Math.*, **2**, 331.

Green, A.E.S., and Stolarski, R.S. (1972) Analytic models of electron impact excitation cross sections, *J. Atmos. Terr. Phys.*, **34**, 1703.

Greenberg, J.M. (1982) What are comets made of? A model based on interstellar dust, in: *Comets* (L. L. Wilkening ed.), Univ. of Arizona Press, Tucson, 131.

Greenberg, J.M., Zhao, N.S., and Hager, J.I. (1989) The interstellar dust model of comet dust constrained by 3.4 mm and 10 mm emission, *Adv. Space Res.*, **9**, 3.

Grewing, M., Praderie F., and Reinhard R. (Eds.) (1988) *Exploration of Halley's comet*, Springer-Verlag, Berlin.

Grim, G. (1969) *Spectroscopy of plasma*. Atomizdat, Moscow, (in russian).

Grishin, A.M., and Fomin, V.M. (1984) *Conjugate and Nonsteady Problems of the Mechanics of Reactive Media*. Nauka, Novosibirsk.

Groth, N., Kley, D. and Schurath, U. (1971) Rate constant for the infrared emission of the NO $(C^2\Pi - A^2 \sum^{+})$ transition, *J. Quant. Spectrosc. Radiat. Transfert*, **11**, 1475.

Grunbaum, F.A. (1971) Propagation of chaos for the Boltzmann equation, *Arch. Ration. Mech. and Anal.*, **42**, 323.

Hanner, M.S. (1986) A preliminary look at the dust in comet Halley, *Adv. Space Res.*, **5**, 325.

Hanner, S., Aitken, D.K., Knacke, R., *et al.* (1985) Infrared spectrophotometry of comet IRAS-Araki-Alcock (1983 d): a bare nucleus revealed?, *Icarus*, **62**, 97.

Hartle, R.E., Donahue, T.M., Grebovsky, J.M., and Mayr, H.G. (1996) Hydrogen and deuterium in the thermosphere of Venus: Solar cycle variations and escape, *J. Geophys. Res.*, **101**, 4525.

Haviland, J.K., and Lavin, M.L. (1962) Applications of the Monte Carlo method to heat trahsfer in a rarefied gas, *Phys. Fluids*, **5**, 1399.

Hays, P.B., Abreu, V.J., Solomon, S.C. and Yee, J.H. (1988) The visible airglow experiment - a review, *Planet. Space Sci.*, **36**, 21.

Hedin, A.E. (1983) A revised thermospheric model based on mass spectrometer and incoherent scatter data: MSIS–83, *J. Geophys. Res.*, **88**, 10,170.

Hedin, A.H. (1987) MSIS-86 thermospheric model, *J. Geophys. Res.*, **92**, 4649.

Hedin, A.E. (1988) A revised thermospheric model based on mass spectrometer and incoherent scatter data: MSIS-83, *J. Geophys. Res.*, **88**, 10170.

Hedin, A.E. (1989) Hot oxygen geocorona as inferred from neutral exospheric models and mass spectrometer measurements, *J. Geophys. Res.*, **94**, 5523.

Hedin, A.E. (1991) Extension of the MSIS thermosphere model into the middle and lower atmosphere, *J. Geophys. Res.*, **96**, 1159.

Hickey, M.P., Richards, P.G., and Torr, D.G. (1995) New sources for the hot oxygen geocorona: Solar cycle, seasonal, latitudinal, and diurnal variations, *J. Geophys. Res.*, **100**, 17377.

Hilbert, D. (1912) Begründung der kinetische Gastheorie, *Mathematische Annalen*, **72**, 562.

Hinteregger, H.E., Fukui, K., and Gibson, B.R. (1981) Observational, reference and model data on solar EUV, from measurements on AE-E, *Geophys. Res. Lett.*, **8**, 1147.

Hirschfelder, J.O., Curtiss, C.F., and Bird, R.B. (1954) *The molecular theory of gases and liquids*. John Willey, New York.

Hoban, S., Renter, D.C., Mumma, M.J., and DiSanti, M.A. (1993) Infrared measurements of methanol in comet P/Swift-Tuttle, in: IAU Symp. No. 160, *Asteroids, Comets, Meteors*, Book of Abstracts, Belgirate (Novara), Italy, 134.

Hochstim, A.R. (Ed.) (1969) *Kinetic processes in gases and plasmas*. Academic Press, New York.

Hodges, R.R. (1990) Monte Carlo simulation of nonadiabatic expansion in cometary atmospheres: Halley, *Icarus*, **83**, 410.

Hodges, R.R. (1993) Isotopic fractionation of hydrogen in planetary exospheres due to ionosphere-exosphere coupling: Implication for Venus, *J. Geophys. Res.*, **98**, 10833.

Hodges, R.R. (1994) Monte Carlo simulation of the terrestrial hydrogen exosphere, *J. Geophys. Res.*, **99**, 23229.

Hodges, R.R., Breig, E.L. (1991) Ionosphere-exosphere coupling through charge exchange and momentum transfer in hydrogen-proton collisions, *J. Geophys. Res.*, **96**, 7697.

Hodges, R.R., Breig, E.L. (1993) Charge transfer and momentum exchange in exospheric $D - H^+$ and $H - D^+$ collisions, *J. Geophys. Res.*, **98**, 1581.

Hodges, R.R., and Tinsley, B.A. (1986) The influence of charge exchange on the velocity distribution of hydrogen in the Venus exosphere, *J. Geophys. Res.*, **91**, 13649.

Hodges, R.R., Tinsley, B.A., and Rohrbaugh, R.P. (1981) The effect of the charge exchange source on the velocity and temperature distributions and their anisotropies in the Earth's exospheres, *J. Geophys. Res.*, **86**, 6917.

Hubert, B., Gérard, J.C., Shematovich, V.I., and Bisikalo, D.V. (1996) High rotational excitation of NO infrared thermospheric airglow: a signature of superthermal nitrogen atoms?, *Geophys. Res. Lett.,*, **23**, 2215.

Huebner, W.F. (1985) The photochemistry of comets, in: *The Photochemistry of Atmospheres* (J. S. Levine ed.), Academic, Orlando, FL.

Huebner, W.F. (Ed) (1990) *Physics and chemistry of comets*. Springer-Verlag, Heidelberg.

Huebner, W.F., and Keady J.J. (1983) Energy balance and photochemical processes in the inner coma, in: *Cometary Exploration*, MTA, Budapest, **I**, 165.

Huebner, W.F., Boice, D.C., and Korth, A. (1989) Halley's polymetric organic molecules, *Adv. Space Res.*, **9**, 29.

Huebner, W.F., Giguere P.T., and Slattery W.E. (1983) Photochemical processes in the inner coma, in: *Cometary Exploration*, MTA, Budapest, **I**, 496.

Hunten, D.M. (1973) The escape of light gases from planetary atmospheres, *J. Atmos. Sci.*, **30**, 1481.

Hunten, D.M. (1982) Thermal and nonthermal escape mechanisms for terrestrial bodies, *Planet. Space Sci.*, **30**, 773.

Hunten, D.M., and Donahue, T.M. (1976) Hydrogen loss from the terrestrial planet, *Rev. Earth Planet. Sci.*, **4**, 265.

Ibadinov, Kh. I. and Aliev, S (1989) Temperature of cometary nuclei made of H_2O and CO_2 ices, *Comety Meteory*, No. 39, 9.

Ip, W.-H. (1988) On hot oxygen corona on Mars, *Icarus*, **76**, 145.

Ip, W.-H. (1989) Photochemical heating of cometary comae. III. The radial variations of the expansion velocity of CN shells in comet Halley, *Astrophys. J.*, **346**, 475.

Ip, W.-H. (1990a) The fast atomic oxygen corona extent of Mars, *Geophys. Res. Lett.*, **17**, 2289.

Ip, W.-H. (1990b) Energetic neutral atoms in cometary comae, *Astrophys. J.*, **353**, 290.

Irvine, W.M., Abraham, Z., A'Hearn, M.F., et al.(1984) Radioastronomical observations of comets IRAS-Araki-Alcock (1983 d) and Sugano-Saigusa-Fujikawa (1983 e), *Icarus*, **60**, 215.

Ishimoto, M., Romick, G.J., and Meng, C.-I. (1992) Energy distribution of energetic O^+ precipitation into the atmosphere, *J. Geophys. Res.*, **97**, 8619.

Ishimoto, M., Torr, M.R., Richards, P.G., and Torr, D.G. (1986) The role of energetic O^+ precipitation in a mid-latitude aurora, *J. Geophys. Res.*, **91**, 5793.

Ivanov, M.S., and Rogasinskii, S.V. (1988a) Comparative analysis of algorithms of direct statistical simulation method in rarefied gas dynamics, *Zhurn. Vych. Mat. Mat. Fiz.*, **23**, 1058, (in russian).

Ivanov, M.S., and Rogasinskii, S.V. (1988b) Analysis of numerical techniques of the direct simulation Monte Carlo method in the rarefied gas dynamics, *Sov. J. Numer. Anal. Math. Modelling*, **3**, 453, (in russian).

Ivanov, M.S., and Rogazinskii, S.V. (1991) Theoretical analysis of traditional and modern schemes of the DSMC method, in: *Rarefield Gas Dynamics*, 629, VCH, New York.

Ivanov, M.S., Cheremissine, F.G., and Yanitskii, V.E. (1987) Statistical modeling of kinetic equations with physical and chemical processes, in: *Modeling in Mechanics*, Novosibirsk, **1**, 62, (in russian).

Ivanovskiy, A.I., Repnev, A.P., and Schvidkovskiy, E.G. (1967) *Kinetic theory of the upper atmosphere*. Hidrometeoizdat, Leningrad, (in russian).

Izakov, M.N. (1967) The structure of neutral upper atmosphere, *Space Sci. Rev.*, **7**, 579, (in russian).

Jacchia, L.G. (1977) Thermospheric temperature, density, and composition: new models, SAO Sp. Rep. 375, Cambridge, Massachussets.

Jackman, C.H., Garvey, R.H., and Green, A.E.S. (1977) Electron impact on atmospheric gases I. Updated cross sections, *J. Geophys. Res.*, **82**, 5081.

Jackson, W.M. (1982) Laboratory studies of photochemical and spectroscopic phenomena related to comets, in: *Comets* (L.L.Willkening ed.), Univ. of Arizona Press, Tucson, 496.

Jackson, W.M., Cody, R.J., and Sabety-Dzvonik, M. (1976) Laser induced photoluminescence spectroscopy of cometary radicals, in: *The Study of Comets*, NASA Spec. Publ. No. 393, 706.

Janev, R.K., Langer W.D., Evans Jr. K. and Post Jr. D.E. (1987) *Elementary processes in Hydrogen-Helium plasmas: cross sections and reactio rate coefficients.* Springer-Verlag, Heidelberg.

Jeans, J.H. (1923) *The dynamical theory of gases.* Cambridge Univ. Press, New York.

Jeisberger, E.K., Kissle, J., and Rabe, J. (1989) The composition of comets, in: *Origin and Evolution of Planetary and Satellite Atmospheres*, Univ. of Arizona Press, Tucson, 167.

Jokosky, B.M., Pepin, R.O., Johnston, R.E., and Fox, J.L. (1994) Mars atmospheric loss and isotopic fractionation by solar-wind-induced sputtering and photochemical escape, *Icarus*, **111**, 271.

Jusinski, L.E., Black, G. and Slanger, T. G. (1988) Resonance–enhanced multiphoton ionization measurements of $N(^2D)$ quenching by $O(^3P)$, *J. Phys. Chem.*, **92**, 5977.

Kac, M. (1963) *Probability and related problems in physics.* Mir, Moscow, (in russian).

Kac, M. (1973) Some probabilistic aspects of the Boltzmann equation, in: *The Boltzmann equation theory and application*, Springer, New York, 379.

Kennealy, J.P., Del Greco, F.P., Caledonia, G.E., and Green, B.D. (1978) Nitric oxide chemiexcitation occurring in the reaction between metastable nitrogen atoms and oxygen molecules, *J. Chem. Phys.*, **64**, 1574.

Khare, B.N., Thompson, W.R., Murray, B.G., *et al.* (1989) Solid organic residues produced by irradiation of hydrocarbon containing H_2O and H_2O/NH_3 ices: Infrared spectroscopy and astronomical implication, *Icarus*, **79**, 350.

Kim, S.J., and A'Hearn, M.F. (1991) Upper limits of SO and SO_2 in comets, *Icarus*, **90**, 79.

Kim, S.J., Drossart P., Caldwell J., and Maillard J.P. (1990) Temperatures of the Jovian auroral zone inferred from 2-μm H_2 quadrupole line observations, *Icarus*, **64**, 233.

Kirby, K., Constantinides, E.R., Babeu, S., Oppenheimer, M., and Victor, G.A. (1979) Photoionization and photoabsorption cross sections of thermospheric species: He, O, N_2 and O_2, *At. Data Nucl. Data Tables*, **23**, 63.

Klimontovich, Yu.L. (1980) *Kinetic theory of electromagnetic processes.* Nauka, Moscow, (in russian).

Knudsen, W.C. (1973) Escape of ^{4}He and fast O atoms from Mars and inferences on the ^{4}He mixing ratio, *J. Geophys. Res.*, **78**, 8049.

Kockarts, G. (1980) Nitric oxide cooling in the terrestrial atmosphere, *Geophys. Res. Lett.*, **7**, 137.

Kolesnichenko, A.V., and Marov, M.Ya. (1985) Modeling of gas dynamic processes in the gas-dust atmosphere of a comet, *Preprint No. 61*, Keldysh Inst., Moscow.

Kolesnichenko, A.V., and Skorov, Yu.V. (1987) Hydrodynamic model of the outflow of a gas suspension from a cometary nucleus (slightly dusty coma), *Preprint No. 113*, Keldysh Inst., Moscow.

Kolmogorov, A.N. (1938) Analytical methods in probability theory, *Advances in Mathematics*, **5**, 51, (in russian).

Konno, I., and S. Wyckoff, S. (1989) Atomic and molecular abundances in comet Giacobini-Zinner, *Adv. Space Res.*, **9**, 163.

Korolev, A.E., and Yanitskii, V.E. (1983) Direct stochastic simulation of collisional relaxation in gas mixtures, *J. Comp. Math. and Math. Phys.*, **23**, 674, (in russian).

Korolev, A.E., and Yanitskii, V.E. (1985) Development of statistical particle-in-cell method for relaxation problems of chemically reacting gas mixtures, *J. Comp. Math. and Math. Phys.*, **25**, 431, (in russian).

Koura, K. (1970) Transient Couette flow of rarefied binary gas mixtures, *Phys. Fluids*, **13**, 1457.

Kozyra, J.V., Cravens, T.E., and Nagy, A.F. (1982) Energetic O^+ precipitation, *J. Geophys. Res.*, **87**, 2481.

Krasnobaev, K.V. (1983) Axisymmetric gas outflow from a cometary nucleus, *Pis 'ma Astron. Zh.*, **9**, 637, (in russian).

Krinberg, I.A. (1978) *Kinetics of electrons in the Earth's ionosphere and plasmosphere.* Nauka, Moscow, (in russian).

Krinberg, I.A., and Taschilin, A.V. (1984) *Ionosphere and plasmosphere.* Nauka, Moscow, (in russian).

Krishna Swamy, K.S. (1984) Chemical composition of comets and its relation to chemical evolution, *J. Sci. Ind. Res.*, **43**, 17.

Krishna Swamy, S., and Wallis, M.K. (1987) Sulfur compounds in cometary IUE spectra, *Mon. Not. R. Astron. Soc.*, **228**, 305.

Kuze, A. and Ogawa, T. (1988) Solar cycle variation of thermospheric NO: a model sensitivity study, *J. Geomagn. Geoelectr.*, **40**, 1053.

LaBudde, R.A., and Bernstein, R.B. (1973) Classical study of rotational excitation of a rigid rotor: Li^++H2. II. Correspondence with quantal results, *J. Chem. Phys.*, **59**, 3687.

Laganà, A., Garcia, E., and Ciccarelli, L. (1987) Deactivation of vibrationally excited nitrogen molecules by collision with nitrogen atoms, *J. Phys. Chem.*, **91**, 312.

Larson, H.P., Davis, D.S., Mumma, M.J., *et al.* (1986) Velocity resolved observations of water in comet Halley, *Astrophys. J.*, **309**, 95.

Larson, S., Sekanina, Z., Levy, D., *et al.* (1987) Comet P/Halley near-nucleus phenomena in 1986, *Astron. Astrophys.*, **187**, 639.

Larson, H.P., Weaver, H.A., Mumma, M.J., and Drapatz, S. (1989) Airborne infrared spectroscopy of comet Wilson (1986 I) and comparison with comet Halley, *Astrophys. J.*, **338**, 1106.

Larson, H.P., Hong-Yao Hu, Mumma, M.J., and Weaver, H.A. (1990) Outbursts of H_2O in Comet P/Halley, *Icarus*, **86**, 129.

Larson, H.P., Hong-Yao Hu, Hsieh, K.S., *et al.* (1991) Descriptions of the neutral gas outflow in comets P/Halley and Wilson (1987 VII) from analyses of velocity resolved H_2O line profiles, *Icarus*, **91**, 251.

Lawler, M.E., Brownlee, D.E., Temple, S., and Wheelock, M.M. (1989) Iron, magnesium, and silicon in dust from comet Halley, *Icarus*, **80**, 225.

Lean, J.L. (1987) Solar ultraviolet irradiance variations: a review, *J. Geophys. Res.*, **92**, 839.

Lebedinets, V. N. (1990) Influx of organic particles of space dust into the atmosphere, *Dokl. Akad. Nauk SSSR*, **314**, 1363 (in russian).

Lebedinets, V. N. (1991) Earth's dust cloud and atmospheric oxygen, *Astron. Herald (Solar System Research)*, **25**, 325.

Lemaire, P., Charra J., Jouchoux A., Vidal-Madjar A., Artzner G.E., Vial J.C., Bonnet R.M. and Skumanich A. (1978) Calibrated full disk solar HI Lyman-α and Lyman-β profiles, *Astrophys. J.*, **223**, L55.

Leontovich, M.A. (1936) General equations of gas kinetic theory from point of view of random processes, *J. Exp. Theor. Phys.*, **5**, 211, (in russian).

Libby, W. F. (1947) Chemistry of energetic atoms produced by nuclear reactions, *J. Am. Chem. Soc.*, **69**, 2523.

Lichtenegger, H., and Komle, N. (1991) Heating and evaporation of icy particles in the vicinity of comets, *Icarus*, **90**, 319.

Lindenfeld, M.J., and Shizgal, B. (1979) Non-Maxwellian effects associated with the thermal escape of the planetary atmosphere, *Planet. Space Sci.*, **27**, 739.

Lipson S.J., P.S. Armstrong, J.A. Dodd, J.R. Lowell, W.A.M. Blumberg, and Nadile R.M. (1994) Subthermal nitric oxide spin-orbit distributions in the thermosphere, *Geophys. Res. Lett.*, **22**, 2421.

Livengood, T.A, Moos H.W. , Ballester G.E., and Prangé R. (1992) Jovian auroral activity, 1981-1991, *Icarus*, **97**, 2645.

Liu, C.Y., and Lees, L. (1961) Kinetic theory description of plane compressible Couette flow, in *Rarefied gas dynamics* (ed. L.Talbot), Academic Press, New York, 391.

Locht, R., Denzer, W., Ruhl, E., and Baumgartel, H. (1992) Photoionization mass spectrometry of kinetic energy-selected ions. The translational energy distribution of N^+/N_2 in the inner-shell ionization energy range, *Chem. Phys.*, **160**, 477.

Logan, D.A., and McElroy, M.B. (1977) Distribution functions for energetic oxygen atoms in the Earth's lower atmosphere, *Planet. Space Sci.*, **25**, 117.

Magee-Sauer, K., Scherb, F., Roesler, F.L., *et al.* (1989) Fabry-Perot observations of NH2 emission from comet Halley, *Icarus*, **82**, 50.

Mahajan, K.K., Ghosh, S., Sethi, N.K., and Kohli, R. (1992) Ionospheric evidence of hot oxygen in the upper atmosphere of Venus, *Geophys. Res. Lett.*, **19**, 1627.

Marconi, H.L., and Mendis, D.A. (1982a) The photochemistry and dynamics of a dusty cometary atmosphere, *Moon and Planets*, **27**, 27.

Marconi, H.L., and Mendis, D.A. (1982b) A multi-fluid model of an H_2O dominated dusty cometary atmosphere, *Moon and Planets*, **27**, 431.

Marconi, H.L., and Mendis, D.A. (1983) The atmosphere of a dirty-clathrate cometary nucleus: A two-phase, multified model, *Astrophys. J.*, **273**, 381.

Marov, M.Ya., and Kolesnichenko, A.V. (1987) *Introduction to planetary aeronomy.* Nauka, Moscow, (in russian).

Marov, M.Ya., and Shematovich, V.I. (1985) Numerical stochastic simulation of photochemistry in cometary atmospheres, *Preprint 176*, Keldysh Inst. of Appl. Math, Moscow, (in russian).

Marov, M.Ya., and Shematovich, V.I. (1986) Numerical simulation of solar short-wave radiation influence on Earth's upper atmosphere, in: *Observations of artificial satellites*, Moscow, **24**, 84, (in russian).

Marov, M.Ya., and Shematovich, V.I. (1987) Numerical investigation of photochemistry of H_2O dominant cometary atmosphere, *Preprint 90*, Keldysh Inst. of Appl. Math, Moscow, (in russian).

Marov, M.Ya., and Shematovich, V.I. (1988) Mathematical description of kinetics and dynamics of rarefied atmospheric gas, in: *Observations of artificial celestial bodies*, Moscow, **84**, 144, (in russian).

Marov, M.Ya., Shematovich, V.I., and Bisikalo, D.V. (1990) *Kinetic simulation of rarefied gas in the aeronomy problems*. Keldysh Institute of Applied Mathematics, Moscow, (in russian).

Marov, M.Ya., Shematovich, V.I., and Bisikalo, D.V. (1996) Nonequilibrium aeronomic processes: a kinetic approach to the mathematical modeling, *Space Science Reviews*, **76**, Nos. 1/2, 1.

Marov, M.Ya., Kolesnichenko, A.V., and Skorov, Yu.V. (1987) Thermal photometric model of a cometary nucleus, *Astron. Herald (Solar System Research)*, **21**, 47.

Marov, M.Ya., Kolesnichenko, A.V., and Skorov, Yu.V. (1993) Modeling of the outflow of a gas suspension from the porous surface layer during sublimation of a cometary nucleus, *Preprint No. 80*, Keldysh Inst., Moscow.

Mason, E.A. and Marrero T.R. (1970) The diffusion of atoms and molecules, *Adv. Mol. Phys.*, **6**, 156.

Massey, H.S.W., and Mohr, C.B.O. (1934) Free paths and transport phenomena in gases and the quantum theory of collisions. II. The determination of the laws of force between atoms and molecules,*Proc. Roy. Soc. Lond. A*, **144**, 188.

Mauersberger, K., Engebretson, M.J., Potter, W.E., Kayser, D.C. and Nier, A.O. (1975) Atomic nitrogen measurements in the upper atmosphere, *Geophys. Res. Lett.*, **2**, 337.

McCoy, R.P. (1983)T hermospheric odd nitrogen, 1, NO, N(^{4}S), O(^{3}P) densities from rocket measurements of the NO δ and γ bands and the O_2 Herzberg I bands, *J. Geophys. Res.*, **88**, 3197.

McElroy, M.B. (1972) Mars : an evolving atmosphere, *Science*, **175**, 443.

McEwan, M.J., and Phillips, L.F. (1975) *Chemistry of the atmosphere*. John Wiley, New York.

McKean, H.P. (1966) A class of Markov processes associated with nonlinear para- bolic equation, *Proc. Nat. Acad. Sci. USA*, **56**, 1907.

McNeal, R.J., Whitson, M.E., and Cook, G.R. (1974) Temperature dependence of the quenching of vibrationally excited nitrogen by atomic oxygen, *J. Geophys. Res.*, **79**, 1527.

Meier, R.R. (1991) Ultraviolet spectroscopy and remote sensing of the upper atmosphere, *Space Science Reviews*, **58**, 1.

Mendis, D.A. (1988) Exploration of Halley's comet: Symposium summary, in: *Exploration of Halley's Comet* (M. Grewing, F. Praderie, and R. Reinhard eds.), Springer-Verlag, Berlin-Heidelberg, 939.

Mendis, D.A., Houpis, H.L.E., and Marconi, M.L. (1985) The physics of comets, *Fundam. Cosmic Phys.*, **10**, 1.

Meriwether, J.W., Atreya, S.K., Donahue, T.M., and Burnside, R.G. (1980) Measurements of Balmer alpha emission from the hydrogen geocorona, *Geophys. Res. Lett.*, **7**, 967.

Miroshin, R.N. (1967) Improved analysis of distribution functions and kinetic equations for rarefied gas, in: *Aerodynamics of Rarefied Gases*, Leningrad State Univ., Leningrad, (in russian).

Mitchell, D.L., Lin, R.P., Carlson, C.W., *et al.* (1992) The origin of complex organic ions in the coma of comet Halley, *Icarus*, **98**, 125.

Mitchner, M., and Kruger, C. (1976) *Partially ionized gases*. Mir, Moscow, (in russian).

Moddeman, W.E., Carlson, T.A., Krause, M.O., Pullen, B.P., Bull, W.E., and Schweitzer, G.K. (1971) Determination of the K-LL Auger spectra of N_2, O_2, CO, NO, H_2O and CO_2, *Chem. Phys.*, **55** , 2317.

Moffet, R.J. (1988) Rates of electron cooling in the upper atmosphere, *Planet. Space Sci.*, **36**, 65.

Moller, G., and Jackson, W.M. (1990) Laboratory studies of polyoxymethylene: Applications to comets, *Icarus*, **86**, 189.

Montroll, A., and Lebowitz, G. (Eds.) (1983) *Nonequilibrium phenomena: Boltzmann equation*. Springer-Verlag, New York.

Morgan, J.E., and Schiff, H.J. (1964) Diffusion coefficients of O and N atoms in inert gases, *Can. J. Phys.*, **47** , 300.

Moroz, V.I., *et al.* (1987) Detection of parent molecules in comet P/Halley from KS-Vega experiments, *Astron. Astrophys.*, **187**, 513.

Mott-Smith, H.M. (1951) The solution of the Boltzmann equation for a shock wave, *Phys. Rev*, **82**, 885.

Mumma, M.J., Hoban, S, Renter, D. C., and DiSanti, M. (1993) Methanol in recent comets: Evidence for two distinct cometary populations, in: IAU Symp. No. 160, *Asteroids, Comets, Meteors*, Book of Abstracts, Belgirate (Novara), Italy, 227.

Nagy, A.F., and Banks, P.M. (1970) Photoelectron fluxes in the ionosphere, *J. Geophys. Res.*, **75**, 6260.

Nagy, A.F., and Cravens, T.E.(1988) Hot oxygen atoms in the upper atmospheres of Venus and Mars, *Geophys. Res. Lett.*, **15**, 433.

Nagy, A.F., Cravens, T.E., Yee, J.-H., and Stewart, A.I. (1981) Hot oxygen atoms in the upper atmospheres of Venus, *Geophys. Res. Lett.*, **8**, 629.

Nagy, A.F., Kim, J., and Cravens, T.E.(1990) Hot hydrogen and oxygen atoms in the upper atmospheres of Venus and Mars, *Ann. Geophys.*, **8**, 251.

Nanbu, K. (1980) Direct simulation scheme derived from the Boltzmann equation, *J. Phys. Soc. Japan*, **49**, 2042.

Newburn, R.L., and Spinrad, H. (1984) Spectrophotometry of 17 comets. I. The emission features, *Astron. J.*, **89**, 289.

Newburn, R.L., and Spinrad, H. (1985) Spectrophotometry of 17 comets. II. The continuum, *Astron. J.*, **90**, 2591.

Newburn, R.L., and Spinrad, H. (1988) Spectrophotometry of 17 comets. III. Post-Halley updates plus eight additional comets, *Bull. Am. Astron. Soc.*, **20**, 841.

Nicolet, M. (1945) Contribution à l'étude de l'ionosphère, *Inst. Météorol. Belg. Mem.*, **19** , 162.

Nicolet, M. and Aikin, A.C. (1960) The formation of the D region of the ionosphere, *J. Geophys. Res.*, **65**, 1469.

Nier, A.O., and McElroy, M.B. (1977) Composition and structure of Mars' upper atmosphere: Results from the neutral mass spectrometers on Viking 1 and 2, *J. Geophys. Res.*, **82**, 4341.

Nordsiek, A., and Hicks, B. (1967) Monte Carlo evaluation of the Boltzmann collision integral, in: *Rarefied gas dynamics* (ed. C.L.Brundin), Academic Press, New York, 695.

Norton, R.B. and Barth, C.A. (1970) Theory of nitric oxide in the Earth's atmosphere, *J. Geophys. Res.*, **75**, 3903.

Oliver, W.L. (1997) Hot oxygen and the ion energy budget, *J. Geophys. Res.*, **102**, 2503.

Opik, J. (1963) Selective escape of gases, *Geophys. J.*, **7**, 490.

Oran, E.S., Julienne, P.S. and Strobel, D.F.(1975) The aeronomy of odd nitrogen in the thermosphere, *J. Geophys. Res.*, **80** , 3903.

Pattengill, M.D. (1975) A comparison of classical trajectory and exact quantal cross sections for rotationally inelastic Ar-N2 collisions, *Chem. Phys. Letters*, **36**, 25.

Peterson, L.R., Sawada, T., Bass, J.N., and Green, A.E.S. (1973) Electron energy deposition in a gaseous mixture, *Comp. Phys. Comm.*, **5**, 239.

Phelps A.V. (1990) Cross sections and Swarm coefficients for H^+, H_2^+, H_3^+, H, H_2 and H^- in H_2 for energies from O.1 eV to 10 keV, *J. Phys. Chem. Ref. Data*, **19**, 653.

Piper, L.G. (1989) The rate coefficient for quenching of $N(^2D)$ by $O(^3P)$, *J. Chem. Phys.*, **91**, 3516.

Polak, L.S. (1979) *Nonequilibrium chemical kinetics and its applications*. Nauka, Moscow, (in russian).

Polak, L.S., Goldenberg, M.Y., and Levitsky, A.A. (1984) *Computational methods in chemical kinetics*. Nauka, Moscow.

Preston, G.W. (1967) The spectrum of comet Ikeya-Seki (1965f), *Astrophys. J.*,**147**,718.

Prialnik, D., and Bar-Nun A. (1990) Gas rease in comet nuclei, *Astrophys. J.*, **363**, 274.

Prigogine, I. (1962) *Nonequilibrium statistical mechanics*. Mir, Moscow, (in russian).

Probstein, R.F. (1969) The dusty gas dynanaics of comet heads, in: *Problems of the Hydrodynamics and Mechanics of a Continuous Medium*, Nauka, Moscow, 397 (in russian).

Pyarnpuu, A.A., and Shematovich, V.I. (1985) Structural stochastic simulation of collisional processes in rarefied gaseous media, in: *Reports in Applied Math.*, Computing Center, Moscow, (in russian).

Pyarnpuu, A.A., and Shematovich, V.I. (1987) Construction of structural stochastic models of collisional processes in rarefied gaseous media, in: *Modeling in Mechanics*, Novosibirsk, **1**, 109, (in russian).

Pyarnpuu, A.A., Shematovich, V.I., and Zmievskaya, G.I. (1981) Development of constructive physical-probabilistic analogue of collisional processes in rarefied gas, *Reports of USSR Ac. of Sci.*,**258**, 815, (in russian).

Pyarnpuu, A.A., Tsvetkov, G.A., and Shematovich, V.I. (1986) Structural stochastic simulation of relaxation problems, in: *Reports in Applied Math.*, Computing Center, Moscow, (in russian).

Rees, M.H. (1989) *Physics and chemistry of the upper atmosphere*. Cambridge University Press, Cambridge.

Rego D., Prangé R. and Gérard J.C. (1994) Auroral Lyman α and H_2 bands from the giant planets(1- Excitation by proton precipitation in the Jovian atmosphere, *J. Geophys. Res.*, **99**, 17075.

Richards, P.G., and Torr, D.G. (1983) A simple theoretical model for calculating and parameterising the ionospheric photoelectron flux, *J. Geophys. Res.*, **88**, 2155.

Richards, P.G., and Torr, D.G. (1984) An investigation of the consistency of the ionospheric measurements of the photoelectron flux and solar EUV flux, *J. Geophys. Res.*, **89**, 5625.

Richards, P.G., and Torr, D.G. (1985) The altitude dependence of the ionospheric photoelectron flux: a comparison of the theory and measurements, *J. Geophys. Res.*, **90**, 2877.

Richards, P.G., and Torr, D.G. (1988) Ratio of photoelectron to EUV ionization rates for aeronomic studies, *J. Geophys. Res.*, **93**, 4060.

Richards, P.G., Fennelly, J.A., and Torr, D.G. (1994b) EUVAC : A solar EUV flux model for aeronomic calculations, *J. Geophys. Res.*, **99**, 8981.

Richards, P.G., Hickey, M.P., and Torr, D.G. (1994b) New sources for the hot oxygen geocorona, *Geophys. Res. Lett.*,**21**657.

Roble, R.G. (1995) Energetics of the mesosphere and thermosphere, in: *The Upper Mesosphere and Lower Thermosphere : a review of experiment and theory*, *Geophys. Monograph* **87**, 234, American Geophysical Union, Washngton.

Roble , R.G. and Emery, B.A. (1983) On the global ı eun temperature of the thermosphere, *Planet. Space Sci.*, **31**, 597.

Roble, R.G. and Kasting, J.F. (1984) The zonally averaged circulation, temperature, and compositional structure of the lower thermosphere and variations with geomagnetic activity, *J. Geophys. Res.*, **89**, 1711.

Roble, R.G. and Ridley, E.C. (1994) A thermosphere-ionosphere-mesosphere-electrodynamics general circulation model (TIME-GCM) : Equinox solar minimum simulation, 30-500 km, *Geophys. Res. Lett.*, **21**, 417.

Roble, R.G., Ridley, E.C., and Dickinson, R.E. (1987) On the mean global structure of the thermosphere, *J. Geophys. Res.*, **92**, 8745.

Roble, R.G., Ridley, E.C., and Richmond, A.D. (1988) A coupled thermosphere/ionosphere general circulation model, *Geophys. Res. Lett.*, **15**, 1325.

Roettger, E.F., Feldman P.D., A'Hearn M.F., *et al.* (1989) IUE observations of the evolution of the comet Wilson (1986 1): Comparison with P/Halley, *Icarus*, **80**, 303.

Rohrbaugh, R.P., and Nisbet, J.S. (1973) Effect of energetic oxygen atoms on neutral density models, *J. Geophys. Res.*, **78** , 6768.

Rusch, D.W. (1973) Satellite ultraviolet measurements of nitric oxide fluorescence with a diffusive transport model, *J. Geophys. Res.*, **78**, 5676.

Rusch, D.W. and Barth, C.A. (1975) Satellite measurements of nitric oxide in the polar region, *J. Geophys. Res.*, **80**, 3719.

Rusch, D.W., and Gérard, J.C. (1980) Satellite studied of $N(^2D)$ emission and ion chemistry in aurorae, *J. Geophys. Res.*, **85**, 1285.

Rusch, D.W., and Sharp, W.E. (1981) Nitric oxide delta band emission in the Earth's atmosphere: comparison of a measurement and theory, *J. Geophys. Res.*, **86**, 10111.

Rusch, D.W., Gérard, J.C., and Fesen, C.G. (1991) The diurnal variation of NO, $N(^2D)$ and ions in the thermosphere : a comparison of satellite measurements to a model, *J. Geophys. Res.*, **96**, 11331.

Rutherford, J.A., and Vroom, D.A. (1974) The reaction of atomic oxygen with several atmospheric ions, *J. Chem. Phys.*, **61**, 2514.

Samarskii, A.A.(1971) *Introduction to the theory of difference schemes*. Nauka, Moscow, (in russian).

Samarskii, A.A., and Popov, Yu.P. (1981) *Difference Methods of Solving Problems of Gas Dynamics*. Nauka, Moscow, (in russian).

Sampson, D.H. (1965) *Radiative contributions to energy and momentum transport in a gas*. Interscience Publishers, New York.

Schafer, D.A., Newman, J.H., Smith, K.A., and Stebbings, R.F. (1987) Differential cross sections for scattering of 0.5-, 1.5-, and 5.0-keV oxygen atoms by He, N_2, and O_2, *J. Geophys. Res.*, **92**, 6107.

Scherb, F., Magee-Sauer K., Roesler F.L., and Harlander J. (1990) Fabry-Perot observations of comet Halley H_2O^+, *Icarus*, **86**, 172.

Schmitt, G.A., Abreu V.J. and Hays P.B. (1981) Non-thermal $O(^1D)$ produced by dissociative recombination of O_2^+: a theoretical model and observational results, *Planet. Space Sci.*, **29**, 1095.

Schunk, R.W., and Nagy, A.F. (1978) Electron temperatures in the F-region of the ionosphere: theory and observations, *Rev. Geophys.*, **16**, 355.

Schunk, R.W., and Nagy, A.F. (1980) Ionospheres of the t errestrial planets, *Rev. Geophys.*, **18**, 813.

Seiff A., Kirk D.B., Knight T.C.D., Mihalov J.D., Blanchard R.C., Young R.E., Schubert G., von Zahn U., Lehmacher G., Milos F.S., and Wuang J. (1996) Structure of the atmosphere of Jupiter : Galileo probe measurements, *Science*, **272**, 844.

Sharma, R.D., Sun, Y., and Dalgarno, A. (1993) Highly rotationally excited nitric oxide in the terrestrial thermosphere, *Geophys. Res. Lett.*, **20**, 2043.

Sharma, R.D., Kharchenko, V.A., Sun, Y., and Dalgarno, A. (1996) Energy distribution of fast nitrogen atoms in the nighttime terrestrial thermosphere, *J. Geophys. Res.*, **101**, 275.

Sharp, R.D., Johnson, R.E., and Shelley, E.G. (1974) The morphology of energetic O^+ in the magnetosphere, *J. Geophys. Res.*, **79**, 144.

Sharp, R.D., Johnson, R.E., and Shelley, E.G. (1976a) The morphology of energetic O^+ ions during two magnetic storms: Temporal variations, *J. Geophys. Res.*, **81**, 3283.

Sharp, R.D., Johnson, R.E., and Shelley, E.G. (1976b) The morphology of energetic O^+ ions during two magnetic storms: Latitudinal variations, *J. Geophys. Res.*, **81**, 3292.

Shelley, E.G., Johnson, R.E., and Sharp, R.D. (1972) Satellite observation of energetic heavy ions during a geomagnetic storm, *J. Geophys. Res.*, **77**, 6104.

Shelley, E.G., Johnson, R.E., and Sharp, R.D. (1974) The morphology of energetic O^+ in the magnetosphere, in: *Magnetospheric Physics*, 135, D.Reidel, Hingham, Massachusets.

Shematovich, V.I. (1979) Numerical algorithms for non-stationary statistical model of gas mixtures with inner degrees of freedom, in: *Reports in Applied Math.*, Computing Center, Moscow, (in russian).

Shematovich, V.I. (1980) *Non-stationary statistical imulation of collisional physical and chemical processes in rarefied gas*, Ph. D. Thesis, Computing Center, Moscow, (in russian).

Shematovich, V.I. (1982) Numerical investigation of photochemical processes in the upper atmosphere: numerical stochastic model of photoprocesses, in: *Scientific Reports of Institute of Astronomy*, Moscow, **55**, 160, (in russian).

Shematovich, V.I. (1987) Numerical stochastic simulation of kinetics of atmospheric photochemistry, in: *Mathematical problems of applied aeronomy*, Keldysh Inst. of Applied Math., Moscow, 199, (in russian).

Shematovich, V.I., Bisikalo, D.V., and Marov, M.Ya. (1991a) Kinetic approach to the mathematical modeling of collisional physical and chemical processes in planetary atmospheres, in: *Rarefield Gas Dynamics*, 345, VCH, New York.

Shematovich, V.I., Bisikalo, D.V., and Gérard, J.-C. (1991b) Non thermal nitrogen atoms in the Earth's thermosphere. I. Kinetics of "hot" $N(^4S)$, *Geophys. Res. Lett.*, **18**, 1691.

Shematovich, V.I., Bisikalo, D.V., and Gérard, J.-C. (1992) The thermospheric odd nitrogen photochemistry : role of non-thermal $N(^4S)$ atoms, *Ann. Geophys.*, **10**, 792.

Shematovich, V.I., Bisikalo, D.V., and Gérard, J.-C. (1994) A kinetic model of the formation of the hot oxygen geocorona. I. Quiet geomagnetic conditions, *J. Geophys. Res.*, **99**, 217.

Shizgal, B., and Arkos, G.G. (1996) Nonthermal escape of the atmospheres of Venus, Earth, and Mars, *Rev. Geophys.*, **34**, 483.

Shizgal, B., and Lindenfeld, M.J. (1980) Further studies of non-maxwellian effects associated with the thermal escape of a planetary atmosphere, *Planet. Space Sci.*, **28**, 159.

Shul'man, L.M. (1972) *Dynamics of cometary atmosphere. Neutral gas.* Naukova Dumka, Kiev, (in russian).

Shul'man, L.M. (1982) Ion-molecular clusters in cometary nuclei, *Preprint ITF-81-141R*, Inst. Teor. Fiz. Acad. Nauk Ukr. SSR, Kiev.

Shul'man, L.M. (1987) *Nuclei of comets.* Nauka, Moscow, (in russian).

Singh, P.D., de Almeida A.A., and Huebner W.F. (1991) The states of carbon and nitrogen atoms after photodissociation of CN, CH, CH^+, C_2, C_3, and CO in comets, *Icarus*, **90**, 74.

Siskind, D.E., Barth, C.A. and Cleary, D.D. (1990) The possible effect of solar soft X rays on thermospheric nitric oxide, *J. Geophys. Res.*, **95**, 4311.

Siskind, D.E., Barth, C.A. and Roble, R.G. (1989) The response of thermospheric nitric oxide to an auroral storm, 1., Low and mid latitudes, *J. Geophys. Res.*, **94**, 16885.

Siskind, D.E., Strickland, D.J., Meier, R.R., Majeed, T., and Eparvier, F.G. (1995) On the relationship between the solar soft X ray flux and thermospheric nitric oxide: an update with an improved photoelectron model, *J. Geophys. Res.*, **100**, 19687.

Skinner, T.E., Durrance S.T., Feldman P.D., and Moos H.W. (1984) IUE observations of latitudinal and temporal variations in the Jovian auroral emission, *Astrophys. J.*, **278**, 441.

Skorochod, A.V. (1983) *Stochastic equations for complex systems.* Nauka, Moscow, (in russian).

Smith, D.R., and Ahmadjian, M. (1993) Observation of nitric oxide rovibrational bandhead emissions in the quiescent airglow during the CIRRIS-1A space shuttle experiment, *Geophys. Res., Lett.* **20**, 2679.

Smith, D.R., Adler-Golden, S. and Roth, C. (1993) Empirical c orrelations in thermospheric NO density measured from rocket and satellites, *J. Geophys. Res.*, **98**, 9453.

Smyth, W.H., and Schempp W.V. (1989) [OI] in comet Halley, *Icarus*, **82**, 61.

Smyth, W.H., Marconi M.L., and Combi M.R. (1991) Analysis of hydrogen Lyaman-α observations of the coma of comet P/Halley near perihelion, *Icarus*, **113**, 119.

Snyder, L.E., Palmer P., and DePater I. (1989a) Radio detaction of formaldegyde emission from comet Halley, *Astron. J.*, **97**, 246.

Snyder, L.E., Palmer P., and DePater I. (1989b) VLA observations of formaldegyde emission from comets Halley and Machholtz(1988 j), *Publ. Astron. Soc. Pac.*, **101**, 882.

Solomon, S. (1983) The possible effects of translationally excited nitrogen atoms on lower thermospheric odd nitrogen, *Planet. Space Sci.*, **33**, 135.

Spitzer, L., Jr. (1952) The terrestrial atmosphere above 300 km, in: *The Atmospheres of the Earth and Planets*, Second Edition, Univ. Chicago Press, Chicago.

Spohn, H. (1980) Kinetic equations from hamiltonian dynamics: Markovian limits, *Rev. Modern Phys.*, **53**, 569.

Stakhanov, I.P. (1979) *The physical Nature of Ball Lightning.* Atomizdat, Moscow (in Russian).

Stamnes, K. (1980) Analytic approach to electron transport and energy degradation, *Planet. Space Sci.*, **28**, 427.

Stamnes, K., and Rees, M.H. (1983) Inelastic scattering effects on photoelectron spectra and ionospheric electron temperature, *J. Geophys. Res.*, **88**, 6301.

Stebbings, R.F., Smith, A.C.H., and Ehrhardt, H. (1964) Charge transfer between oxygen atoms and O^+ and H^+ ions, *J. Geophys. Res.*, **69**, 2349.

Strathdee S., and Browning R. (1979) Dissociative photoionisation of H_2 : proton kinetic energy spectra, *J. Phys: Atom. Molec. Phys.*, **12**, 1789.

Strickland, D.J., Book, D.L., Coffey, T.P., and Fedder, J.A. (1976) Transport equation techniques for the deposition of auroral electrons, *J. Geophys. Res.*, **81**, 2755.

Strobel, D.F. (1982) Chemistry and evolution of Titan's atmosphere, *Planet. Space Sci.*, **30**, 839.

Strobel, D.F., Hunten, D.M., and McElroy, M.B. (1970) Production and diffusion of nitric oxide, *J. Geophys. Res.*, **75**, 4307.

Struminskii, V.V. (1982)Kinetic theory of gas mixtures, in: *Molecular Gas Dynamics*, Nauka, Moscow, 132, (in russian).

Takacs, P.Z., Broadfoot, A.L., Smith, G.R., and Kumar, S. (1980) Mariner 10 observations of hydrogen Lyman alpha emission from the Venus exosphere: Evidence of complex structure, *Planet. Space Sci.*, **28**, 687.

Tan, Z., Chen, Y-K., Varghese, P.L., and Howell, J.R. (1989) New numerical strategy to evaluate the collision integral of the Boltzmann equation, *Prog. in Astro. and Aero.*, **118**, 359.

Tcheremissine, F.G. (1985) Numerical methods for the direct solution of the kinetic Boltzmann equation, *USSR Comput. Math. Math. Phys.*, **25**, 156, (in russian).

Tcheremissine, F.G. (1991) Fast solution of the Boltzmann equation, in: *Rarefied gas dynamics* (ed. A.E.Beylich), VCH, Weinheim, 273.

Tennyson, P.D., Feldman, P.D., Hartig, J.F. and Henry, R.C. (1986) Near-midnight observations of nitric oxide δ- and γ-band chemiluminescence, *J. Geophys. Res.*, **91**, 10141.

Tharamel, J., Kharchenko, V., and Dalgarno, A. (1995) Thermalization of fast atoms in an oxygen atmosphere: quantum calculations of energy transition rates in elastic collisions, *Eos Trans. AGU*, **76**, 438.

Tinsley, B.A., Hodges, R.R., and Rohrbaugh, R.P. (1986) Monte Carlo models for the terrestrial exosphere over a solar cycle, *J. Geophys. Res.*, **91**, 13631.

Titheridge, J.E. (1996) Direct allowance for the effect of photoelectrons in ionospheric modeling, *J. Geophys. Res.*, **101**, 357.

Tobiska, W.K. (1991) Revised solar extreme ultraviolet flux model, *J. Atmos. Terr. Phys.*,

53, 1005.

Tobiska, W.K., and Barth, C.A. (1990) A solar EUV flux model, *J. Geophys. Res.*, **95**, 8243.

Tokunaga, A.T., and Booke T.Y. (1990) Did comet form from unaltered interstellar dust and ices? The evidance from infrared spectroscopy, *Icarus*, **86**, 208.

Torr D.G. and Torr, M.R. (1979) Chemistry of the thermosphere and ionosphere, *J. Atmos. Terr. Phys.*, **41**, 797.

Torr, M.R., and Torr, D.G. (1982) The role of metastable species in the thermosphere, *Rev. Geophys. Space Phys.*, **20**, 91.

Torr, M.R., and Torr, D.G. (1985) Ionization frequencies for solar cycle 21 : revised, *J. Geophys. Res.*, **90**, 6675.

Torr, M.R., Richards, P.G., and Torr, D.G. (1980) A new determination of the ultraviolet heating efficiency of the thermosphere, *J. Geophys. Res.*, **85**, 6819.

Torr, M.R., Torr, D.G., Chang, T., Richards, P., Swift, W., and Ni, N. (1995) Thermospheric nitric oxide from the ATLAS 1 and spacelab 1 missions, *J. Geophys. Res.*, **100**, 17,389.

Torr, M.R., Torr, D.G., Ong, R.A., and Hinteregger, H.E. (1979) Ionization frequencies for major thermospheric constituents as a function of solar cycle, *Geophys. Res. Lett.*, **6**, 771.

Torr, M.R., Torr, D.G., Roble, R.G., and Ridley, E.C. (1982) The dynamic response of the thermosphere to the energy influx resulting from energetic O^+ ions, *J. Geophys. Res.*, **87**, 5290.

Torr, M.R., Torr, D.G., Richards, P.G., and Yung, S.P. (1990) Mid- and low-latitude model of thermospheric emissions, 1. $O^+(^2P)$ 7320 A and $N_2(2P)$ 3371 A, *J. Geophys. Res.*, **95**, 21147.

Torr, M.R., Walker, J.C.G., and Torr, D.G. (1974) Escape of fast oxygen from the atmosphere during geomagnetic storms, *J. Geophys. Res.*, **79**, 5267.

Trafton L., Gérard J.C., Munhoven G. and Waite J.H. (1994) High resolution spectra of Jupiter's northern auroral ultraviolet emission with the Hubble Space Telescope, *Astrophys. J.*, **421**, 816.

Truhlar, D.G., and Muckerman, J.T. (1979) Reactive scattering cross sections III: quasiclassical and semiclassical methods, in: *Atom-Molecule Collision Theory*, edited by R. B. Bernstein, pp. 505, Plenum, New York.

Van Dishoeck, E.F., and Dalgarno A. (1984) The dissociation of OH and OD in comets by solar radiation, *Icarus*, **59**, 305.

Van Zyl, B., and Stephen, T.M. (1994) Dissociative ionization of H_2, N_2, and O_2 by electron impact, *Phys. Rev. A*, **50** , 3164.

Vallander, S.V. (1967) Probabilistic interpretation of rarefied gas kinetics problems, in: *Aerodynamics of Rarefied Gases*, Leningrad State Univ., Leningrad, **3**, 5, (in russian).

Uhlenbeck, G.E. (1963) Boltzmann equation, Appendix I, in: Kac M. *Probability and related problems in physics*, Mir, Moscow, (in russian).

Ulrich, B.L., and Conclin E.J. (1975) Discovery of CH_3CN in comets, *Nature*, **248**, 121.

Waite, J.H. Jr., Cravens T.E. , Kozyra J., Nagy A.F., Atreya S.K. and Chen R.H. (1983) Electron precipitation and related aeronomy of the Jovian thermosphere and ionospher'e, *J. Geophys. Res.,*, **88**, 6143.

Wallis, M.K., and Krishna Swamy K.S. (1987) Some diatomic molecules from comet P/Halley's UV spectra near spacecraft flybys, *Astron. Astropys.*, **187**, 329.

Weaver, H.A., Mumma M.J., and Larson H.P. (1987) Infrred investigation of water in comet P/Halley, *Astron. Astrophys.*, **187**, 411.

Weaver, H.A., Feldman P.H., McPhate J.B., *et al.* (1993) Deriving abundances from cometary spectra taken with the HST, in: IAU Symp. No. 160, *Asteroids, Comets, Meteors*, Book of Abstracts, Belgirate (Novara), Italy, 309.

Wegmann, R., Schmidt U., Huebner W.F., and Boise D.C. (1987) Cometary MHD and chemistry, *Astron. Astrophys.*, **187**, 339.

Whipple, E.G. (Jr) (1974) Theory of reaction product velocity distributions, *J. of Chem.*

Phys., **60** , 1345.

Whipple, F.L. (1950) A comet model. II. Physical relations for comets and meteors, *Astrophys. J.*, **13**, 464.

Whipple, F.L. (1984) Present status of the icy conglomerate model, *Smithson. Astrophys. Obs. Rep.*, No. 1966.

Whipple, F.L. (1989a) The forest and the trees, *Smithson. Astrophys. Obs. Rep.*, No. 2978.

Whipple, F.L. (1989b) Comets in the space age, *Astrophys. J.*, **241**, 1.

Whipple, F.L. (1989c) The cometary nucleus: current concepts, in: *Proceedings of the 20th ESLAB Symposium on the Exploration of the Halley's comet*, **2**, (B.B.Battrick, E.J.Rolfe, and R.Reillard eds.), ESA SP-250.

Wickramasinghe, D.T., and Allen D.A. (1986) Discovery of organic grains in comet Halley, *Nature*, **323**, 44.

Wilson, J.K., and Scheider, N.M. (1994) Io's Fast Sodium: implications for nuclear and atomic atmospheric escape, *Icarus*, **111**, 31.

Wise, J.O, Carovillano, R.L. , Carlson, H.C., Roble, R.G., Adler-Golden, S., Nadile, R.M. and Ahmadjian, M. (1995) CIRRIS 1A global observations of 15-μ NO limb radiance in the lower thermosphere during moderate to active geomagnetic activity, *J. Geophys. Res.*, **100** , 21357.

Yanitskii, V.E. (1975) Statistical model of ideal gas flow and some its features, in: *Numerical Methods of Fluid Mechanics*, Novosibirsk, **6**, 139, (in russian).

Yanitskii, V.E. (1988) The stochastical models of perfect gas with some finite number of particles, Computer Center of the USSR Ac. Sci., preprint, (in russian).

Yanitskii, V.E. (1991) Operator approach to the direct Monte-Carlo simulation theory in rarefied gas dynamics, in: *Rarefield Gas Dynamics*, 770, VCH, New York.

Yee, J.H., and Dalgarno, A. (1985) Energy transfer of $O(^1S)$ atoms in collision with $O(^3P)$ atoms, *Planet. Space Sci.*, **33**, 825.

Yee, J.H., and Hays, P.B. (1980) The oxygen polar corona, *J. Geophys. Res.*, **85**, 1795.

Yee, J.-H., Guberman, S.L., and Dalgarno, A. (1990) Collisional quenching of $O(^1D)$ by $O(^3P)$, *Planet. Space Sci.*, **38**, 647.

Yee, J.H., Meriwether, J.W., and Hays, P.B. (1980) Detection of a corona of fast oxygen atoms during solar maximum, *J. Geophys. Res.*, **85**, 3396.

Yen, S.M., Hicks, B., and Osteen, R.M. (1974) Further development of a Monte Carlo method for the evaluation of the Boltzmann collision integral, in: *Rarefied gas dynamics* (eds. M.Becker, M.Fiebig), DFVLR-Press, Porz-Wahn, A.12-1.

Ytrehus, T. (1977) Theory and experiments on gas kinetics in evaporation, in: *Rarefied Gas Dynamics*, Springer-Verlag, New York,**2**, 1197.

Ytrehus, T. (1994) Moment solutions in the kinetic theory of strong evaporation and condensation: Application in cometary dust-gas dynamics, *Prog. in Astro. and Aero.*, **158**.

Zahnle, K., and Grinspoon D. (1990) Comet dust as a source of aminoacids at the Cretaceous/Tertiary boundary, *Nature*, **348**, 157.

Zeippen, C.J. (1982) Transition probabilities for forbidden lines in the 2 p^3 configuration, *Mon. Not. R. Astron. Soc.* **198**, 111.

Zipf, E.C., and McLaughlin, R.W. (1978) On the dissociation of nitrogen by electron impact and by EUV photo–absorption, *Planet. Space Sci.*, **26**, 449.

Zipf, E.C. , Espy, P.S. and Boyle, C.F. (1980) The excitation and collisional deactivation of metastable $N(^2P)$ atoms in auroras, *J. Geophys. Res.*, **85**, 687.

Zmievskaya, G.I., Pyarnpuu, A.A., and Shematovich, V.I. (1979) Simulation of physical and chemical processes in gas mixtures, *Reports of USSR Academy of Sci.*, **247**, 561, (in russian).

Zmievskaya, G.I., Pyarnpuu, A.A., and Shematovich, V.I. (1980) Mathematical base for physical-probabilistic analogue development for physical and chemical kinetics, in: *Reports in Applied Math.*, Computing Center, Moscow, (in russian).

Zmievskaya, G.I., Marov, M.Ya., and Shematovich, V.I. (1982) Numerical investigation of photochemical processes in the upper atmosphere: stochastic simulation of UV solar radiation influence on rarefied multicomponent gas, in: *Scientific Reports of Institute of Astronomy*, Moscow, **55**, 144, (in russian).

Zmievskaya, G.I., Pyarnpuu, A.A., and Shematovich, V.I. (1983) Stochastic model of discrete rarefied medium as a mean for investigation of physical and chemical kinetics problems, in: *Numerical Methods of Fluid Mechanics*, Novosibirsk, **18**, 74, (in russian).

Zmievskaya, G.I., Marov, M.Ya., and Shematovich, V.I. (1984) Stochastic simulation of interaction between the solar radiation and rarefied gas of the Earth's upper atmosphere, in: *Observations of artificial satellites*, Moscow, **21**, 249.